Plug-in Hybrid Electric Vehicle (PHEV)

Plug-in Hybrid Electric Vehicle (PHEV)

Special Issue Editor

Joeri Van Mierlo

MDPI • Basel • Beijing • Wuhan • Barcelona • Belgrade

Special Issue Editor
Joeri Van Mierlo
Vrije Universiteit Brussels
Belgium

Editorial Office
MDPI
St. Alban-Anlage 66
4052 Basel, Switzerland

This is a reprint of articles from the Special Issue published online in the open access journal *Applied Sciences* (ISSN 2076-3417) from 2018 to 2019 (available at: https://www.mdpi.com/journal/applsci/special_issues/Plug_in_Hybrid_Electric_Vehicle)

For citation purposes, cite each article independently as indicated on the article page online and as indicated below:

LastName, A.A.; LastName, B.B.; LastName, C.C. Article Title. *Journal Name* **Year**, *Article Number*, Page Range.

ISBN 978-3-03921-453-2 (Pbk)
ISBN 978-3-03921-454-9 (PDF)

Cover image courtesy of MOBI Mobility, Logistics & Automotive Technology Research Center.

Contents

About the Special Issue Editor

Joeri Van Mierlo is a key player in the Electromobility field. He is Professor at Vrije Universiteit Brussel, one of the top universities in this field. Prof. Dr. ir. Joeri Van Mierlo leads MOBI—Mobility, Logistics and Automotive Technology Research Centre (https://mobi.research.vub.be) with a multidisciplinary and growing team of over 100 staff members. Prof. Van Mierlo was Visiting Professor at Chalmers University of Technology, Sweden (2012). He is expert in the field of Electric and Hybrid vehicles (batteries, power converters, energy management simulations) as well as to the environmental and economical comparison of vehicles with different drive trains and fuels (LCA, TCO). Prof. Van Mierlo is Vice President of AVERE (www.avere.org), the European Electric Vehicle Association, and also Vice President of its Belgian section ASBE (www.asbe.be). He chairs the EPE chapter "Hybrid and Electric vehicles" (www.epe-association.org). He is an active member of EARPA (European Automotive Research Partner Association) and member of EGVIA (European Green Vehicle Initiative Association). He is IEEE Senior Member and member of IEEE Power Electronics Society (PELS), IEEE Vehicular Technology Society (VTS) and IEEE Transportation Electrification Community. He is the co-author of more than 500 scientific publications and Editor-in-Chief of World Electric Vehicle Journal.

Editorial

Special Issue "Plug-In Hybrid Electric Vehicle (PHEV)"

Joeri Van Mierlo

Director of MOBI-Mobility, Logistics and Automotive Technology Research Centre, Vrije Universiteit Brussel, Faculty of Engineering, ETEC—Department of Electrical Engineering and Energy Technology, Core Lab of Flanders Make, 1050 Brussels, Belgium; Joeri.van.mierlo@vub.be

Received: 11 July 2019; Accepted: 11 July 2019; Published: 16 July 2019

Abstract: Climate change, urban air quality, and dependency on crude oil are important societal challenges. In the transportation sector especially, clean and energy-efficient technologies must be developed. Electric vehicles (EVs) and plug-in hybrid electric vehicles (PHEVs) have gained a growing interest in the vehicle industry. Nowadays, the commercialization of EVs and PHEVs has been possible in different applications (i.e., light duty, medium duty, and heavy duty vehicles) thanks to the advances in energy-storage systems, power electronics converters (including DC/DC converters, DC/AC inverters, and battery charging systems), electric machines, and energy efficient power flow control strategies. This Special Issue is focused on the recent advances in electric vehicles and (plug-in) hybrid vehicles that address the new powertrain developments and go beyond the state-of-the-art (SOTA).

Keywords: novel propulsion systems; emerging power electronics; including wide bandgap (WBG) technology; emerging electric machines; efficient energy management strategies for hybrid propulsion systems; energy storage systems; life-cycle assessment (LCA)

1. Introduction

In light of the current challenges of climate change, urban air quality, and dependency on crude oil [1–3], this special issue was introduced to collect the latest research on plug-in hybrid electric vehicles. There were 21 papers submitted to this special issue, of which 11 papers were accepted. When looking back to this special issue, various topics have been addressed, mainly on drive trains and energy management (four papers), batteries (five papers), and environmental assessments (two papers).

2. Drive Trains and Energy Management

The first paper, authored by Zheng Chen [4], proposes an energy management strategy for a power-split plug-in hybrid electric vehicle (PHEV) based on reinforcement learning (RL). Firstly, a control-oriented power-split PHEV model was built, and then the RL method was employed based on the Markov decision process (MDP) to find the optimal solution according to the built model. During the strategy search, several different standard driving schedules were chosen, and the transfer probability of the power demand was derived based on the Markov chain. Accordingly, the optimal control strategy was found by the Q-learning (QL) algorithm, which can decide suitable energy allocation between the gasoline engine and the battery pack. Simulation results indicate that the RL-based control strategy could not only lessen fuel consumption under different driving cycles but also limit the maximum discharge power of the battery, compared with the charging depletion/charging sustaining (CD/CS) method and the equivalent consumption minimization strategy (ECMS) [4].

Renxin Xiao and his co-authors compared different energy management methods in their paper [5]. This paper proposes a comparison study of energy management methods for a parallel plug-in hybrid electric vehicle (PHEV). Based on detailed analysis of the vehicle driveline, quadratic convex functions are presented to describe the nonlinear relationship between engine fuel-rate and battery charging power at different vehicle speeds and driveline power demand. The engine-on power threshold is estimated by the simulated annealing (SA) algorithm, and the battery power command is achieved by convex optimization with target of improving fuel economy, compared with the dynamic programming (DP)-based method and the charging depleting-charging sustaining (CD/CS) method. In addition, the proposed control methods are discussed at different initial battery state of charge (SOC) values to extend the application. Simulation results validate that the proposed strategy based on convex optimization can save fuel consumption and reduce the computation burden noticeably [5].

Duong Tran describes in his paper the development of DC/DC multiport converters (MPC) [6]. These converters are gaining interest in the field of hybrid electric drivetrains (i.e., vehicles or machines), where multiple sources are combined to enhance their capabilities and performances in terms of efficiency, integrated design, and reliability. This hybridization will lead to more complexity and high development/design time. Therefore, a proper design approach is needed to optimize the design of the MPC as well as its performance and to reduce development time. In this research article, a new design methodology based on a multi-objective genetic algorithm (MOGA) for non-isolated interleaved MPCs is developed to minimize the weight, losses, and input current ripples that have a significant impact on the lifetime of the energy sources. The inductor parameters obtained from the optimization framework are verified by the finite element method (FEM) COMSOL software, which shows that inductor weight of optimized design is lower than that of the conventional design. The comparison of input current ripples and losses distribution between optimized and conventional designs are also analyzed in detail, which validates the perspective of the proposed optimization method, taking into account emerging technologies, such as wide-bandgap semiconductors (SiC, GaN) [6].

The last paper in the domain of drive trains and energy management is from Yi-Fan Jia et al. [7]. A drive system with an open-end winding permanent magnet synchronous motor (OW-PMSM) fed by a dual inverter and powered by two independent power sources is suitable for electric vehicles. By using an energy conversion device as primary power source and an energy storage element as secondary power source, this configuration can not only lower the DC-bus voltage and extend the driving range but also handle the power sharing between two power sources without a DC/DC (direct current to direct current) converter. Based on a drive system model with voltage vector distribution, this paper proposes a desired power-sharing calculation method and three different voltage vector distribution methods. By their selection strategy, the optimal voltage vector distribution method can be selected according to the operating conditions. On the basis of the integral synthesizing of the desired voltage vector, the proposed voltage vector distribution method can reduce the inverter switching frequency while making the primary power source follow its desired output power. Simulation results confirm the validity of the proposed methods, which improve the primary power source's energy efficiency by regulating its output power and lessening inverter switching loss by reducing the switching frequency. This system also provides an approach to the energy management function of electric vehicles [7].

3. Energy Storage Systems for Electric and Hybrid Vehicles

Insu Cho introduces an accurate state of charge (SOC) approach [8]. Current optimization strategy for a parallel hybrid requires much computational time and relies heavily on the drive cycle to accurately represent driving conditions in the future. With increasing application of the lithium-ion battery technology in the automotive industry, development processes and validation methods for the battery management system (BMS) have attracted attention. This paper proposes an algorithm to analyze charging characteristics and improve accuracy for determining state of charge (SOC), the equivalent of a fuel gauge for the battery pack, during the regenerative braking period of a Transmission-mounted electrical device (TMED)-type parallel hybrid electric vehicle [8].

Another SOC estimation method is proposed by Chi Zhang [9]. Accurate battery modeling is essential for the state-of-charge (SOC) estimation of electric vehicles, especially when vehicles are operated in dynamic processes. Temperature is a significant factor for battery characteristics, especially for the hysteresis phenomenon. A lack of existing literatures on the consideration of temperature influence in hysteresis voltage can result in errors in SOC estimation. Therefore, this paper gives an insight to the equivalent circuit modeling, considering the hysteresis and temperature effects. A modified one-state hysteresis equivalent circuit model is proposed for battery modeling. The characterization of hysteresis voltage versus SOC at various temperatures was acquired by experimental tests to form a static look-up table. In addition, a strong tracking filter (STF) was applied for SOC estimation. Numerical simulations and experimental tests were performed in a commercial 18650 type Li(Ni1/3Co1/3Mn1/3)O2 battery. The results were systematically compared with extended Kalman filter (EKF) and unscented Kalman filter (UKF). The results of comparison showed the following: (1) the modified model has more voltage tracking capability than the original model and (2) the modified model with STF algorithm has better accuracy, robustness against initial SOC error, voltage measurement drift, and convergence behavior than EKF and UKF [9].

In the paper of Omid Rahbari et al. [10], two techniques that are congruous with the principle of control theory are utilized to estimate the state of health (SOH) of real-life plug-in hybrid electric vehicles (PHEVs) accurately, which is of vital importance to battery management systems. The relation between the battery terminal voltage curve properties and the battery state of health is modelled via an adaptive neuron-fuzzy inference system and a group method of data handling. The comparison of the results demonstrates the capability of the proposed techniques for accurate SOH estimation. Moreover, the estimated results are compared with the direct actual measured SOH indicators using standard tests. The results indicate that the adaptive neuron-fuzzy inference system with 15 rules based on an SOH estimator has better performances over the other technique, with a 1.5% maximum error in comparison to the experimental data [10].

The impact of ageing when using various state of charge (SOC) levels for an electrified vehicle is investigated in the paper of Evelina Wikner [11]. An extensive test series is conducted on Li-ion cells, based on graphite and NMC/LMO electrode materials. Lifetime cycling tests are conducted during a period of three years in various 10% SOC intervals, during which the degradation as function of number of cycles is established. An empirical battery model is designed from the degradation trajectories of the test result. An electric vehicle model is used to derive the load profiles for the ageing model. The result showed that, when only considering ageing from different types of driving in small depth of discharges (DODs), using a reduced charge level of 50% SOC increased the lifetime expectancy of the vehicle battery by 44% to 130%. When accounting for the calendar ageing as well, this proved to be a large part of the total ageing. By keeping the battery at 15% SOC during parking and limiting the time at high SOC, the contribution from the calendar ageing could be substantially reduced [11].

The aim of this paper of Mahdi Soltani et al. [12] is to investigate the effectiveness of a hybrid energy storage system in heavy duty applications, in protecting the battery from damage due to the high-power rates during charging and discharging. Public transportation based on electric vehicles has attracted significant attention in recent years due to its lower overall emissions. Fewer charging facilities in comparison to gas stations, limited battery lifetime, and extra costs associated with its replacement present some barriers to achieving wider acceptance. A practical solution to improve the battery lifetime and driving range is to eliminate the large-magnitude pulse current flow from and to the battery during acceleration and deceleration. Hybrid energy storage systems that combine high-power (HP) and high-energy (HE) storage units can be used for this purpose. Lithium-ion capacitors (LiC) can be used as a HP storage unit, which is similar to a supercapacitor cell but with a higher rate capability, a higher energy density, and better cyclability. In this design, the LiC can provide the excess power required while the battery fails to do so. Moreover, hybridization enables a downsizing of the overall energy storage system and decreases the total cost as a consequence of lifetime, performance, and efficiency improvement. The procedure followed and presented in this paper demonstrates the good performance of the evaluated hybrid storage system in reducing the negative consequences of the power peaks associated with urban driving cycles and its ability to improve the lifespan by 16% [12].

4. Environmental Assessments of Electrified Vehicles

Benedetta Marmiroli presents a review on vehicle life-cycle assessment (LCA) studies [13]. LCAs on electric mobility are providing a plethora of diverging results. Forty-four articles published from 2008 to 2018 have been investigated in this review in order to find the extent and the reason behind this deviation. The first hurdle can be found in the goal definition followed by the modelling choice as both are generally incomplete and inconsistent. These gaps influence the choices made in the life cycle inventory (LCI) stage, particularly in regards to the selection of the electricity mix. A statistical regression is made with results available in the literature. It emerges that, despite the wide-ranging scopes and the numerous variables present in the assessments, the electricity mix's carbon intensity can explain 70% of the variability of the results. This encourages a shared framework to drive practitioners in the execution of the assessment and policy makers in the interpretation of the results [13].

Nils Hooftman et al. [14] compare the environmental impact of the combination of a 40 kWh EV and a trailer options with a range of conventional cars and EVs, differentiated per battery capacity. In this paper, they distinguish plug-in hybrid electric vehicles (PHEVs), electric vehicles (EVs) with a modest battery capacity of 40 kWh, and long-range EVs with 90 kWh installed. Given that the average motorist only rarely performs long-distance trips, both the PHEV and the 90 kWh EV are considered to be over-dimensioned for their purpose, although consumers tend to perceive the 40 kWh EV range as too limiting. Therefore, in-life range modularity by means of occasionally using a range-extender trailer for a 40 kWh EV is proposed, based on either a petrol generator as a short-term solution or a 50 kWh battery pack. A life-cycle assessment (LCA) is presented for comparing the different powertrains for their environmental impact, with the emphasis on local air quality and climate change. Therefore, the combination of a 40 kWh EV and the trailer options is benchmarked with a range of conventional cars and EVs, differentiated per battery capacity. Next, the local impact per technology is discussed on a well-to-wheel base for the specific situation in Belgium, with specific attention given to the contribution of non-exhaust emissions of particulate matter (PM) due to brake, tyre, and road wear. From a life cycle point of view, the trailer concepts outperform the 90 kWh EV for the discussed midpoint indicators as the latter is characterized by a high manufacturing impact and by a mass penalty resulting in higher contributions to non-exhaust PM formation. Compared to a petrol PHEV, both trailers are found to have higher contributions to diminished local air quality, given the relatively low use phase impact of petrol combustion. Concerning human toxicity, the impact is proportional to battery size, although the battery trailer performs better than the 90 kWh EV due to its occasional application rather than carrying along such high capacity all the time. For climate change, we see a clear advantage of both the petrol and the battery trailer, with reductions ranging from one-third to nearly 60%, respectively [14].

Funding: This editorial received no external funding.

Acknowledgments: This issue would not be possible without the contributions of various talented authors, hardworking and professional reviewers, and dedicated editorial team of Applied Sciences. Congratulations to all authors—no matter what the final decisions of the submitted manuscripts were, the feedback, comments, and suggestions from the reviewers and editors helped the authors to improve their papers.

Conflicts of Interest: The author declares no conflict of interest.

References

1. Messagie, M.; Boureima, F.S.; Coosemans, T.; Macharis, C.; Van Mierlo, J.; Mierlo, J. A Range-Based Vehicle Life Cycle Assessment Incorporating Variability in the Environmental Assessment of Different Vehicle Technologies and Fuels. *Energies* **2014**, *7*, 1467–1482. [CrossRef]
2. Berckmans, G.; Messagie, M.; Smekens, J.; Omar, N.; Vanhaverbeke, L.; Van Mierlo, J. Cost Projection of State of the Art Lithium-Ion Batteries for Electric Vehicles Up to 2030. *Energies* **2017**, *10*, 1314. [CrossRef]

3. Mierlo, J.V. The World Electric Vehicle Journal, The Open Access Journal for the e-Mobility Scene. *World Electr. Veh. J.* **2018**, *9*, 1. [CrossRef]
4. Chen, Z.; Hu, H.; Wu, Y.; Xiao, R.; Shen, J.; Liu, Y. Energy Management for a Power-Split Plug-In Hybrid Electric Vehicle Based on Reinforcement Learning. *Appl. Sci.* **2018**, *8*, 2494. [CrossRef]
5. Xiao, R.; Liu, B.; Shen, J.; Guo, N.; Yan, W.; Chen, Z. Comparisons of Energy Management Methods for a Parallel Plug-In Hybrid Electric Vehicle between the Convex Optimization and Dynamic Programming. *Appl. Sci.* **2018**, *8*, 218. [CrossRef]
6. Tran, D.; Chakraborty, S.; Lan, Y.; Van Mierlo, J.; Hegazy, O. Optimized Multiport DC/DC Converter for Vehicle Drivetrains: Topology and Design Optimization. *Appl. Sci.* **2018**, *8*, 1351. [CrossRef]
7. Jia, Y.F.; Chu, L.; Xu, N.; Li, Y.K.; Zhao, D.; Tang, X. Power Sharing and Voltage Vector Distribution Model of a Dual Inverter Open-End Winding Motor Drive System for Electric Vehicles. *Appl. Sci.* **2018**, *8*, 254. [CrossRef]
8. Cho, I.; Bae, J.; Park, J.; Lee, J. Experimental Evaluation and Prediction Algorithm Suggestion for Determining SOC of Lithium Polymer Battery in a Parallel Hybrid Electric Vehicle. *Appl. Sci.* **2018**, *8*, 1641. [CrossRef]
9. Zhang, C.; Yan, F.; Du, C.; Rizzoni, G. An Improved Model-Based Self-Adaptive Filter for Online State-of-Charge Estimation of Li-Ion Batteries. *Appl. Sci.* **2018**, *8*, 2084. [CrossRef]
10. Rahbari, O.; Mayet, C.; Omar, N.; Van Mierlo, J. Battery Aging Prediction Using Input-Time-Delayed Based on an Adaptive Neuro-Fuzzy Inference System and a Group Method of Data Handling Techniques. *Appl. Sci.* **2018**, *8*, 1301. [CrossRef]
11. Wikner, E.; Thiringer, T. Extending Battery Lifetime by Avoiding High SOC. *Appl. Sci.* **2018**, *8*, 1825. [CrossRef]
12. Soltani, M.; Ronsmans, J.; Kakihara, S.; Jaguemont, J.; Bossche, P.V.D.; Van Mierlo, J.; Omar, N. Hybrid Battery/Lithium-Ion Capacitor Energy Storage System for a Pure Electric Bus for an Urban Transportation Application. *Appl. Sci.* **2018**, *8*, 1176. [CrossRef]
13. Marmiroli, B.; Messagie, M.; Dotelli, G.; Van Mierlo, J. Electricity Generation in LCA of Electric Vehicles: A Review. *Appl. Sci.* **2018**, *8*, 1384. [CrossRef]
14. Hooftman, N.; Messagie, M.; Joint, F.; Segard, J.-B.; Coosemans, T. In-Life Range Modularity for Electric Vehicles: The Environmental Impact of a Range-Extender Trailer System. *Appl. Sci.* **2018**, *8*, 1016. [CrossRef]

Article

Energy Management for a Power-Split Plug-In Hybrid Electric Vehicle Based on Reinforcement Learning

Zheng Chen [1], Hengjie Hu [1], Yitao Wu [1], Renxin Xiao [1], Jiangwei Shen [1,*] and Yonggang Liu [2,3,*]

[1] Faculty of Transportation Engineering, Kunming University of Science and Technology, Kunming 650500, China; chen@kmust.edu.cn (Z.C.); huhengjie1995@163.com (H.H.); yitaowumail@gmail.com (Y.W.); xrx1127@foxmail.com (R.X.)

[2] State Key Laboratory of Mechanical Transmissions, Chongqing University, Chongqing 400044, China

[3] School of Automotive Engineering, Chongqing University, Chongqing 400044, China

* Correspondence: shenjiangwei6@163.com (J.S.); andyliuyg@cqu.edu.cn (Y.L.)

Received: 24 October 2018; Accepted: 30 November 2018; Published: 4 December 2018

Abstract: This paper proposes an energy management strategy for a power-split plug-in hybrid electric vehicle (PHEV) based on reinforcement learning (RL). Firstly, a control-oriented power-split PHEV model is built, and then the RL method is employed based on the Markov Decision Process (MDP) to find the optimal solution according to the built model. During the strategy search, several different standard driving schedules are chosen, and the transfer probability of the power demand is derived based on the Markov chain. Accordingly, the optimal control strategy is found by the Q-learning (QL) algorithm, which can decide suitable energy allocation between the gasoline engine and the battery pack. Simulation results indicate that the RL-based control strategy could not only lessen fuel consumption under different driving cycles, but also limit the maximum discharge power of battery, compared with the charging depletion/charging sustaining (CD/CS) method and the equivalent consumption minimization strategy (ECMS).

Keywords: energy management strategy; Markov decision process (MDP); plug-in hybrid electric vehicles (PHEVs); Q-learning (QL); reinforcement learning (RL)

---

1. Introduction

In recent years, as the greenhouse effect and air pollution have become increasingly severe, green energy attracts more attention in all walks of life. In automotive industry, exhaust emission from conventional fuel vehicles is an important factor that causes the environmental pollution. Developing new energy vehicles (NEVs) has shown its significance in reducing emission and lessening induced air pollution. Currently, NEVs can be mainly classified into three types, i.e., fuel cell vehicles, battery electric vehicles (BEVs) and hybrid electric vehicles (HEVs), and they are usually equipped with an energy storage system, such as a battery pack or a super-capacitor [1,2]. For BEVs, it can be powered purely by the battery pack or the super-capacitor. Plug-in hybrid electric vehicles (PHEVs) are considered to combine advantages of both BEVs and HEVs [3]. Compared with HEVs, the prominent advantage of PHEVs is that the battery pack can be recharged by the external charging plug, thereby supplying certain all electric range (AER). Compared with BEVs, the controller of PHEVs can start the engine to sustain the battery when a certain battery state of charge (SOC) threshold is reached and meanwhile supply the extended driving range. Consequently, it is critical to manage the power distribution between the battery and the engine properly in PHEVs.

Energy management strategy (EMS) of PHEVs is responsible for power and energy distribution among different energy storage systems, such as gasoline engine and electromotor. Different control tradeoff of energy management target is mentioned in related literatures [4,5] including fuel economy improvement [6], and tailpipe emission reduction [7]. Rule based and optimization based methods are mostly considered, as discussed by the authors of [8]. Rule based methods are relatively easier to exploit and are widely employed in practice [9,10]. In [9], a classified rule based EMS is designed, which emphasizes on different operating modes of PHEVs, and simulation results yields satisfied emission reduction. However, these rule based strategies highly depend on design process and engineering experience, thus leading to longer design time [11]. On the contrary, modern real-time and global optimization based algorithms can be applied with provable optimal guarantee. In particular, dynamic programming (DP), adopted by many researchers, is generally treated as an emblematic algorithm among all the optimal methods [12–15]. In [12], the investigators proposed an intelligent EMS based on DP, by which numerical simulation results manifest the improved fuel economy dramatically. Quadratic programming (QP) is also a mature algorithm to search for the optimal result with affordable operational budgets [16], compared with DP. Pontryagin minimum principle (PMP) [17] and equivalent consumption minimization strategy (ECMS) [18] are also widely adopted in EMS of PHEVs. In addition, model predictive control (MPC) [19], is extensively investigated as a real-time optimization manner applying to EMS of PHEVs. Furthermore, intelligent algorithms such as simulated annealing (SA) optimization [17], neural network (NN) [20], genetic algorithm (GA) [21] are also employed for EMS of PHEVs in recent years.

Nowadays, with development of artificial intelligence (AI) technology, reinforcement learning (RL) is becoming more and more popular in various fields including robotic control, intelligent system, and energy management of power grids. In [22], a parallel control architecture based on the RL technology is applied for robotic manipulation, thereby enabling robots to easily adapt to the environment variation. RL is also introduced in the field of energy management of PHEVs in [23–30]. In [23], the investigators find that the RL based EMS cannot only guarantee the vehicle dynamic performance, but also improve the fuel economy, and as a result, can outperform stochastic dynamic program (SDP) in terms of adaptability and learning ability. In [24], the Kullback–Leibler (KL) divergence technique is applied to calculate the power transition probability matrices of the RL algorithm to find the optimal power distribution ratio between the battery and the super-capacitor. Simulation results show that this kind of control policy cannot only effectively decrease the battery charging frequency and control the maximum discharging current, but also maximize the energy efficiency to cut down the overall cost under diverse conditions. In [25], a novel RL based method is proposed combining with the remaining travel distance estimation, and the controller could continuously search for the optimal strategy and learn from the previous process. In [26], a RL method called TD (λ)-learning is employed for the HEV, and simulation results manifest that the RL based policy can improve the fuel economy by 42%. In [27], a blended real-time control strategy is proposed based on the Q-learning (QL) method to balance the overall performance and optimality. A bi-level control strategy is proposed in [28], in which the fuzzy encoding predictor and the KL divergence rate are employed to predict the driver's power demand in the higher level, and the lower level is mainly focused on employing the RL algorithm to find the optimal solution.

Based on the above discussion, it is imperative to further apply the RL technique for energy management of power-split PHEVs. Hence, the main motivation of the energy management strategy is to further refine the battery power based on the RL by selecting proper state and action variables. As a result, the objectives for both optimal fuel economy and battery power restriction can be met at the same time, thereby prolonging the battery life potentially. For the sake of achieving the target, the powertrain of a power-split PHEV is modeled and analyzed first. Subsequently, considering that the proposed method should be applicable in most driving conditions, the Markov chain is adopted to estimate the transition probability matrix regarding demanded power under different driving cycles. Finally, the QL algorithm is conducted to develop and finally form the EMS towards reaching the optimal target. Furthermore, the proposed EMS is compared with the CD/CS strategy to validate the optimality under different driving cycles by simulations. The rest of this article is structured as follows: Section 2 describes the simplified vehicle structure and the fuel consumption model. In Section 3, the

RL based framework is proposed to realize the optimal EMS. In Section 4, corresponding simulations prove the proposed method is superior to the CD/CS algorithm. Section 5 concludes the article.

2. PHEV Powertrain Model

In this paper, the model under study is a power-split PHEV derived from Autonomie. A typical power-split PHEV model is the Toyota Prius PHEV. The powertrain structure of the vehicle is shown in Figure 1, which consists of a 39 ampere-hour (Ah) traction battery pack, a gasoline engine, a final drive, a planetary transmission and two electric motors, i.e., Motor 1 and Motor 2. The engine, Motor 1 and Motor 2 connect with the planet carrier, the ring gear and the sun gear, respectively. As can be seen in Figure 1, motor 2 is employed to provide a significant portion of the electric power, and motor 1 is mainly used as a generator. The main parameters are listed in Table 1.

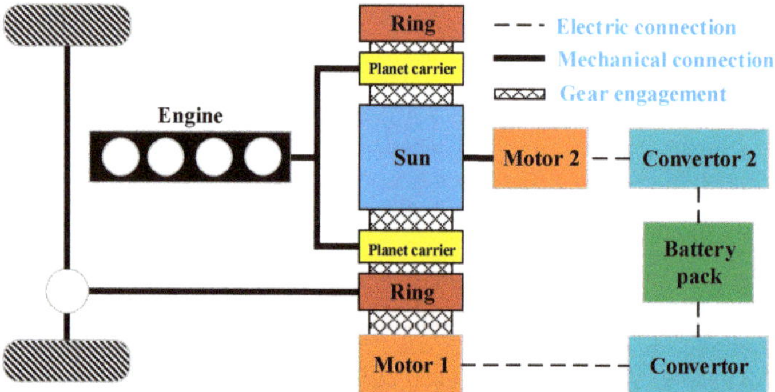

Figure 1. Power-Split plug-in hybrid electric vehicle (PHEV) powertrain structure.

Table 1. Main parameters of power-split PHEV.

Parts	Parameters	Value
Vehicle	Mass	1801 kg
Battery	Rated capacity	39 Ah
Motor 1	Peak power	50 kW
	Rated power	25 kW
Motor 2	Peak power	30 kW
	Rated power	15 kW
Engine	Rated power	57 kW
Planetary gear set	Sun gear	30
	Ring gear	78

2.1. Energy Management Problem

This paper focuses on minimizing the total fuel consumption. Hence, the fuel index β can be established as,

$$\beta = \min F_{total} = \min \int_0^T F_{rate} dt \tag{1}$$

where F_{total} is the total fuel consumption, F_{rate} donotes the fuel rate. T is the total driving time. For the sake of calculating the fuel rate by appropriate simplification, F_{rate} can be determined as,

$$F_{rate} = f(T_{eng}, \omega_{eng}) \tag{2}$$

where ω_{eng}, T_{eng} denote the speed and the torque of engine, respectively. To minimize the fuel consumption, the relationship between the vehicle power request and the fuel consumption needs to be analyzed in detail.

2.2. Power Request Model

Given a certain driving cycle, the power required to drive the vehicle powertrain can be calculated as,

$$P_{req} = (F_f + F_w + F_i)v \tag{3}$$

where P_{req} is the vehicle request power, F_f, F_w, and F_i represent the resistance derived from the road, air drag and vehicle inertial, respectively. v denotes the driving velocity. The resistances, that merely associated with vehicle and environment parameters, can be expressed as,

$$\begin{cases} F_f = mgf \\ F_w = C_d A v^2 / 21.15 \\ F_i = \delta mg \end{cases} \tag{4}$$

where m is the total mass, f denotes the road resistance coefficient, g is the gravity coefficient, A is the frontal area of the vehicle, C_d is the aerodynamic drag coefficient, and δ is the rotational mass coefficient. As shown in Figure 1, the power flow equations can be formulated to describe the corresponding power flow, as:

$$\begin{cases} P_{req} = P_{final} \cdot \eta_{final} \\ P_{final} = (P_{mot1} + P_{mot2} + P_{eng}) \cdot \eta_{gear} \\ P_{bat} = (P_{mot1}/\eta_{c1} + P_{mot2}/\eta_{c2}) + P_{acc} \\ P_{eng} = f_{eng}(T_{eng}, \omega_{eng}) \end{cases} \tag{5}$$

where P_{final} is the driveline power, P_{mot1}, P_{mot2}, and P_{eng} are the output power of motor 1, motor 2 and engine, respectively. P_{acc} denotes the power of electric accessories and is assumed to be a constant value, i.e., 220 W. η_{gear}, η_{final} and η_c are the transmission efficiency factor of gear, final drive and electric convertor, respectively. As seen in Figure 1, the planetary gear set works as the coupling device that connects the engine and the motors, and the corresponding dynamic equations are expressed as follows:

$$\begin{cases} \omega_{eng} = \frac{1}{1+i_{gear}}\omega_{mot2} + \frac{i_{gear}}{1+i_{gear}}\omega_{mot1} \\ T_{eng} = -(1+i_{gear})T_{mot2} = -\frac{1+i_{gear}}{i_{gear}}T_{mot1} \\ \omega_{ring} = \omega_{mot1} = \frac{v}{r_{whl}}r_{final} \end{cases} \tag{6}$$

where i_{gear} is the transmission ratio of the planetary gear, ω_{mot1}, ω_{mot2}, and ω_{ring} are the speed of motor 1, motor 2 and ring gear, respectively; T_{mot1} and T_{mot2} are the torque of two motors; r_{whl} denotes the radius of the wheel and r_{final} is the final driveline ratio. In this article, we choose to ignore the inertial of planet gear, sun gear and ring gear for ease of managing the energy distribution.

Based on the above descriptions, the instantaneous fuel consumption F_{rate} can be redefined as:

$$F_{rate} = f(T_{eng}, \omega_{eng}) = f(P_{bat}, P_{req}, v) \tag{7}$$

Now we can find that F_{rate} can be directly determined by P_{bat}, thus it is necessary to model the battery and analyze its power relationship.

2.3. Battery Model

To analyze the power relationship of the battery, a simplified battery model is presented here, which consists of an internal resistor and an open circuit voltage source, and the corresponding calculation equations of the battery model can be described as:

$$
\begin{cases}
P_{bat} = OCV \cdot i_{bat} - i_{bat}^2 R_{int} \\
i_{bat} = \dfrac{OCV - \sqrt{OCV^2 - 4R_{int}P_{bat}}}{2R_{int}} \\
SOC = SOC_{init} - \dfrac{1}{C_{bat}} \int_0^t i_{bat} dt
\end{cases}
\tag{8}
$$

where OCV denotes the battery open circuit voltage, i_{bat} is the battery current, R_{int} is the battery internal resistance, C_{bat} is the battery capacity, SOC is the battery SOC and SOC_{init} is its initial value. Detailed battery parameters varying with SOC are shown in Figure 2. It can be found that R_{int} decreases from 0.1403 ohm to 0.09 ohm and OCV ranges from 165 V to 219.7 V.

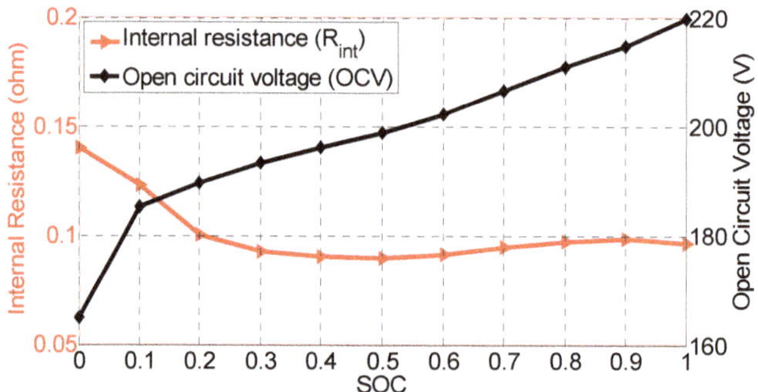

Figure 2. OCV and R_{int} variation with state of charge (SOC).

From the above analysis, we can find that if the battery power is predetermined, the energy distribution strategy inside the vehicle can be achieved. By this manner, the control strategy distributions can be ascertained by the battery power. In order to ensure safety of all components and consider their power limitations and performance extension, some constraint conditions are imposed:

$$
\begin{cases}
P_{bat\_min} \leq P_{bat} \leq P_{bat\_max} \\
P_{mot1\_min} \leq P_{mot1} \leq P_{mot1\_max} \\
P_{mot2\_min} \leq P_{mot2} \leq P_{mot2\_max} \\
P_{eng\_min} \leq P_{eng} \leq P_{eng\_max} \\
P_{req\_min} \leq P_{req} \leq P_{req\_max} \\
SOC_{min} \leq SOC \leq SOC_{max}
\end{cases}
\tag{9}
$$

where parameters with subscripts min and max mean their corresponding minimum and maximum values, respectively. In the next step, the RL based strategy is introduced to achieve the energy management of the PHEV.

3. Reinforcement Learning for Energy Management

To apply the RL for energy management of PHEVs, we need to build the vehicle power transition probability model first.

3.1. Transition Probablity Model

Markov chain model is a discrete time and state stochastic process with Markov property, of which the state is a sequence with multiple finite random variables. In this process, the selection of the next state is related to the current state and the current action, and does not show any relationship with the previous historical state. In addition, the change of state is independent of time, but is transferred by probability. According to the finite-state Markov chain driver model introduced in [31], the actual driving cycle can be considered as the stochastic Markov chain. The request power is treated as a stochastic variable and can be modeled by the Markov chain. To obtain the transition probability matrix, several standard driving cycles shown in Figure 3 are recorded and analyzed to estimate the transition probability matrix of the demanded power. The selected driving cycles not only include urban, suburban and highway driving conditions, but also involve some intense speed profiles, of which the velocity scale, the acceleration and deceleration frequency can cover most of the driving conditions. According to speed profiles of partially selected driving cycles depicted in Figure 3, the transition probability of the demand power can be calculated based on the maximum likelihood estimation, as:

$$\begin{cases} p_{s,s'} = \dfrac{n_{s,s'}}{n_s} \\ n_s = \sum\limits_{k=1}^{K} p_{s,s'k} \end{cases} \tag{10}$$

where $n_{s,s'}$ represents the counted number transiting from s to s', and n_s is the total number for all transitions of s. $p_{s,s'}$ means the transition probability of the driver's power demand transferred from the current moment to the next moment at each velocity state.

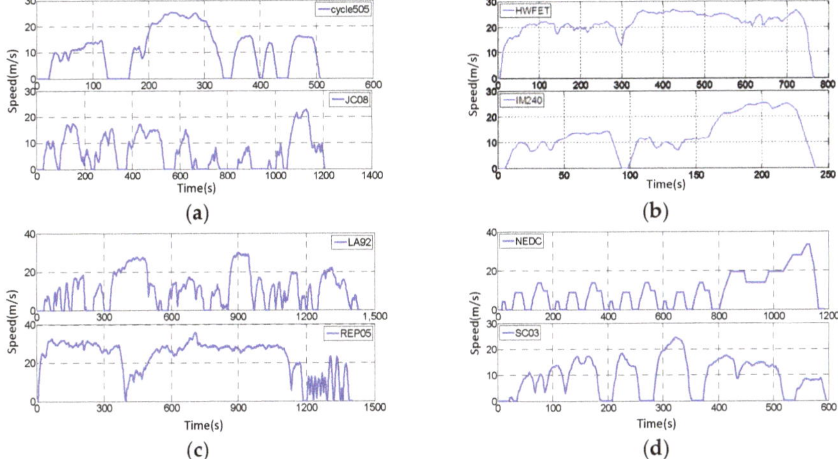

Figure 3. Drive cycle curves: (**a**) Cycle505 and JC08 cycles; (**b**) Highway Fuel Economy Test (HWFET) and IM240 cycles; (**c**) LA92 and REP05 cycles; and (**d**) New European Driving Cycle (NEDC) and SC03 cycles.

According to calculation based on the Markov chain, the transition probability matrix for vehicle speed of 30 km/h and 80 km/h are shown in Figure 4. It can be found that the request power scope is from −40 kW to 40 kW at speed of 30 km/h and the request power scope is from −80 kW to 80 kW at speed of 80 km/h. The transition probability is limited within 0.1 to 0.7, and most of the distribution is concentrated on a diagonal. In addition, it can be clearly seen from Figure 4 that the transiting probability of power request moving from the current state to the next state with different speed values is obviously different.

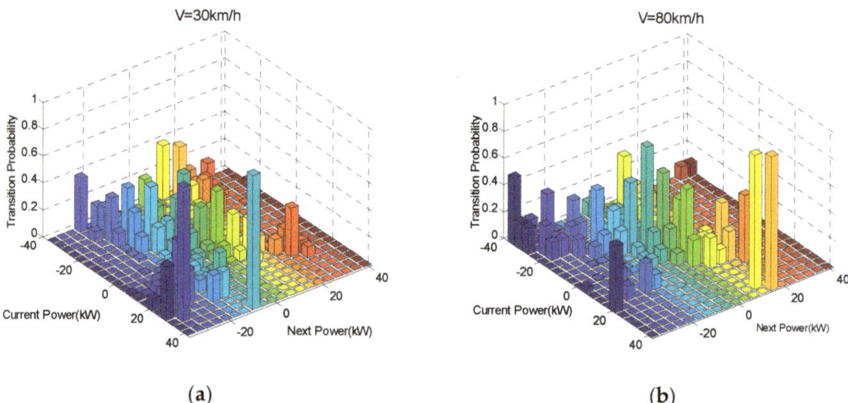

Figure 4. The transition probability map. (**a**) The transition probability map at V = 30 km/h; (**b**) The transition probability map at V = 80 km/h.

3.2. Reinforcement Learning Algorithm

RL, as a significant machine learning method, can conduct repeated explorations in which the agent takes a series of actions in its environment to maximize its designated benefits. The agent-environment interaction for RL is illustrated in Figure 5.

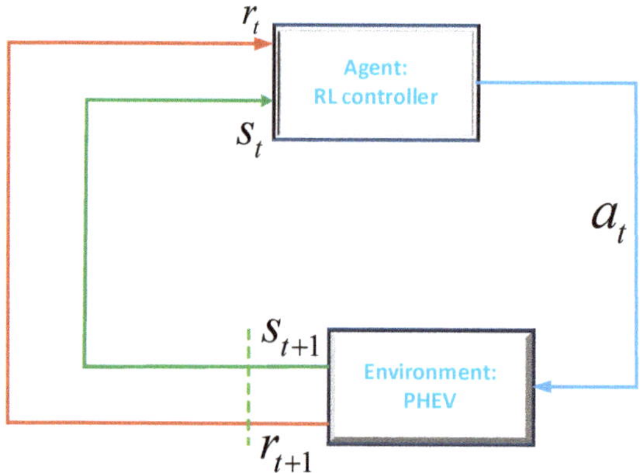

Figure 5. The agent-environment interaction.

The agent-environment interaction can be regarded as a Markov decision process, and the RL mainly focuses on solving the Markov decision process based on a series of iteration. In this paper, the state variable $s \in S$ includes the power request, SOC and the vehicle speed and the action variable $a \in A$ is the battery power. The reward function r, which evaluates the current action, is defined as the immediate fuel consumption of the engine.

The object function could be written as the total reward for the finite future at each state, which can be described as:

$$V^*(s) = E\left(\sum_{t=0}^{\infty} \gamma^t r_t\right) \tag{11}$$

where $\gamma \in [0,1]$ is the discount factor to guarantee convergence of the agent during the learning process. Since any state is different and each state is unique, the object function can be reformulated as:

$$V^*(s) = \min_{a \in A}(r(s,a) + \gamma \sum_{s' \in S} p_{sa,s'} V^*(s')) \tag{12}$$

where $p_{sa,s'}$ indicates the transition probability of state variables that change from s to s' based on action a, and $r(s,a)$ indicates the reward of applying action a to transfer from s to s'.

The optimal control strategy is determined by Bellman's principle:

$$\pi^*(s) = \operatorname*{argmin}_{a}(r(s,a) + \gamma \sum_{s' \in S} p_{sa,s'} V^*(s')) \tag{13}$$

As a popular candidate of RL algorithms, the QL algorithm is simple and easy to implement [32], and has been widely employed to solve the optimal value function of MDP. The QL algorithm can obtain a strategy to maximize the sum of expected discounted rewards by directly optimizing an iterated value function Q. According to the updated Q value, the agent needs to examine every action in each iteration to make sure that the learning process can converge. In terms of these merits, we employ the QL algorithm as the kernel algorithm to train, learn and finally achieve the energy management of PHEVs. In the QL algorithm, the Q value, i.e., the state-action value, can be written as:

$$Q^*(s,a) = r(s,a) + \gamma \sum_{s' \in S} p_{sa,s'} \min_{a} Q^*(s',a') \tag{14}$$

Furthermore, the updated rule of Q value can be described as:

$$Q(s,a) \leftarrow Q(s,a) + \eta(r + \gamma \min_{a'} Q(s',a') - Q(s,a)) \tag{15}$$

where $\eta \in [0,1]$ is a decaying factor.

According to the above discussion, the proposed method consists of a simplified vehicle model, a transition probability matrix, a reward matrix and the QL control strategy, where the reward matrix is computed via the simplified vehicle model and the control strategy is calculated according to the power transition matrix, the reward matrix and the QL algorithm feedback. Table 2 lists the pseudocode of the QL algorithm, and it can clearly illustrate the iterative process of QL algorithm. The optimal control strategy is derived through the iterative process shown in Table 2. Figure 6 summarized the detailed procedures of QL in Matlab [19]. First, the QL algorithm and the MDP as well as the related parameters are combined and discretized. Then, the power transition matrix is calculated based on the driver model. Based on the discrete variables and the simplified PHEV model, the reward matrix R is calculated. After iteration, the QL algorithm can be applied successfully to find the optimal energy management solution.

Table 2. The pseudocode of Q–Learning (QL) algorithm.

The QL Algorithm Framework
1. Arbitrarily initialize Q(s,a), S
2. Repeat each step
3. According to the Q(s,a) (ε-greedy policy), choose A
4. Take action A, observe R, S'
5. Update the Q(s,a), $S \leftarrow s'$
6. Until S is terminal

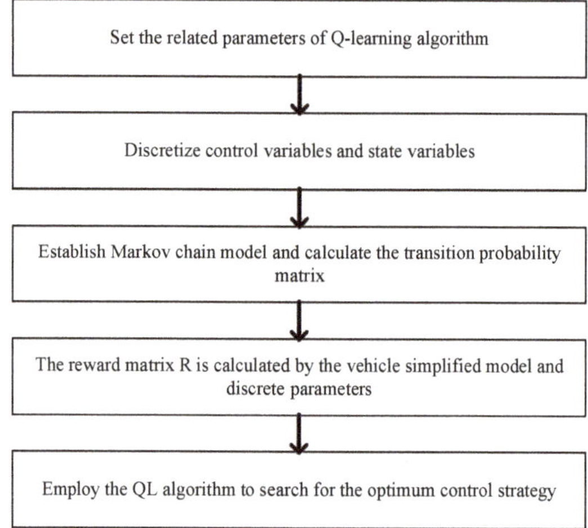

Figure 6. Procedures of the QL calculation.

The optimal control strategy based on the RL algorithm is shown in Figure 7. The battery power ranges from −12 kW to 12 kW, the required power range is limited within −45 kW to 45 kW, and the SOC ranges from 0.3 to 0.9. It can be found that the optimal battery power can be determined by state variables, i.e., the required power, SOC and the vehicle speed. Figure 8 shows the convergence process of the QL algorithm, where the mean discrepancy is applied to measure the difference of the Q values. We can find that with increase of the iterations, the mean discrepancy gradually decreases to 0. From this point, the effectiveness and convergence of the QL algorithm can be proved.

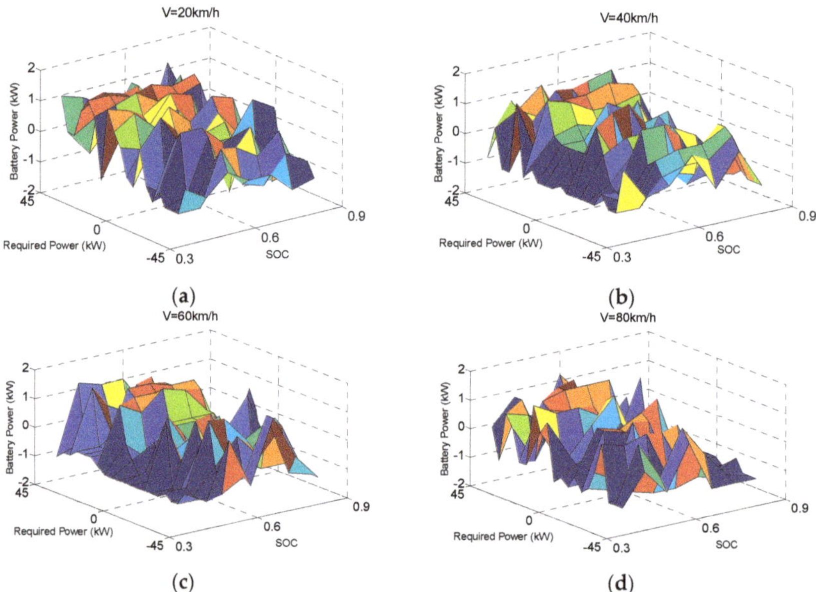

Figure 7. Optimal control strategy based on RL algorithm with different speeds. (**a**) The optimal control action variable at V = 20 km/h; (**b**) The optimal control action variable at V = 40 km/h; (**c**) The optimal control action variable at V = 60 km/h; and (**d**) The optimal control action variable at V = 80 km/h.

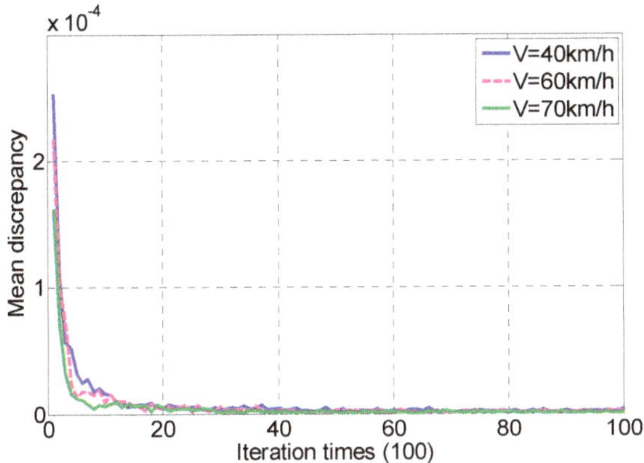

Figure 8. Mean discrepancy of the Q-values.

4. Simulation and Result

In this article, simulations are conducted based on the Autonomie and Matlab/Simulink. New European Driving Cycle (NEDC), Highway Fuel Economy Test (HWFET) and Urban Dynamometer Driving Schedule (UDDS), shown in Figure 9, are employed to verify the proposed strategy. The selected driving cycles can represent most of the driving pattern under different driving conditions.

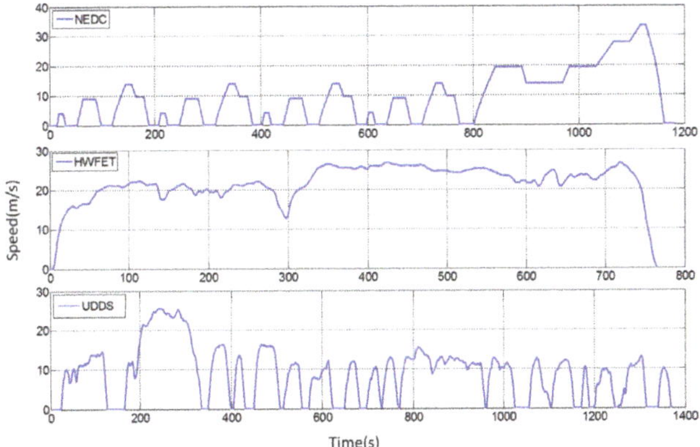

Figure 9. Profile of simulation driving cycles.

To compare the performance of the proposed method, the charge depletion/charge sustaining (CD/CS) algorithm is introduced as a benchmark, which is widely employed in actual applications. In addition, the ECMS is also employed to compare the performance of the proposed algorithm. For the CD/DS algorithm, the power distribution of the vehicle can be easily achieved by setting a series of control parameters without any pre-known information of driving conditions. During the CD stage, except for some specific situation, the engine generally remains shut down, and the tractive power is mainly provided by the battery until the SOC drops to a specified lower threshold (e.g., 30%). Then, the vehicle is powered by both the engine and the battery to remain SOC near the specified value under the CS stage. The detailed CD/CS control scheme can be described [12] as:

$$P_{bat} = \begin{cases} P_{req} & SOC > 36\% \\ \min(27804.9, P_{req}) & 33\% \leq SOC \leq 36\% \\ \min(27804.9 \cdot (SOC - 0.3)/0.03, P_{req}) & 30\% \leq SOC \leq 33\% \\ \max(-28157.5 \cdot (SOC - 0.3)/0.03, P_{req}) & P_{req} < 0, 27\% \leq SOC \leq 30\% \\ \max(-28157.5 \cdot (SOC - 0.3)/0.03, P_{req} - P_{eng\_max}) & P_{req} > 0, 27\% \leq SOC \leq 30\% \\ \max(-28157.5, P_{req}) & P_{req} < 0, SOC < 27\% \\ \max(-28157.5, P_{req} - P_{eng\_max}) & P_{req} > 0, SOC < 27\% \end{cases} \tag{16}$$

where P_{eng_max} represents the maximum power of engine.

The ECMS algorithm, as a classical real-time optimization algorithm, transfers the electric consumption of the battery to the equivalent fuel consumption and then tries to minimize the fuel consumption. During each time constant, the vehicle power request is distributed to the battery and the engine according to the minimum principle. By this way, the whole fuel consumption can be reduced and the fuel economy can be improved simultaneously. A typical solution of the ECMS can be formulated based on the Hamilton function, as:

$$H(x(t), u(t), \lambda(t), t) = F_{rate\_eng}(u(t), t) + \lambda \cdot f(x(t), u(t), t) \tag{17}$$

where λ is an equivalent factor that can be adjusted dynamically or can be fixed as a constant value. $x(t)$ and $u(t)$ are state variables and control variables, respectively. In this paper, $x(t)$ includes the battery SOC, the vehicle power demand, and the vehicle speed. Similar to before, $u(t)$ is the battery power. By solving (17), the optimal solution can be found and the final fuel consumption can be obtained.

In simulation validation, three standard cycles are selected to splice multifarious and verifiable conditions. Cycle 1 is consisted of two NEDC cycles, one UDDS cycle and two HWFET cycles, Cycle 2 is comprised of two UDDS cycles, two NEDC cycles and two HWFET cycles, and Cycles 3 and 4 includes five and six HWFET cycles. Cycles 5 and 6 are consisted of six and seven UDDS cycles, respectively. The fuel consumption results with the SOC correction [33] are listed in Table 3. It can be found that compared with the CD/CS scheme, the RL based control strategy can effectively reduce the fuel consumption by 10.1%, 9.31%, 4.84%, 4.49%, 5.95% and 5.13% under different driving cycles. Compared with the ECMS, the RL algorithm can gain similar fuel consumption savings. Thus, the validity of RL based algorithm can be proved. More intuitively, Figure 10 shows the battery power comparison with respect to the proposed algorithm, the ECMS and the CD/CS scheme. The power range of the battery based on the RL algorithm is from -12 kW to 12 kW, while the battery power based on the CD/CS algorithm ranges from -30 kW to 5 kW. It can be recognized that the EMS based on the RL algorithm is capable of controlling the range of the battery power variation smaller than that of the CD/CS method, and the RL method can restrict the maximum battery discharge power. Here we can conclude that the EMS based on the RL control strategy can protect the battery and extend the battery life to some extent.

Figure 11 shows the SOC curve under different driving cycles. The initial SOC is supposed to be 90%, and the minimum SOC threshold is 30%. Compared with the results of the CD/CS scheme, the SOC downward trend based on the RL method is more smoothly. Figure 12 illustrates the fuel consumption under four driving cycles. According to Figures 11 and 12, we can find that the optimized control strategy does not take effect completely in the entire cycle, and works before the battery SOC drops to a certain value. Even so, the proposed algorithm can still effectively reduce the fuel consumption.

To further discover improvements of the RL based strategy, the engine operating points for both RL based method and the CD/CS method under four driving cycles are depicted in Figure 13. It can be obviously found that by implementing the RL based algorithm, the engine working efficiency is higher than 30% in most cases. Compared with the CD/CS strategy, the proposed method can make the engine working points more densely in the high efficiency area. Moreover, it can be noticed that based on the RL based method, the majority of engine working points gather near the optimal operating line, not like that by the CD/CS algorithm. Therefore, it can explain that why the fuel consumption based on the proposed method is less than that based on the CD/CS method.

Table 3. Fuel economy comparison.

Driving Cycle	Strategy	Ending SOC (%)	Fuel Consumption (kg)	Saving (%)
	CD/CS	30.57	1.3205	-
Cycle 1	ECMS	30.21	1.2061	8.39
	RL method	30.45	1.1851	10.1
	CD/CS	30.57	1.7374	-
Cycle 2	ECMS	30.21	1.6067	7.15
	RL method	30.25	1.5702	9.31
	CD/CS	30.57	1.9951	-
Cycle 3	ECMS	30.21	1.8734	5.78
	RL method	30.25	1.8930	4.84
	CD/CS	30.57	2.6360	-
Cycle 4	ECMS	30.21	2.4819	5.60
	RL method	30.25	2.5121	4.49
	CD/CS	30.14	1.2574	-
Cycle 5	ECMS	29.31	1.1681	5.91
	RL method	29.33	1.1688	5.95
	CD/CS	30.14	1.6551	-
Cycle 6	ECMS	2931	1.5488	5.52
	RL method	29.32	1.5563	5.13

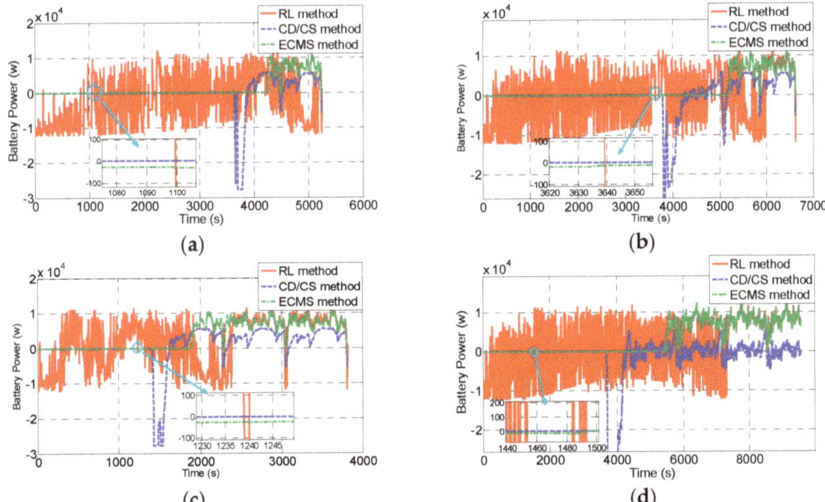

Figure 10. Battery power comparison under driving cycles. (**a**) Cycle 1; (**b**) Cycle 2; (**c**) Cycle 3; and (**d**) Cycle 6.

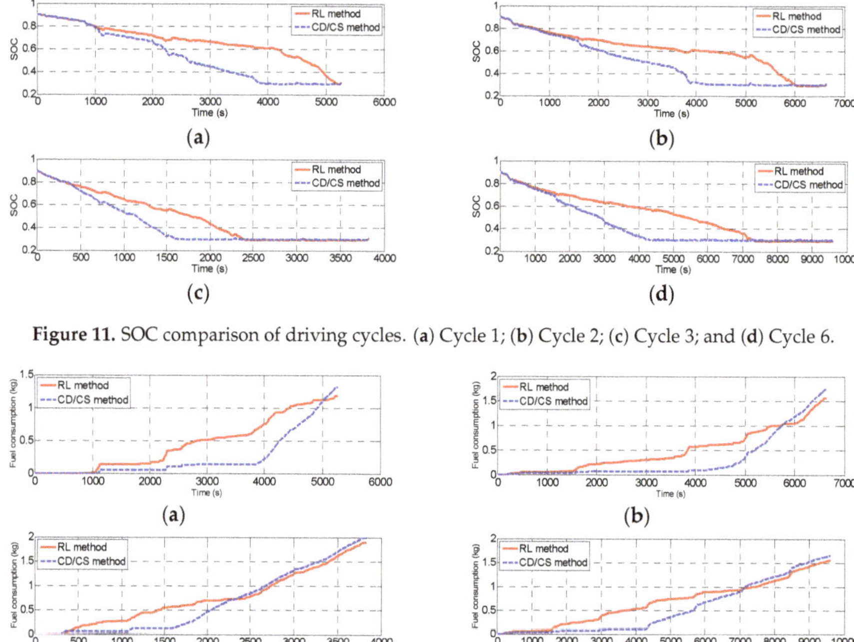

Figure 11. SOC comparison of driving cycles. (**a**) Cycle 1; (**b**) Cycle 2; (**c**) Cycle 3; and (**d**) Cycle 6.

Figure 12. Fuel consumption results. (**a**) Cycle 1; (**b**) Cycle 2; (**c**) Cycle 3; and (**d**) Cycle 6.

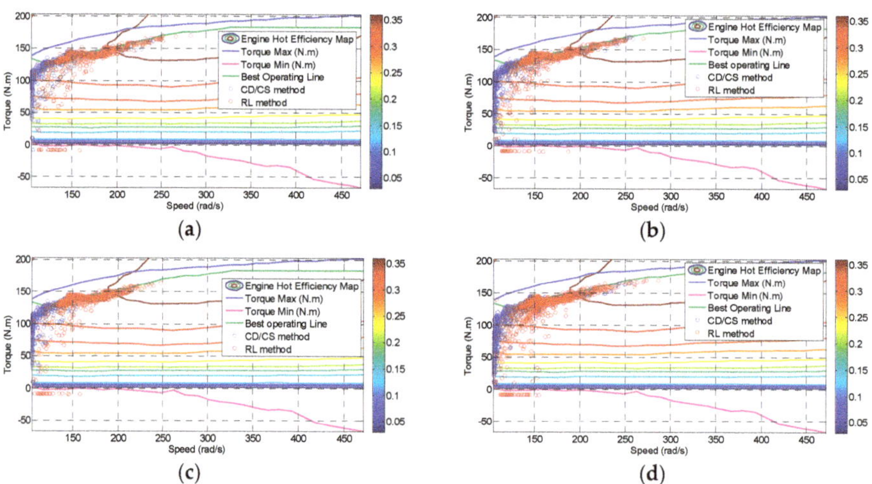

Figure 13. Engine hot efficiency results. (**a**) Cycle 1; (**b**) Cycle 2; (**c**) Cycle 3; and (**d**) Cycle 6.

5. Conclusions

In this paper, the Q-learning RL algorithm has been employed for the energy management of a power-split PHEV. The mathematical vehicle model is built after detailed powertrain analysis. By combining Q-learning method with MDP, the RL model of PHEV is constructed and the optimal result based on RL is obtained where the battery power is optimized. Three standard driving cycles are chosen for simulation verification. Simulation results manifest that the proposed RL algorithm can guarantee a preferable fuel consumption and show more effectiveness than the CD/CS algorithm. In addition, the proposed algorithm can restrict the battery current within a narrower range, thus extending the battery life cycle to some extent.

Our next step work will focus on exploring a more stable Markov chain model and more advanced optimization algorithm. In addition, the proposed algorithm will be further investigated to update the transition probability matrix of the Markov driver chain in real time, and hardware-in-the-loop and actual vehicle validation will be conducted to verify the real control performance of the proposed method.

Author Contributions: Z.C. and H.H. drafted this paper, discussed combine reinforcement learning with Markov decision process. Y.W. and R.X. provided some energy management strategy suggestions. J.S. oversaw the research. Y.L. revised the paper and provide some technical help.

Funding: This work is supported by the National Science Foundation of China (Grant No. 61763021 and 51775063) in part and the National Key Research and Design Program of China (Grant No. 2018YFB0104000 and 2018YFB0104900) in part. Most importantly, the authors would also like to thank the anonymous reviewers for their valuable comments and suggestions.

Conflicts of Interest: The authors declare no conflict of interest.

References

1. Repp, S.; Harputlu, E.; Gurgen, S.; Castellano, M.; Kremer, N.; Pompe, N.; Woerner, J.; Hoffmann, A.; Thomann, R.; Emen, F.M. Synergetic effects of Fe^{3+} doped spinel $Li_4Ti_5O_{12}$ nanoparticles on reduced graphene oxide for high surface electrode hybrid supercapacitors. *Nanoscale* **2018**, *10*, 1877–1884. [CrossRef] [PubMed]

2. Genc, R.; Alas, M.O.; Harputlu, E.; Repp, S.; Kremer, N.; Castellano, M.; Colak, S.G.; Ocakoglu, K.; Erdem, E.J.S.R. High-Capacitance Hybrid Supercapacitor Based on Multi-Colored Fluorescent Carbon-Dots. *Sci. Rep.* **2017**, *7*, 11222. [CrossRef] [PubMed]

3. Martinez, C.M.; Hu, X.; Cao, D.; Velenis, E.; Gao, B.; Wellers, M. Energy Management in Plug-in Hybrid Electric Vehicles: Recent Progress and a Connected Vehicles Perspective. *IEEE Trans. Veh. Technol.* **2017**, *66*, 4534–4549. [CrossRef]

4. Trovao, J.P.F.; Santos, V.D.N.; Antunes, C.H.; Pereirinha, P.G.; Jorge, H.M. A Real-Time Energy Management Architecture for Multisource Electric Vehicles. *IEEE Trans. Ind. Electron.* **2015**, *62*, 3223–3233. [CrossRef]

5. Lu, X.; Sun, K.; Guerrero, J.M.; Vasquez, J.C.; Huang, L. State-of-Charge Balance Using Adaptive Droop Control for Distributed Energy Storage Systems in DC Microgrid Applications. *IEEE Trans. Ind. Electron.* **2014**, *61*, 2804–2815. [CrossRef]

6. Sabri, M.F.M.; Danapalasingam, K.A.; Rahmat, M.F. A review on hybrid electric vehicles architecture and energy management strategies. *Renew. Sustain. Energy Rev.* **2016**, *53*, 1433–1442. [CrossRef]

7. Hu, X.; Martinez, C.M.; Yang, Y. Charging, power management, and battery degradation mitigation in plug-in hybrid electric vehicles: A unified cost-optimal approach. *Mech. Syst. Signal Process.* **2017**, *87*, 4–16. [CrossRef]

8. Feng, T.; Yang, L.; Gu, Q.; Hu, Y.; Yan, T.; Yan, B. A Supervisory Control Strategy for Plug-In Hybrid Electric Vehicles Based on Energy Demand Prediction and Route Preview. *IEEE Trans. Veh. Technol.* **2015**, *64*, 1691–1700. [CrossRef]

9. Wirasingha, S.G.; Emadi, A. Classification and Review of Control Strategies for Plug-In Hybrid Electric Vehicles. *IEEE Trans. Veh. Technol.* **2011**, *60*, 111–122. [CrossRef]

10. Peng, J.; He, H.; Xiong, R. Rule based energy management strategy for a series-parallel plug-in hybrid electric bus optimized by dynamic programming. *Appl. Energy* **2017**, *185*, 1633–1643. [CrossRef]

11. Gao, Y.; Ehsani, M. Design and Control Methodology of Plug-in Hybrid Electric Vehicles. *IEEE Trans. Ind. Electron.* **2010**, *57*, 633–640. [CrossRef]

12. Chen, Z.; Mi, C.C.; Xu, J.; Gong, X.; You, C. Energy Management for a Power-Split Plug-in Hybrid Electric Vehicle Based on Dynamic Programming and Neural Networks. *IEEE Trans. Veh. Technol.* **2014**, *63*, 1567–1580. [CrossRef]

13. Wu, J.; He, H.; Peng, J.; Li, Y.; Li, Z. Continuous reinforcement learning of energy management with deep Q network for a power split hybrid electric bus. *Appl. Energy* **2018**, *222*, 799–811. [CrossRef]

14. Zhang, S.; Xiong, R. Adaptive energy management of a plug-in hybrid electric vehicle based on driving pattern recognition and dynamic programming. *Appl. Energy* **2015**, *155*, 68–78. [CrossRef]
15. Xie, S.; Sun, F.; He, H.; Peng, J. Plug-In Hybrid Electric Bus Energy Management Based on Dynamic Programming. *Energy Procedia* **2016**, *104*, 378–383.
16. Chen, Z.; Mi, C.C.; Xiong, R.; Xu, J.; You, C. Energy management of a power-split plug-in hybrid electric vehicle based on genetic algorithm and quadratic programming. *J. Power Sources* **2014**, *248*, 416–426. [CrossRef]
17. Chen, Z.; Mi, C.C.; Xia, B.; You, C. Energy management of power-split plug-in hybrid electric vehicles based on simulated annealing and Pontryagin's minimum principle. *J. Power Sources* **2014**, *272*, 160–168. [CrossRef]
18. Chen, Z.; Xia, B.; You, C.; Mi, C.C. A novel energy management method for series plug-in hybrid electric vehicles. *Appl. Energy* **2015**, *145*, 172–179. [CrossRef]
19. Li, G.; Zhang, J.; He, H. Battery SOC constraint comparison for predictive energy management of plug-in hybrid electric bus. *Appl. Energy* **2017**, *194*, 578–587. [CrossRef]
20. Murphey, Y.L.; Park, J.; Chen, Z.; Kuang, M.L.; Masrur, M.A.; Phillips, A.M. Intelligent Hybrid Vehicle Power Control-Part I: Machine Learning of Optimal Vehicle Power. *IEEE Trans. Veh. Technol.* **2012**, *61*, 3519–3530. [CrossRef]
21. Chen, Z.; Xiong, R.; Wang, C.; Cao, J. An on-line predictive energy management strategy for plug-in hybrid electric vehicles to counter the uncertain prediction of the driving cycle. *Appl. Energy* **2017**, *185*, 1663–1672. [CrossRef]
22. Hester, T.; Quinlan, M.; Stone, P. RTMBA: A Real-Time Model-Based Reinforcement Learning Architecture for Robot Control. In Proceedings of the 2012 IEEE International Conference on Robotics and Automation, Saint Paul, MN, USA, 14–18 May 2012; pp. 85–90.
23. Liu, T.; Zou, Y.; Liu, D.; Sun, F. Reinforcement Learning of Adaptive Energy Management With Transition Probability for a Hybrid Electric Tracked Vehicle. *IEEE Trans. Ind. Electron.* **2015**, *62*, 7837–7846. [CrossRef]
24. Xiong, R.; Cao, J.; Yu, Q. Reinforcement learning-based real-time power management for hybrid energy storage system in the plug-in hybrid electric vehicle. *Appl. Energy* **2018**, *211*, 538–548. [CrossRef]
25. Liu, C.; Murphey, Y.L. Power Management for Plug-in Hybrid Electric Vehicles using Reinforcement Learning with Trip Information. In Proceedings of the 2014 IEEE Transportation Electrification Conference and Expo, Dearborn, MI, USA, 15–18 June 2014.
26. Lin, X.; Wang, Y.; Bogdan, P.; Chang, N.; Pedram, M.; IEEE. Reinforcement Learning Based Power Management for Hybrid Electric Vehicles. In Proceedings of the 2014 IEEE/Acm International Conference on Computer-Aided Design, Dearborn, MI, USA, 15–18 June 2014; pp. 32–38.
27. Qi, X.; Wu, G.; Boriboonsomsin, K.; Barth, M.J. A Novel Blended Real-time Energy Management Strategy for Plugin Hybrid Electric Vehicle Commute Trips. In Proceedings of the 2015 IEEE 18th International Conference on Intelligent Transportation Systems, Las Palmas, Spain, 15–18 September 2015; pp. 1002–1007. [CrossRef]
28. Liu, T.; Hu, X. A Bi-Level Control for Energy Efficiency Improvement of a Hybrid Tracked Vehicle. *IEEE Trans. Ind. Inform.* **2018**, *14*, 1616–1625. [CrossRef]
29. Liu, T.; Hu, X.; Li, S.E.; Cao, D. Reinforcement Learning Optimized Look-Ahead Energy Management of a Parallel Hybrid Electric Vehicle. *IEEE-Asme Trans. Mechatron.* **2017**, *22*, 1497–1507. [CrossRef]
30. Zou, Y.; Liu, T.; Liu, D.; Sun, F. Reinforcement learning-based real-time energy management for a hybrid tracked vehicle. *Appl. Energy* **2016**, *171*, 372–382. [CrossRef]
31. Hong, Y.Y.; Chang, W.C.; Chang, Y.R.; Lee, Y.D.; Ouyang, D.C. Optimal Sizing of Renewable Energy Generations in a Community Microgrid Using Markov Model. *Energy* **2017**, *135*, 68–74. [CrossRef]
32. Holland, O.; Snaith, M. Extending the Adaptive Heuristic Critic and Q-Learning—From Facts to Implications. *Artif. Neural Netw.* **1992**, *2*, 599–602.
33. Hou, C.; Ouyang, M.; Xu, L.; Wang, H. Approximate Pontryagin's minimum principle applied to the energy management of plug-in hybrid electric vehicles. *Appl. Energy* **2014**, *115*, 174–189. [CrossRef]

 *applied sciences*

Article

An Improved Model-Based Self-Adaptive Filter for Online State-of-Charge Estimation of Li-Ion Batteries

Chi Zhang [1,2], Fuwu Yan [1,2], Changqing Du [1,2,*] and Giorgio Rizzoni [3]

1   Hubei Key Laboratory of Advanced Technology for Automotive Components, Wuhan 430070, China; Zhangchi_whut@163.com (C.Z.); Yanfuwu@vip.sina.com (F.Y.)
2   Hubei Collaborative Innovation Center for Automotive Components Technology, Wuhan 430070, China
3   Department of Mechanical and Aerospace Engineering, The Ohio State University, 201 W 19th Ave., Columbus, OH 43210, USA; rizzoni.1@osu.edu
*   Correspondence: cq_du@whut.edu.cn; Tel.: +86-027-87850553

Received: 28 September 2018; Accepted: 23 October 2018; Published: 28 October 2018

Abstract: Accurate battery modeling is essential for the state-of-charge (SOC) estimation of electric vehicles, especially when vehicles are operated in dynamic processes. Temperature is a significant factor for battery characteristics, especially for the hysteresis phenomenon. Lack of existing literatures on the consideration of temperature influence in hysteresis voltage can result in errors in SOC estimation. Therefore, this study gives an insight to the equivalent circuit modeling, considering the hysteresis and temperature effects. A modified one-state hysteresis equivalent circuit model was proposed for battery modeling. The characterization of hysteresis voltage versus SOC at various temperatures was acquired by experimental tests to form a static look-up table. In addition, a strong tracking filter (STF) was applied for SOC estimation. Numerical simulations and experimental tests were performed in commercial 18650 type $Li(Ni_{1/3}Co_{1/3}Mn_{1/3})O_2$ battery. The results were systematically compared with extended Kalman filter (EKF) and unscented Kalman filter (UKF). The results of comparison showed the following: (1) the modified model has more voltage tracking capability than the original model; and (2) the modified model with STF algorithm has better accuracy, robustness against initial SOC error, voltage measurement drift, and convergence behavior than EKF and UKF.

Keywords: state of charge; strong track filter; modified one-state hysteresis model; $Li(Ni_{1/3}Co_{1/3}Mn_{1/3})O_2$ battery

---

1. Introduction

The concerns in energy crisis and global warming have driven the development of alternative energy vehicles rapidly. The Electric Vehicles (EVs), which are among the ultimate solutions for sustainable transportation, have attracted attention in aspects such as rechargeable power batteries and Battery Management System (BMS).

A key estimative parameter, state of charge (SOC) of a battery, indicates its residual capacity and reflects the remaining range of an electric vehicle. An accurate SOC estimation can not only predict the remaining range for the EVs to relieve the "range anxiety" for the drivers, but it can also help to determine an effective management strategy to avoid cell damage from over-charging and over-discharging. However, due to the complexity of the chemical and physical processes involved, characteristics of batteries present a distinct nonlinear feature, which makes their online monitoring a challenging task. Therefore, special algorithms for SOC online estimation are required.

Previous papers coining the term 'SOC' can date back to the 1960s [1]. After several decades' efforts, a great variety of approaches have been engaged in targeting monitoring the SOC for EVs. Generally, the SOC estimation algorithms can be divided into three categories.

The first methods are based on the direct measurement, including the residual capacity method, the open-circuit voltage (OCV) method, and the Ampere-hour counting based method [2]. The residual capacity method calculates the SOC by discharging the battery to the lower cut-off voltage in controlled test equipment. It is also the most reliable method under laboratory conditions, but this is obviously not the case in online monitoring for BMS. The OCV-based method requires a long rest period, thereby not practical for EV applications. Moreover, the performance of the OCV-based method becomes severe for which the battery characteristic of voltage platform is flat, such as $LiFePO_4$ (LFP). To the authors' knowledge, only the Ampere-hour counting based method is suitable for online monitoring. The Ampere-hour counting method estimates the SOC by integrating the flow-in and flow-out current of the battery. This method reportedly has several theoretical limitations and is an open loop method that cannot correct the accumulative error caused by current measuring transducers drift. In addition, the estimation accuracy is dependent on the initial SOC. This method has low robustness against the acquired signal quality, as well as initial SOC information.

The second types are the machine learning methods, such as the Artificial Neural Network (ANN), Fuzzy Logic (FL), and Support Vector Machine (SVM). These methods are also called "black box" model by which these do not need the detailed information of the battery system. Mahmoud Ismail et al. [3,4] developed an ANN SOC estimator for commercial Li-ion battery. The network is trained by the input of current and voltage, and the output is the battery SOC. The algorithm is validated by the benchmark driving cycles and can achieve a relatively high degree of accuracy. Claudio Burgos et al. [5] introduced a fuzzy-based model to characterize the discharge behavior of lead-acid batteries. Du Jiani et al. [6] presented a methodology of FL to describe the equivalent circuit model parameters on SOC and temperature effects. The SVM model is also a smart choice for SOC estimation, which has been used in several literatures [7–11]. Although much research focused on the machine learning methods, some of which have shown good performance, the common shortfall of these methods is the heavy computational burden that makes the online implementation unpractical.

The last algorithms are based on the control-oriented battery model. The model-based estimators are used to calculate the SOC by characterizing the battery behavior through measurable signals like current, terminal voltage, and temperature. These model-based estimators include the electrochemical model, the Equivalent Circuit Model (ECM), and the empirical model. The electrochemical model is the most precise model among the others, wherein it describes the electrochemical reaction processes of the battery by adopting a set of partial differential equations. However, it is also the most complicated model for the limited BMS computational resource. Meanwhile, the empirical model usually has a simple model structure with a low computation demand. However, a large number of experiments are required to build a database. Moreover, for most BMS applications, the ECM is used as a solution, because it meets the best compromise between the accuracy and model complexity.

Currently, the combinations of ECM and system filtering theory have drawn continuous attention by scholars and industry developers. Among all the system filtering theories, the Kalman filter (KF) is the most frequently used [12]. Compared with other system filtering algorithms, the KF method does not need an accurate initial value of SOC because the result will gradually approach the optimal value and the current measurement error will be updated during the operational process of the algorithms. Meanwhile, it is a closed-loop observer, and it can achieve accurate and continuous estimation performance during the whole range of battery operations. The abovementioned advantages make KF a promising solution for BMS application implementation. Moreover, the ordinary KF is only suitable for linear systems, whereas the BMS applications require the use of more complex and nonlinear algorithms. Several advanced modifications have been proposed, such as an extended Kalman filter (EKF), unscented Kalman filter (UKF), and Central-Difference Kalman filter (CDKF). Gregory L. Plett was the first to establish EKF for SOC estimation [13–16]. Based on his pioneering work, a great variety of research concerning the applications of EKF to the non-linear system have been reported in several literatures [17–22]. However, EKF is essentially a first-order Taylor series expansion of the state-space equations that have the theoretical limitation of estimation accuracy under the case of high dynamic current change. In addition, the EKF needs to calculate the Jacobi matrix, which leads to the algorithm being inefficient or even numerical instabilities in the implementation of a low-cost microcontroller. As an alternative system linear approach, the UKF was developed to improve the estimation accuracy and to compute reliability Theoretically, all KF variants require the knowledge of process and measurement noise covariance. Inappropriate tuning parameters may lead to low convergence and high oscillation. Therefore, adaptive technology has been introduced to combine with the KFs. Sun et al. [23] presented an adaptive UKF to estimate the SOC of EV applications. The adaptive adjustment of the noise covariance was dealt by covariance matching methods. Xiong et al. [24] proposed a data-driven-based approach for SOC estimation by employing an adaptive EKF algorithm. Their methods achieved good accuracy and convergence for different types of lithium-ion batteries.

To summarize, a modern smart algorithm for online SOC estimation in a BMS application requires the following characteristics: capable of describing the first- or second-order nonlinear behavior of battery system under dynamic excitation; adaptive adjustment for system noise matrices and high converge robustness against the drastic change of current; numerical stability; and ease of implementation in an embedded chip.

To address the abovementioned problems, an optimized model-based algorithm combined with Strong Tracking Filter (STF) was put forward for online SOC monitoring. The one-state hysteresis model was applied to state-space function and the hysteresis voltage was considered in terms of temperature. The proposed algorithm has the following advantages: (1) strong robustness against model uncertainties; (2) strong tracking ability of the mutation status; and (3) a moderate computational burden. Various experimental tests were designed to validate the proposed approach. The comparisons among the EKF, UKF, and STF were carried out to evaluate the performance of the proposed algorithm.

The remainder of this paper was organized as follows: The battery modeling was introduced in Section 2. In Section 3, the experimental setup and identification results were demonstrated in detail. In Section 4, EKF, UKF, and STF were proposed for the implementation of SOC estimation. Section 5 illustrated the experimental results in comparison with EKF and UKF in the aspects of estimation accuracy, robustness, and convergence behavior. Section 6 presents the conclusions.

2. Battery Modeling

For control-oriented battery online monitoring, a precise battery state-space model must be available. As mentioned, the ECM is the most widely used choice in combination with the KFs. The ECM is based on the Thevenin's theorem, which approximates the battery's electrical behavior through a voltage source and some resistances and capacitors. The accuracy of ECM was enhanced by adding extra resistance-capacitance terms (RC network) into the circuits. However, the complication of the model structure could lead to inefficient parameters' identification and low real-time computation. Obviously, a reasonable model must consider simulation accuracy, parameterization efficiency, and computation burden. Additionally, hysteresis is also a very significant variable for lithium-ion batteries [25–28]. However, studies for this variable are lacking as it is rarely being considered. In this paper, the hysteresis model was used. As shown in Figure 1, the proposed model consists of an Electro-Motive Force (EMF) voltage resource, a hysteresis voltage resource, a resistor R_0, and a RC network connected in series. The EMF voltage resource and the hysteresis voltage resource are together to form a controlled voltage source $-U_{oc}$ (OCV). R_0 is the ohmic resistance. It represents the instantaneous voltage variation caused by the electrolyte and the active mass. R_1 and C_1 represent the polarization resistance and polarization capacitance, respectively. These are also used to depict the transient response of the cell caused by double-layer capacity effects.

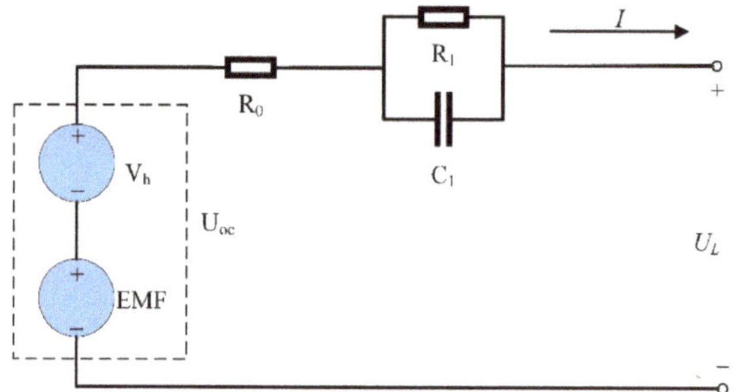

Figure 1. The schematic diagram of the one-state hysteresis equivalent circuit model (ECM).

The hysteresis effect between the charge and discharge boundary curves is illustrated in Figure 2a. The curves are obtained by charging/discharging the cell at C/25 rate (a current corresponding to the manufacturer's rated capacity for a 25 h discharge to low cut-off voltage) at a temperature of 30 °C; 30 °C is the standard temperature according to the latest U. S. Advanced Battery Consortium (USABC) battery test manual [29]. The charge voltage is the upper curve and the discharge voltage is the lower curve. At such low rate, we believed that the voltage drop was caused by ohmic resistance, and polarization is small enough to ignore (less than 1 mV in this case). Thus, the voltage difference at each SOC is caused by the hysteresis phenomenon.

The space function equations corresponding to Figure 1 are derived as follows:

First, according to the Kirchhoff voltage law, the governing equation of the cell model can be derived as follows:

$$\begin{cases} U_L = U_{oc}(SOC, T) + IR_0 + U_1 + h \\ \overline{U_1} = -\frac{1}{R_1 C_1} U_1 + \frac{1}{C_1} I \end{cases} \tag{1}$$

where U_L is the terminal voltage of the cell. EMF is just a theoretical concept, where the EMF is approximated by $U_{oc}(SOC, T)$ after an hour rest, which is the function of SOC and temperature. U_1 is the voltage over the RC network and h is the voltage caused by hysteresis. I is the main circuit current. The charging process is positive, and the discharging process is negative. R_1 is the polarization resistance and C_1 is the polarization capacitance. h is the hysteresis voltage, it is the function of the SOC and temperature.

The SOC of the cell can be expressed as follows:

$$SOC(k) = SOC(0) + \frac{\int_0^k \eta I dt}{C_N} \tag{2}$$

where SOC(0) is the initial SOC value of the cell; SOC(k) is the current SOC value at time k; C_N is the nominal capacity and the function of the discharge rate and temperature; η is the coulomb efficiency, the function of current and temperature [30].

To implement the one-state hysteresis model to KFs, the above equations need to be discretized to the following linear discrete form [31]:

$$U_{L,\,k} = U_{oc,k} + R_0 I_k + U_{1,\,k} + h_k \tag{3}$$

$$U_{1,k} = \exp\left(-\frac{T_s}{R_1 C_1}\right) U_{1,k-1} + \left[1 - \exp\left(-\frac{T_s}{R_1 C_1}\right)\right] R_1 I_{k-1} \tag{4}$$

$$SOC_k = SOC_{k-1} + \frac{\eta T_s}{C_N} I_{k-1} \tag{5}$$

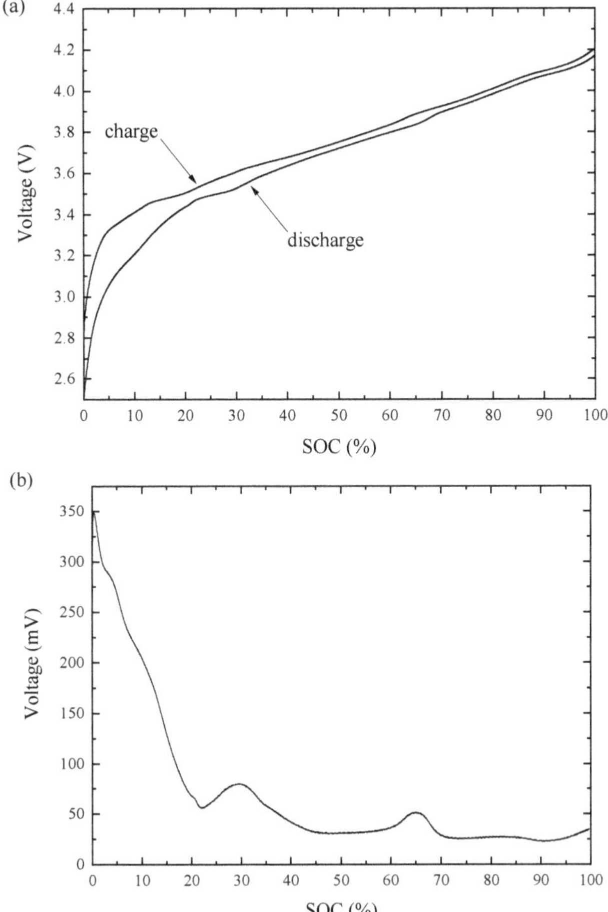

Figure 2. Open circuit voltage (OCV) discharge and charge boundary at 30 °C: (**a**) the discharge and charge curve; (**b**) the hysteresis level.

For the hysteresis state function, let $h(S, t)$ be the hysteresis voltage as the function of SOC and time, and $\hat{S} = dS/dt$.

$$\frac{dh(S, t)}{dS} = \gamma \mathrm{sgn}(\hat{S})(M(S, \hat{S}) - h(S, t)) \tag{6}$$

where $M(S, \hat{S})$ is the maximum hysteresis voltage as a function of SOC and SOC changing rate. γ is a positive constant to tune the rate of decay, and $\mathrm{sgn}(\hat{S})$ is to make the equation stable for charge and discharge process.

Since $dh(S, t)/dt = dh(S, t)/dS \times dS/dt$, the deformation of Equation (6) can be expressed by Equation (7):

$$\frac{dh(S, t)}{dt} = \gamma \mathrm{sgn}(\hat{S})(M(S, \hat{S}) - h(S, t)) \times \frac{dS}{dt} \tag{7}$$

According to Equation (2), it is easy to conclude that $dS/dt = \eta I/C_N$. Meanwhile, $\mathrm{sgn}(\hat{S}) \times dS/dt = |dS/dt|$. Then, Equation (7) is rewritten by:

$$\hat{h}(t) = \left| \frac{\eta I(t)\gamma}{C_N} \right| M(S, \hat{S}) - \left| \frac{\eta I(t)\gamma}{C_N} \right| h(t) \tag{8}$$

And then, discretize Equation (8) to:

$$h_k = \exp\left(-\left|\frac{\eta I(t)\gamma}{C_N}\right|\right) h_{k-1} + \left(1 - \exp\left(-\left|\frac{\eta I(t)\gamma}{C_N}\right|\right)\right) M(S, \hat{S}) \tag{9}$$

3. Battery Experiments

3.1. Test Bench

As shown in Figure 3, an experimental test bench was established to study the characteristics of the lithium-ion battery. The experimental setup consisted of the following: (1) a set of Sony US18650VTC5a type cells (Sony Corp., Tokyo, Japan) with $Li(Ni_{1/3}Co_{1/3}Mn_{1/3})O_2$ (NCM) cathode and graphite anode; (2) a thermal chamber with temperature control; (3) current and voltage sensors; (4) a battery test station; (5) a host computer with software to set up the database; and (6) a Matlab 2017b (MathWorks Inc., Natick, MA, USA) to run the model. The cell is a commercial battery that is used for high specific energy demand application with a nominal voltage of 3.7 V and a nominal capacity of 2.5 Ah. The main characteristics of nominal voltage and nominal capacity are shown in Table 1. For the tests, the temperature chamber was set at $-20\,°C$, $-10\,°C$, $0\,°C$, $10\,°C$, $20\,°C$, $30\,°C$, $40\,°C$, $50\,°C$, respectively. The battery test station (NEWARE BTS4002, Shenzhen, China) was used for test profile control. Each channel of the BTS was capable of ±20 A current and $+5$ V voltage. The accuracy of current and voltage measurement was 50 mA and 10 mV, respectively.

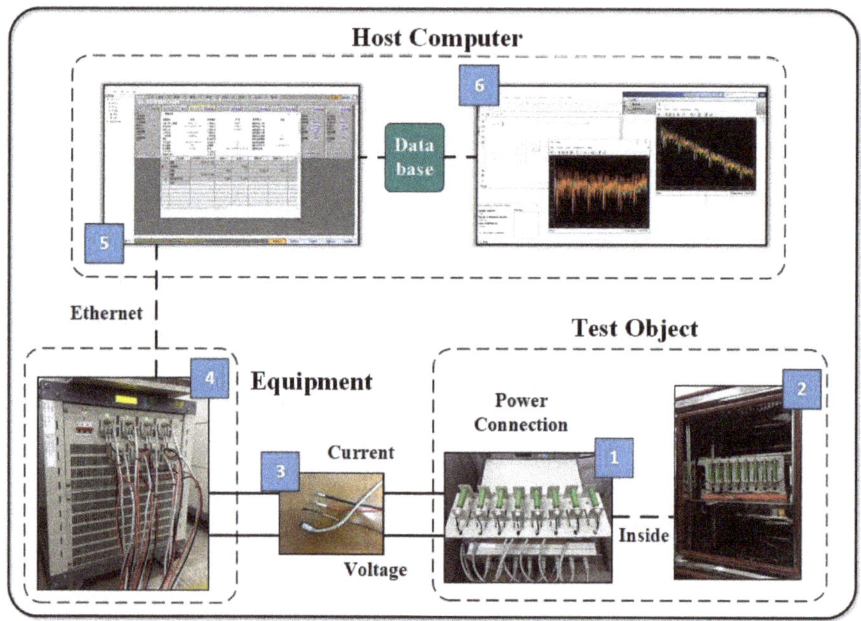

Figure 3. Schematic diagram of the battery test bench.

Table 1. Characteristics of Sony US18650VTC5a cell.

Item	Rating
Capacity	2.5 Ah
End of discharge voltage	2.5 V
Maximum charge voltage	4.2 V
Maximum discharge current rate	12 C

3.2. Test Schedule

Overall, the tests involved in this study were categorized into one of two types, namely, a parameterization and a validation test schedule. The parameterization test includes hysteresis voltage test and a series of HPPC (Hybrid Pulse Power Characterization) tests.

The hysteresis voltage is a considerable factor for battery modeling. Thus, the hysteresis voltage test was conducted at various temperatures to improve the model accuracy. During the hysteresis voltage test, the cells were discharged by 1/25 C constant current, until the low cut-off voltage (2.5 V) was reached. After the 1 h period rest, the cells were recharged by the same 1/25 C rate until the upper cut-off voltage (4.2 V). This test was carried out at temperatures of 0 °C, 15 °C, 30 °C, and 45 °C. There were 4 cycles involved in total, all the discharge and chaege cycles was carried out by constant current (CC) method.

The HPPC test profile as referred to [29], was composed of a series of hybrid pulse power steps with 1 h rest at each SOC point. The current profile and the voltage response of a HPPC microcycle are shown in Figure 4. The HPPC test was carried out with 1 C rate under the temperatures ranging from 0 °C to 50 °C, with an interval of 10 °C. In HPPC test, the model parameters of OCV, R_0, R_1, C_1 were identified online. The identification results were discussed in Section 3.4.

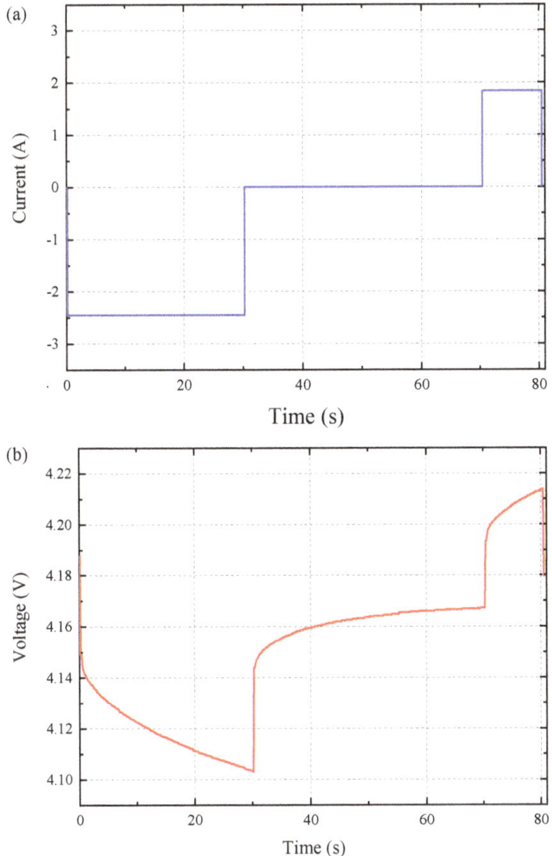

Figure 4. Current profile and voltage response of a hybrid pulse power characterization (HPPC) microcycle.

The purpose of the validation test was to evaluate the performance of the modified model, as well as the SOC approach, which will be introduced in the next section. Therefore, a current sequence derived from the Federal Urban Driving Schedule (FUDS) cycle was used in this study. FUDS is a dynamic current load which is designed to simulate the battery operation case in EV applications. In summary, the entire profile consisted of nine typical FUDS microcycles; a 1/3 C constant current discharge between the FUDS cycles and the constant current discharge was a rest period of one hour. To implement the FUDS to the cells, the battery size factor (BSF) was scaled down to 9.25 Wh. The cells were fully charged before the test and were discharged to the low cut-off voltage of 2.5 V at last. Therefore, the "true value" of SOC was calculated by the accumulative Ampere-hours flowing in and out of the battery. Figure 5 shows the sampled current and voltage of the validation test profile. The validation test was run at the temperature ranges from −20 °C to 50 °C. It is worth pointing out that the test should be terminated immediately when the low cut-off voltage of 2.5 V was reached.

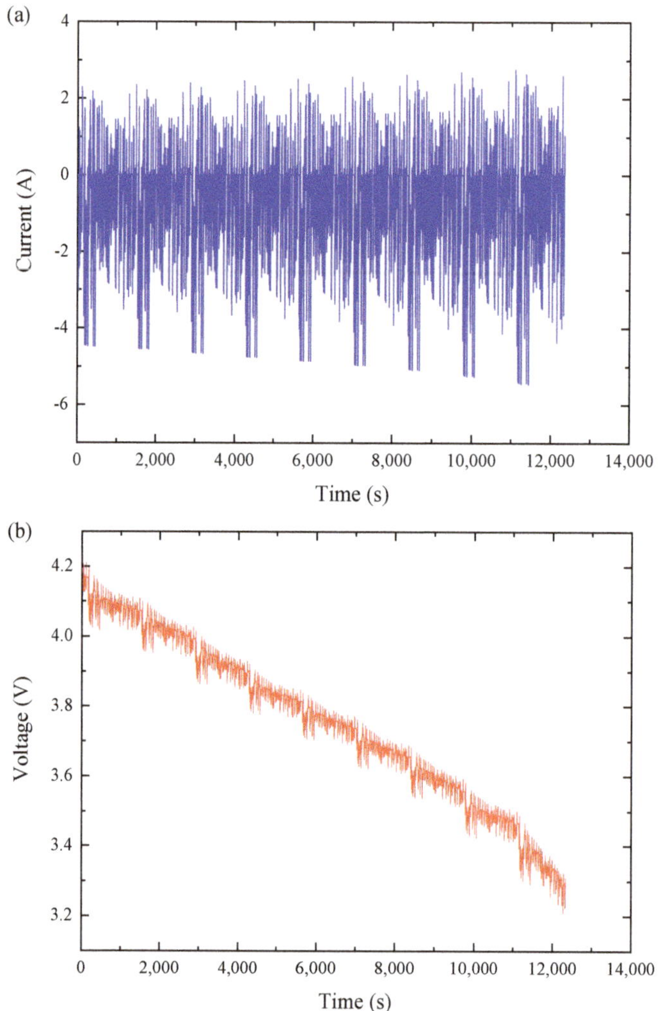

Figure 5. The profile of validation test cycles: (**a**) current profile; (**b**) voltage response.

3.3. Parameters Identification

In this work, two types of battery parameters are needed to be identified, namely the function of $h_v = f(SOC, T)$, and the intrinsic parameters of the ECM which were identified offline and online, respectively. Let $\xi = \left[OCV, R_0^+, R_0^-, R_1, C_1\right]$ be the online identification parameters. The current and voltage response of HPPC data were selected and identified by using recursive least square method (RLS). For online identification, because OCV is the function of SOC, an OCV function is needed for curve fitting. Here, the model proposed by Plett [14] was employed. The OCV at the end of the 1 h rest period of HPPC at each SOC was adopted.

$$OCV = k_0 - \frac{k_1}{SOC} - k_2 SOC + k_3 In(SOC) + k_4 In(1 - SOC) \tag{10}$$

Taking $0\,^{\circ}C$ as an example, the measured OCV and the nonlinear fitted function are displayed in Figure 6a. In Figure 6b, the generalized residual was utilized to describe the degree of the fitting. The data at the rest of the temperatures has the similar results.

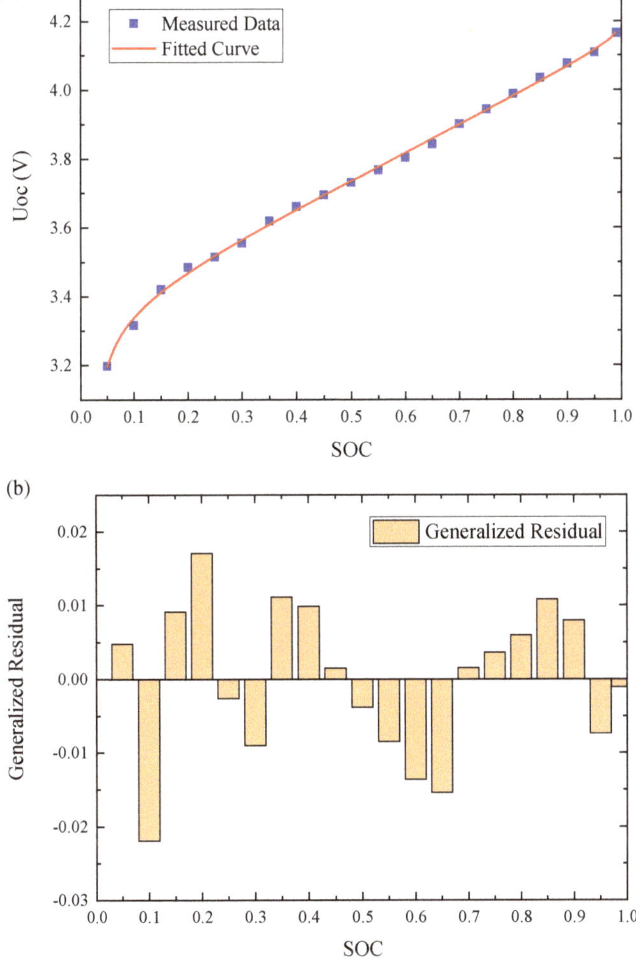

Figure 6. The OCV result at temperature of $0\,^{\circ}C$: (**a**) fitting curve; (**b**) the generalized residual.

To further improve the model accuracy, the OCV was corrected by h_v; the h_v is preprocessed to form an offline h_v-SOC-T look-up table. As shown in Figure 7, the OCV boundaries of discharge and charge process and hysteresis level under various temperatures were depicted.

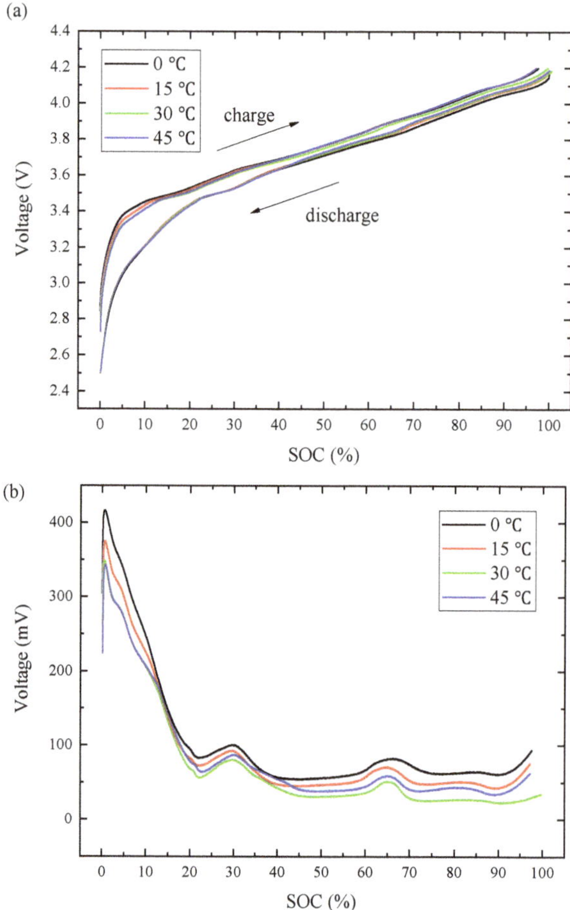

Figure 7. The OCV and hysteresis level under different temperatures: (**a**) the OCV boundaries of discharge and charge process at different temperatures; (**b**) the hysteresis voltage level.

3.4. Identification Results

The identified model was validated by the validation test profile. Here, we defined the original model as lacking hysteresis terms, whereas the improved model as considering hysteresis and temperature effect. The quantitative evaluation of goodness-of-fit of the model, MAE (Mean Absolute Error), and RMSE (Root Mean Squared Error) were introduced by the following equations:

$$\text{MAE} = \left(\frac{\int_{k=0}^{n} |\hat{S}(k) - S(k)|}{n} \right) \times 100\% \tag{11}$$

$$\text{RMSE} = \left(\sqrt{\frac{\sum_{k=0}^{n} (\hat{S}(k) - S(k))}{n}} \right) \times 100\% \tag{12}$$

where $S(k)$ is the experimental data and $\hat{S}(k)$ is the estimated value at step k.

Again, taking 0 °C for example, the simulated voltage by the original model and improved model are shown in Figure 8a and the errors of the two models are depicted in Figure 8b. For the profile at 0 °C, the MAE of the original model and improved model were 47.9 mV and 24.0 mV, respectively; and the RMSE of the original model and improved model were 59.8 mV and 29.9 mV, respectively. The MAE and RMSE at all temperatures are illustrated in Figure 9. Generally, the MAE and RMSE of both the original model and improved model decreased as the temperatures increased. However, as the temperature moves farther away from 30 °C, the performance of the proposed model becomes better than the original model. Unlike the proposed model, the original model's parameters were identified at 0 °C, thus the lack of parameters' correction in terms of the temperatures led the above results.

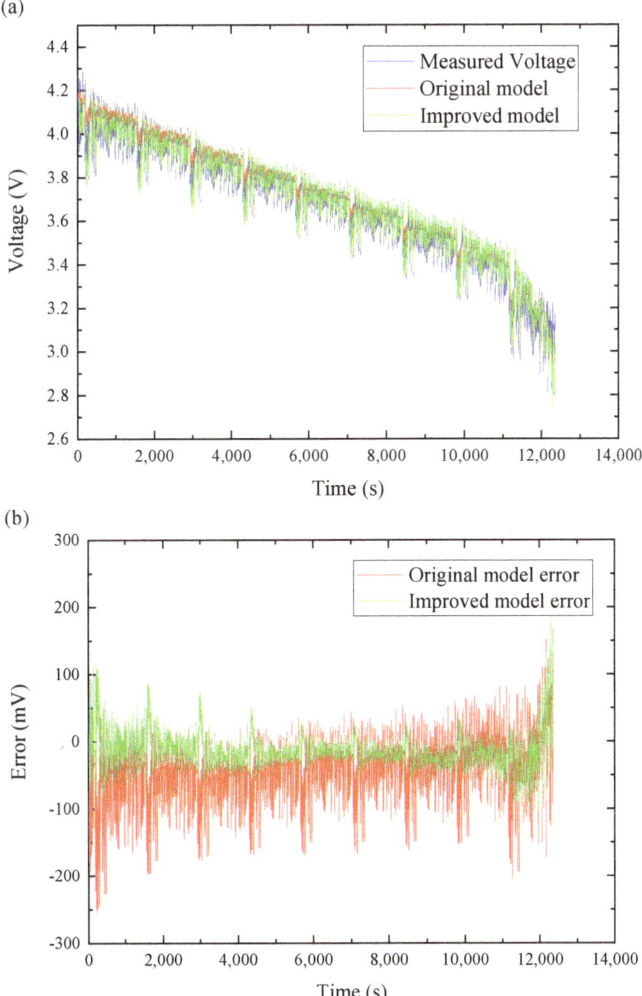

Figure 8. Comparison between measured data and model results at 0 °C: (**a**) voltage of experimental, original model and improved model; (**b**) estimated error of the original and improved model.

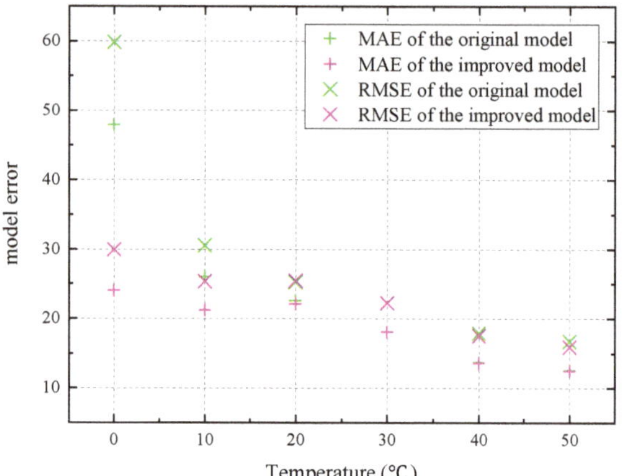

Figure 9. The mean absolute error (MAE) and root mean squared error (RMSE) of the original model and improved model.

4. SOC Estimation Based on EKF and UKF

According to Equations (3)–(5) and (9), the state-space equation is as follows:

$$
\begin{bmatrix} U_{1,k} \\ h_k \\ SOC_k \end{bmatrix}
= \begin{bmatrix} \exp\left(-\frac{T_s}{R_1C_1}\right) & 0 & 0 \\ 0 & \exp\left(-\left|\frac{\eta I_k \gamma T_s}{C_N}\right|\right) & 0 \\ 0 & 0 & 1 \end{bmatrix}
\begin{bmatrix} U_{1,k-1} \\ h_{k-1} \\ SOC_{k-1} \end{bmatrix}
$$
$$
+ \begin{bmatrix} \left[1 - \exp\left(-\frac{T_s}{R_1C_1}\right)\right]R_1 & 0 & 0 \\ 0 & \left(1 - \exp\left(-\left|\frac{\eta I(t)\gamma T_s}{C_N}\right|\right)\right) & 0 \\ 0 & 0 & \frac{\eta T_s}{C_N} \end{bmatrix}
\begin{bmatrix} I_{k-1} \\ M(S,\hat{S}) \\ I_{k-1} \end{bmatrix} + \omega_{k-1}
\tag{13}
$$

$$
U_{L,k} = U_{oc,k} + R_0 I_k + U_{1,k} + h_k + v_k
\tag{14}
$$

where ω_{k-1} is the system process noise and v_k is the measurement noise. Based on the above state-space equations, the proposed model can be implemented on EKF and UKF, respectively

4.1. Extended Kalman Filtering

As there have been plenty of published literatures on the EKF approach, the algorithm principle was shortly introduced, to avoid repetitions. The nonlinear state-space equation and measurement discretized equation can be represented as follows:

$$
x_{k+1} = f(x_k, u_k) + \Gamma_k \omega_k
\tag{15}
$$

$$
y_{k+1} = g(x_k, u_k) + v_k
\tag{16}
$$

where $f(x_k, u_k)$ is the nonlinear state function and $g(x_k, u_k)$ is the nonlinear measurement function. Then, $\hat{A}_k = \left.\frac{\partial f(x_k, u_k)}{\partial x_k}\right|_{x_k = \hat{x}_k}$, $\hat{C}_k = \left.\frac{\partial g(x_k, u_k)}{\partial x_k}\right|_{x_k = \hat{x}_k}$ was defined as the Jacobian matrix of $f(\cdot)$ and $g(\cdot)$, respectively.

The EKF is summarized in Table 2 as referenced in [13].

Table 2. Summary of the extended Kalman filter (EKF) approach for SOC estimation.

Initialization
1. Initialize with:
$\hat{x}_{0\|0} = E(x_0), \, P_{0\|0} = E\left[\left(x_0 - \hat{x}_{0\|0}\right)\left(x_0 - \hat{x}_{0\|0}\right)^T\right]$
State prediction
1. Update the step state vector:
$\hat{x}_{k\|k-1} = A_{k-1}\hat{x}_{k-1\|k-1} + B_{k-1}u_{k-1}$
2. Update the step error covariance:
$P_{k\|k-1} = A_{k-1}P_{k-1\|k-1}A_{k-1}^T + \Gamma_{k-1}Q_{k-1}\Gamma_{k-1}^T$
Measurement update
1. Calculate the Kalman gain:
$K_k = P_{k\|k-1}C_k^T\left(C_kP_{k\|k-1}C_k^T + R_k\right)^{-1}$
2. Update the measurement vector:
$\hat{x}_{k\|k} = \hat{x}_{k\|k-1} + K_k(y_k - g(x_k, u_k))$
3. Update the error covariance:
$P_{k\|k} = (1 - K_kC_k)P_{k\|k-1}$

The key point for implementing the EKF to our proposed model is to deal with the C_k. The detailed derivation process is below:

Assume that the vector of the parameters is $\theta = [R_0^+, \, R_0^-, \, M, \, \lambda]$, where R_0^+ is the ohmic resistance when the cell is charging and R_0^- is the ohmic resistance when the cell is discharging. Then, to calculate C_k^θ:

$$\frac{dg(x_k, \, u_k, \, \theta)}{d\theta} = \frac{\partial g(x_k, \, u_k, \theta)}{\partial \theta} + \frac{\partial g(x_k, \, u_k, \, \theta)}{\partial x_k}\frac{dx_k}{d\theta} \tag{17}$$

$$\frac{dx_k}{d\theta} = \frac{\partial f(x_{k-1}, \, u_{k-1}, \, \theta)}{\partial \theta} + \frac{\partial f(x_{k-1}, \, u_{k-1}, \, \theta)}{\partial x_{k-1}}\frac{dx_{k-1}}{d\theta} \tag{18}$$

The equation can be calculated recursively, and the initial value of $\frac{dx_k}{d\theta}$ ($k = 0$) can be set to zero. In this particular case:

$$\frac{\partial g(x_k, \, u_k, \theta)}{\partial \theta} = \begin{bmatrix} I^+ & I^- & 0 & 0 \end{bmatrix} \tag{19}$$

$$\frac{\partial g(x_k, \, u_k, \theta)}{\partial x_k} = \begin{bmatrix} 1 & 1 & \dfrac{\partial OCV(S_k)}{\partial S_k} \end{bmatrix} \tag{20}$$

$$\frac{\partial f(x_{k-1}, \, u_{k-1}, \, \theta)}{\partial \theta}$$
$$= \begin{bmatrix} 0 & 0 & 0 & 0 \\ 0 & 0 & \left(1 - \exp\left(\left|\frac{\eta i_{k-1}\gamma T_s}{C_N}\right|\right)\right)\mathrm{sgn}(i_{k-1}) & \left(h_{k-1} - M(S, \, \hat{S})\right)\left|\frac{\eta i_{k-1}T_s}{C_N}\right|\exp\left(\left|\frac{\eta i_{k-1}\gamma T_s}{C_N}\right|\right) \\ 0 & 0 & 0 & 0 \end{bmatrix} \tag{21}$$

$$\frac{\partial f(x_{k-1}, \, u_{k-1}, \, \theta)}{\partial x_{k-1}} = \begin{bmatrix} \exp\left(-\frac{T_s}{R_1C_1}\right) & 0 & 0 \\ 0 & \exp\left(-\left|\frac{\eta i_k\gamma T_s}{C_N}\right|\right) & 0 \\ 0 & 0 & 1 \end{bmatrix} \tag{22}$$

With the above equations, EKF can be executed recursively by repeating the steps in Table 2.

4.2. Unscented Kalman Filtering

Kalman filter is an optimum state recursive observer, and the frameworks of both EKF and UKF have a similar prediction-update structure. Alternatively, UKF linearized the system by unscented transform (UT) rather than the Taylor expansion adopted by EKF. UKF not only achieves a higher-order approximation expansion but also avoids the need to compute the Jacobian matrix. Based on the same discretized state-space Equations (13) and (14), the UKF algorithm steps are as follows:

Let χ be a n-dimension vector which obeys $X = N(\chi, P)$ distribution.

Initialization:

$$\overline{x}_0 = E(x_0), \quad P_0 = \left[(x_0 - \overline{x}_0)(x_0 - \overline{x}_0)^T\right] \tag{23}$$

Generate the sigma points via:

$$\chi_{i,k-1} = \begin{cases} \overline{x}_{k-1}, & i = 0 \\ \overline{x}_{k-1} + \left(\sqrt{(n+\lambda)P_{k-1}}\right), & i = 1, 2, \ldots, n \\ \overline{x}_{k-1} - \left(\sqrt{(n+\lambda)P_{k-1}}\right), & i = n+1, \ldots, 2n \end{cases} \tag{24}$$

And their weights are computed via:

$$\begin{cases} W_0^m = \frac{\lambda}{(n+\lambda)} \\ W_i^m = \frac{1}{2(n+\lambda)} & i = 1, \ldots, 2n \end{cases} \tag{25}$$

$$\begin{cases} W_0^c = \frac{\lambda}{(n+\lambda)} + (1 - \alpha^2 + \beta) \\ W_i^c = \frac{1}{2(n+\lambda)} & i = 1, \ldots, 2n \end{cases} \tag{26}$$

where λ is the scaling parameter, which satisfies:

$$\lambda = \alpha^2(n+\kappa) - n \tag{27}$$

α is a small positive number ($10^{-4} \le \alpha \le 1$). β is used for absorbing a priori information for variable χ. For Gaussian distribution, $\beta = 2$ is optimal. κ is a scaling factor that determine the degree of freedom of the sigma points, usually set 0 or $3 - n$ to guarantee the positive definiteness of output variable covariance.

Prediction update:

$$\chi_{i,k|k-1} = f\left(\chi_{i,k-1}, u_k\right) \qquad i = 0, \ldots, 2n \tag{28}$$

$$\overline{x}_{k|k-1} = \sum_{i=0}^{2n} W_i^m \chi_{i,k|k-1} \tag{29}$$

Update the error covariance:

$$P_{k|k-1} = \sum_{i=0}^{2n} W_i^c \left(\chi_{i,k|k-1} - \overline{x}_{k|k-1}\right)\left(\chi_{i,k|k-1} - \overline{x}_{k|k-1}\right)^T + Q_k \tag{30}$$

where Q_k is the covariance matrix of the state noise.

Measurement update:

$$y_{i,\,k|k-1} = g\left(\chi_{i,k-1}, u_k\right) \qquad i = 0, \ldots, 2n \tag{31}$$

$$\overline{y}_{k|k-1} = \sum_{i=0}^{2n} W_i^m y_{i,k-1} \tag{32}$$

Calculate prediction covariance and cross-covariance:

$$P_{yy,\,k} = \sum_{i=0}^{2n} W_i^c \left(y_{i,k|k-1} - \overline{y}_{k|k-1}\right)\left(y_{i,k|k-1} - \overline{y}_{k|k-1}\right)^T + R_k \tag{33}$$

$$P_{xy,\,k} = \sum_{i=0}^{2n} W_i^c \left(\chi_{i,k|k-1} - \overline{x}_{k|k-1} \right) \left(y_{i,k|k-1} - \overline{y}_{k|k-1} \right)^T \tag{34}$$

where R_k is the covariance matrix of the measurement noise.

Calculate the Kalman gain:

$$K_k = \frac{P_{xy,\,k}}{P_{yy,\,k}} \tag{35}$$

State estimate measurement update:

$$\overline{x}_{k|k} = \overline{x}_{k|k-1} + K_k \left(y_k - \overline{y}_{k|k-1} \right) \tag{36}$$

Error covariance measurement update:

$$P_{k|k} = P_{k|k-1} - K_k P_{yy,k} K_k^T \tag{37}$$

The UKF calculation procedure is summarized in Table 3.

Table 3. Summary of the unscented Kalman filter (UKF) approach for SOC estimation.

Initialization
1. Initialize with:
$\overline{x}_0 = E(x_0),\; P_0 = \left[(x_0 - \overline{x}_0)(x_0 - \overline{x}_0)^T \right]$
State prediction
1. Update the step state vector:
$\chi_{i,k\|k-1} = f\left(\chi_{i,k-1},\, u_k \right) \qquad i = 0,\, \ldots,\, 2n$
$\overline{x}_{k\|k-1} = \sum\limits_{i=0}^{2n} W_i^m \chi_{i,k\|k-1}$
2. Update the step error covariance:
$P_{k\|k-1} = \sum\limits_{i=0}^{2n} W_i^c \left(\chi_{i,k\|k-1} - \overline{x}_{k\|k-1} \right)\left(\chi_{i,k\|k-1} - \overline{x}_{k\|k-1} \right)^T + Q_k$
Measurement update
1. Output estimate time update:
$y_{i,\,k\|k-1} = g\left(\chi_{i,k-1}, u_k \right) \qquad i = 0,\, \ldots,\, 2n$
$\overline{y}_{k\|k-1} = \sum\limits_{i=0}^{2n} W_i^m y_{i,k-1}$
2. Calculate prediction covariance:
$P_{yy,\,k} = \sum\limits_{i=0}^{2n} W_i^c \left(y_{i,k\|k-1} - \overline{y}_{k\|k-1} \right)\left(y_{i,k\|k-1} - \overline{y}_{k\|k-1} \right)^T + R_k$
3. Calculate cross-covariance covariance:
$P_{xy,\,k} = \sum\limits_{i=0}^{2n} W_i^c \left(\chi_{i,k\|k-1} - \overline{x}_{k\|k-1} \right)\left(y_{i,k\|k-1} - \overline{y}_{k\|k-1} \right)^T$
Measurement correction
1. Calculate the Kalman gain:
$K_k = \frac{P_{xy,\,k}}{P_{yy,\,k}}$
2. State estimate measurement update:
$\overline{x}_{k\|k} = \overline{x}_{k\|k-1} + K_k \left(y_k - \overline{y}_{k\|k-1} \right)$
3. Error covariance measurement update:
$P_{k\|k} = P_{k\|k-1} - K_k P_{yy,k} K_k^T$

4.3. Strong Tracking Unscented Kalman Filtering

Although UKF can achieve better performance than EKF in estimation accuracy, it should be mentioned that UKF could lose the tracking capability and fail to converge the real value when it comes to the scenario that there is an abnormal change in one of the state vector components. Unfortunately, in real-world conditions, batteries suffer from highly dynamic current load that results in sudden acceleration and deceleration of driver intent. For the algorithm serve in a BMS, the sudden change of state vector, as well as the measurement error caused by the transducer, may lead to poor performance or biased results. Therefore, the STF is introduced in this study to address this problem. The idea of STF is to adjust Kalman gain matrix K_k by introducing the suboptimal multiple fading factors into the covariance matrix $P_{k|k-1}$ [30]. With the adaptively modifying Kalman gain matrix and priori covariance matrix online, STF is able to resist the sudden change of system state vectors. The algorithm for μ_k and $P_{k|k-1}$ are as follows [30,32]:

Firstly, define the residual error ε_k and residual error sequence covariance matrix V_k:

$$\varepsilon_k = y_k - \overline{y}_{k|k-1} \tag{38}$$

$$V_k = \begin{cases} \varepsilon_k \varepsilon_k^T & k = 1 \\ \dfrac{\rho V_{k-1} + \varepsilon_k \varepsilon_k^T}{1+\rho} & k \geq 2 \end{cases} \tag{39}$$

Generally, set $\rho = 0.95$.

Secondly, define the matrixes N_k and M_k as:

$$\begin{cases} N_k = V_k - C_k Q_{k-1} C_k^T - R_k \\ M_k = C_k P_{k|k-1} C_k^T + R_k - V_k + N_k \end{cases} \tag{40}$$

Lastly, the fading factor μ_k is calculated by the following equation:

$$u_k = \begin{cases} \dfrac{tr[N_k]}{tr[M_k]} & u_k \geq 1 \\ 1 & u_k \leq 1 \end{cases} \tag{41}$$

where $tr[\cdot]$ represents the trace of the matrix.

STF enforces the output residuals to be orthogonal or approximately orthogonal at each step to overcome the dynamic errors. The $P_{k|k-1}$ is recalculated by Equation (42) to replace Equation (30). The rest of the algorithm framework is the same as UKF algorithm.

$$P_{k|k-1} = u_k \sum_{i=0}^{2n} W_i^c \left(\chi_{i,k|k-1} - \overline{x}_{k|k-1} \right) \left(\chi_{i,k|k-1} - \overline{x}_{k|k-1} \right)^T + Q_k \tag{42}$$

The process of STF for SOC online estimation is summarized in Figure 10. After the initialization with the state vector $\overline{x}_0$ and error covariance P_0, the data obtained by HPPC were used for model parameters update, then the fading factor was calculated according to the residual error ε_k. Afterward, the typical Kalman steps were performed while the a prior covariance matrix $P_{k|k-1}$ was corrected by fading factor.

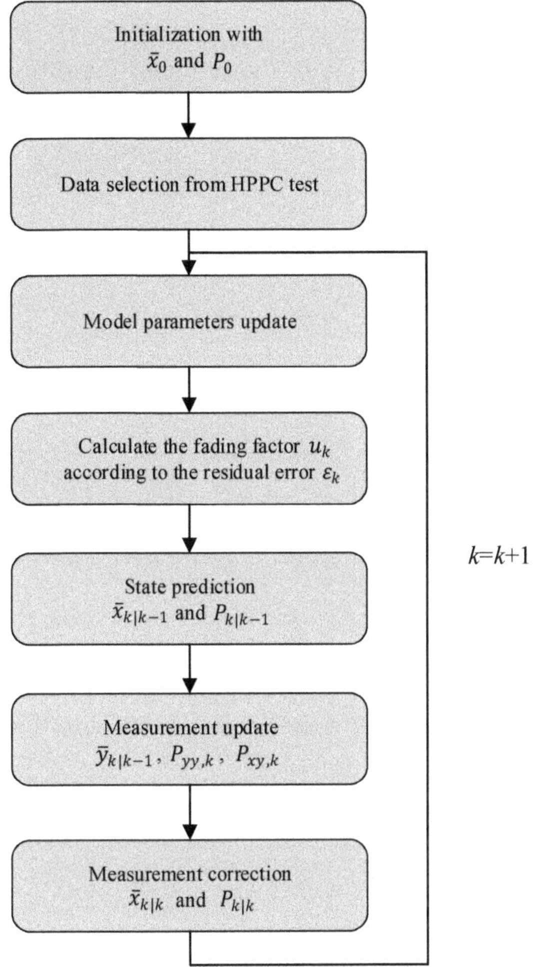

Figure 10. The flow chart of strong tracking filter (STF) algorithms.

5. Results and Discussion

The estimation performance of the three Kalman observers was compared. Since the true value of SOC in the validation test was calculated by the accumulative Ampere-hours flowing in and out of the battery, the test cells were fully charged by the CC–CV method and rested for 1 h at each test temperature to guarantee the initial SOC_0 to be 100%. The block diagram of our proposed SOC observers is illustrated in Figure 11. The validation test profile was loaded into the cells and Matlab/Simulink model, simultaneously. The sampling time was 1 s, and the model parameters were updated according to the current SOC, I, and T. The parameters of each algorithm were configured as follows:

$$Q_k = [0.1\ 0; 0\ 0.00001];\ R_k = 0.001;\ \alpha = 0.002;\ \beta = 2;\ \kappa = 1.$$

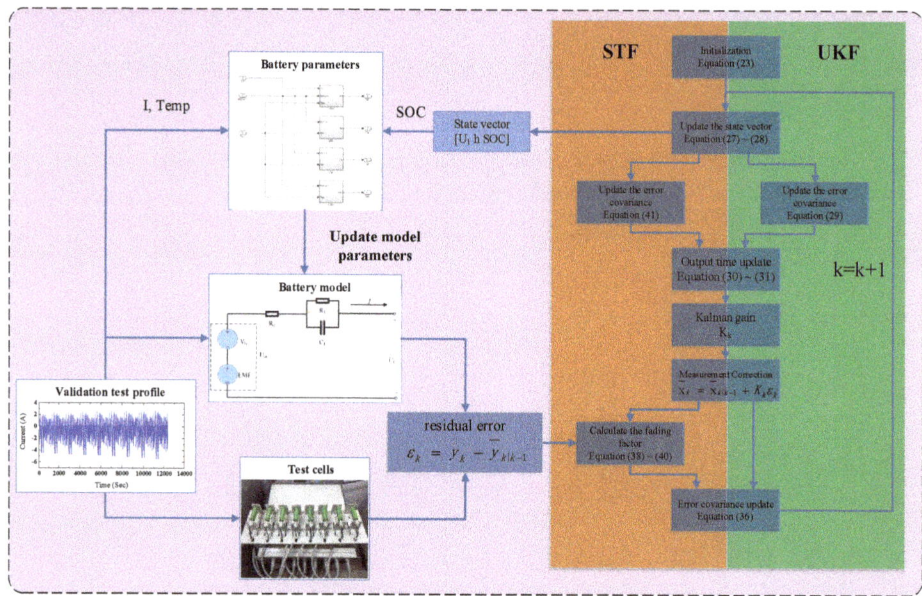

Figure 11. Block diagram of the proposed battery SOC observers.

The prediction accuracy of the three observers under each specific temperature was discussed. The comparative results between the experimental data obtained by the test equipment and the simulation data calculated by the three aforementioned algorithms under 0 °C are shown in Figure 12a. Figure 12b is the zoomed figure for Figure 12a from 4000 s to 6500 s. As a result, it can be seen that the performance of STF was better than UKF, and UKF was better than EKF. As shown in Figure 5a, both charging and discharging processes existed in the FUDS cycles. However, the voltage presented a general descending trend which was corroborated by the SOC result as shown in Figure 12b. Hence, it can be said that EKF has the weakest capacity against the undulation which was caused by the charging and discharging conversion. The weakness in voltage tracking makes EKF curve the most fluctuant one in Figure 12b. Meanwhile, the estimation errors were also demonstrated in Figure 12c, and EKF has the lowest robustness to SOC fluctuation. The max data jitter reached 8% when the battery is working at low temperature. In addition, the MAE and RMSE were employed here again to evaluate the accuracy of EKF, UKF, and STF. Figure 13 shows the MAE and RMSE of the three algorithms at each temperature. For EKF and UKF, the MAE and RMSE decreased as the temperature increased. However, there was an anomaly that occurs in the case when the STF was at 0 °C. We believe that the reason for the anomaly was that when cells were loading at 0 °C, the external performance deteriorated and the voltage fluctuation became more severe when the current direction changed. Consequently, the strong tracking factor μ_k began to step in and corrected the estimation results by adjusting the a prior covariance $P_{k|k-1}$. The statistical data of the MAE and RMSE are summarized in Table 4.

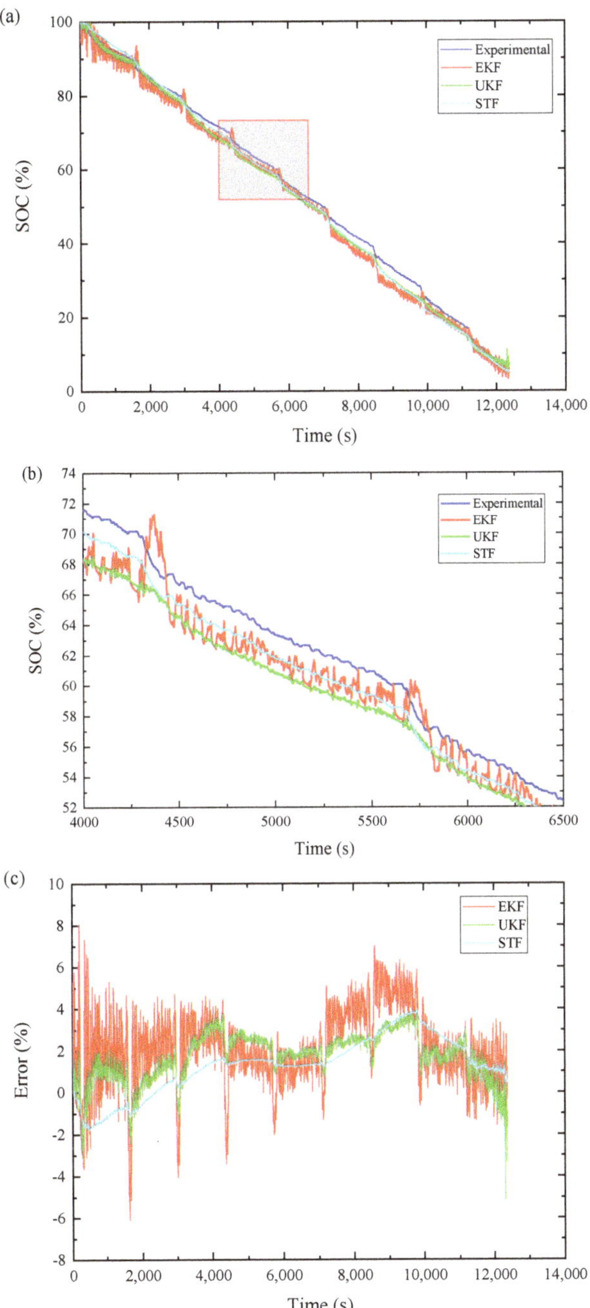

Figure 12. Comparative results of EKF, UKF, and STF for SOC estimation: (**a**) SOC estimation results; (**b**) the enlarged window from 4000 to 6500 s; (**c**) SOC estimation error.

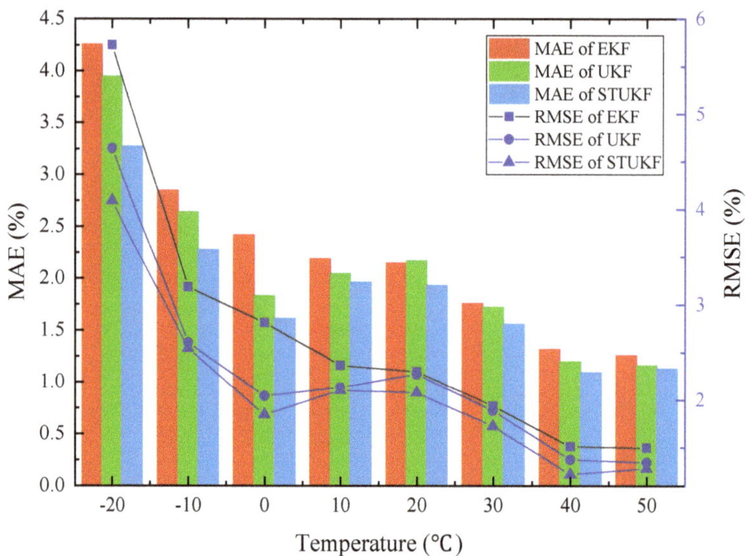

Figure 13. The MAE and RMSE of EKF, UKF, and STF at different temperatures.

Table 4. The statistical data of MAE and RMSE for EKF, UKF, and STF.

				MAE				
	−20 °C	−10 °C	0 °C	10 °C	20 °C	30 °C	40 °C	50 °C
EKF	4.2556	2.8469	2.4179	2.1860	2.1458	1.7587	1.3136	1.2558
UKF	3.9459	2.6408	1.8307	2.0419	2.1688	1.7202	1.1985	1.1593
STF	3.2692	2.2711	1.6102	1.9618	1.9292	1.5580	1.0925	1.1274

				RMSE				
	−20 °C	−10 °C	0 °C	10 °C	20 °C	30 °C	40 °C	50 °C
EKF	5.7263	3.1847	2.8112	2.3636	2.2978	1.9360	1.5081	1.4964
UKF	4.6393	2.6023	2.0419	2.1297	2.2687	1.8886	1.3736	1.3426
STF	4.0924	2.5410	1.8461	2.1017	2.0763	1.7240	1.2160	1.2795

In real applications, it is impossible to obtain the true initial SOC_0 before use. Therefore, the robustness against the unknown initial SOC_0 was also a crucial indicator of the estimation algorithms. Taking the experimental data from 0 °C as examples, Figure 14 shows the comparative results of the three algorithms with a SOC guess of 80% at 0 °C. From Figure 14a, all three algorithms could trace the true trajectory accurately and converge to the true value quickly even with large initial errors. From the enlarged window of 0 to 1500 s as shown in Figure 14b, it was clear that STF and UKF converged faster than EKF. Figure 14c demonstrates the estimation errors; the RMSEs of STF and UKF were 1.98% and 2.26%, respectively, which were 48.28% and 29.91% smaller than those of the EKF. The simulation results were consistent with the fact that UT transformation, which is utilized by UKF and STF, will produce less truncation error than first-order Taylor expansion, which is employed by EKF.

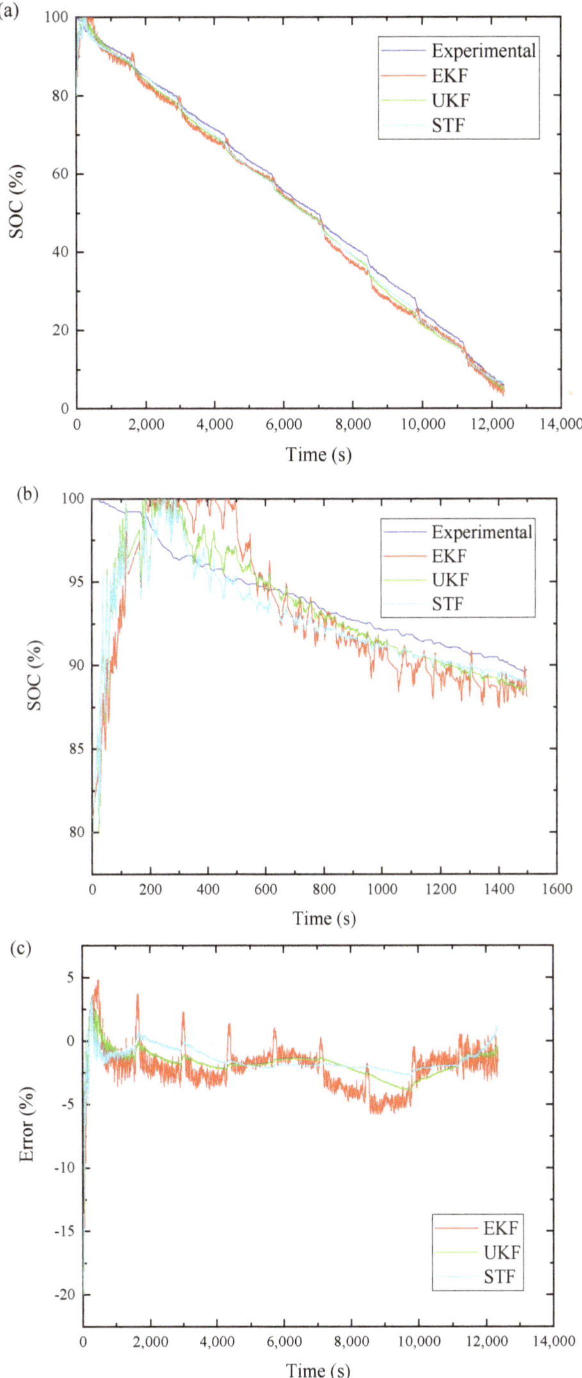

Figure 14. Robustness performance results with a SOC guess of 80% at a temperature of 0 °C: (**a**) SOC estimation results; (**b**) the enlarged window from 0 to 1500 s; (**c**) SOC estimation error.

To further discuss this topic, the initial SOC guess was set to 80%, 50%, and 30%, respectively. Figure 15 shows the part of the estimation results, and all the cases were summarized in Table 5 for comparison. There was an interesting point that could be noted in Figure 15. Regardless of the initial SOC guess values, all the curves converged at a specific point. After that point, the estimated curves present overlapped. We called this occurrence the convergence point, because it represented the convergence time of the algorithm. Judging from the position of the convergence point, the convergence times of UKF and STF were almost the same but a bit less than that of EKF. All the three algorithms were less than 2.5% of the whole operation time. This result is satisfactory for BMS applications.

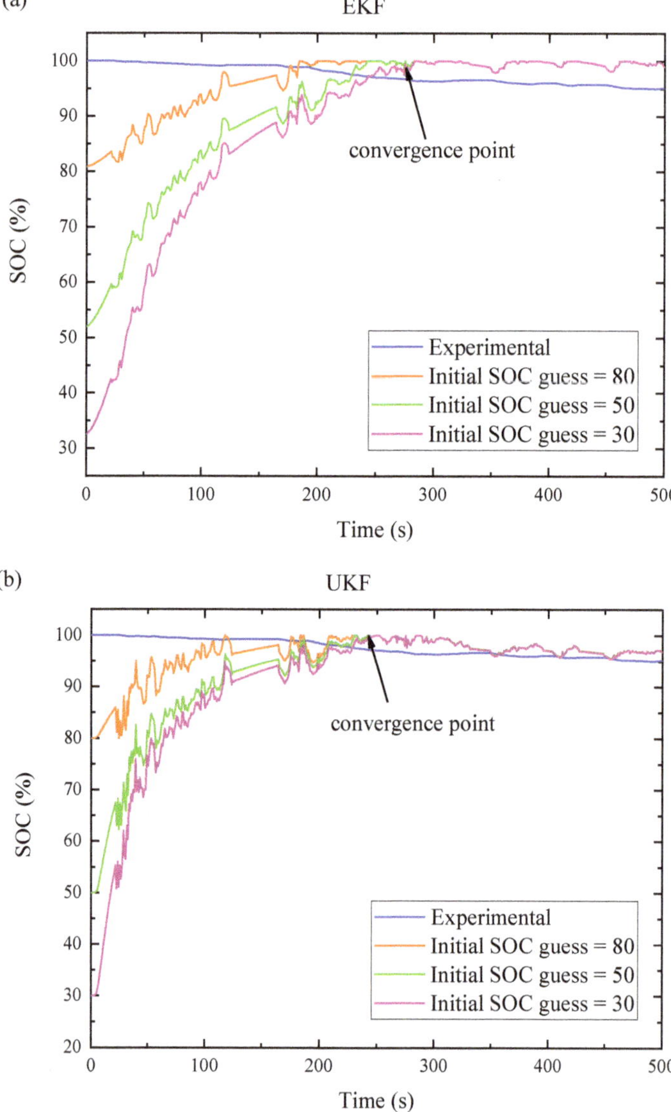

Figure 15. *Cont.*

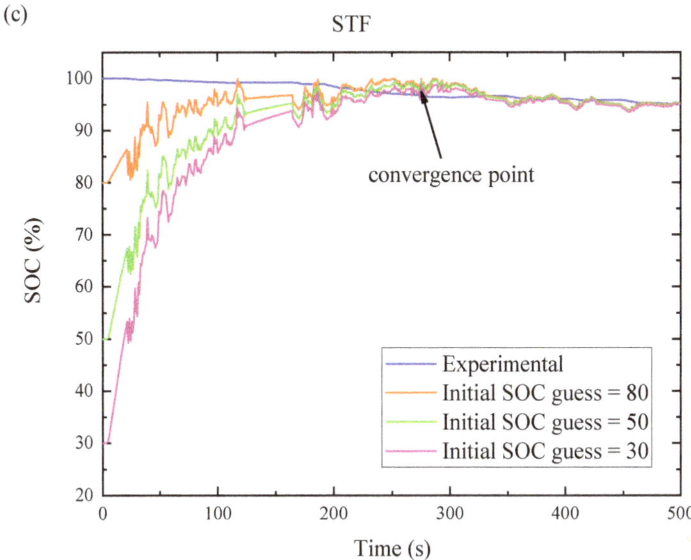

Figure 15. The robustness results for three algorithms: (**a**) EKF; (**b**) UKF; (**c**) STF.

Table 5. The MAE and RMSE by EKF, UKF, and STF at 0 °C.

MAE			
Initial SOC Guess (%)	EKF	UKF	STF
80	2.431	1.906	1.603
50	2.632	2.043	1.732
30	2.761	2.122	1.826
RMSE			
Initial SOC Guess (%)	EKF	UKF	STF
80	2.936	2.260	1.980
50	4.045	3.173	2.967
30	5.032	3.884	3.805

Finally, we were also interested in the robustness against the voltage measurement signal of the three algorithms. The quality of the acquired voltage signals was always a challenge for BMS, especially when the BMS was exposed to the vehicle's high electromagnetic interference environment. Hence, a constant −5 mV voltage measurement offset is designed to simulated the real-world voltage sensor drift in the BMS. We used the test datasheet of 30 °C for validation. The voltage offset data was embedded in the validation test datasheet of 30 °C with the initial SOC at 100%.

As shown in Figure 16, the STF had minimum error and the fastest convergence rate. Moreover, the MAE and RMSE of the three algorithms were tabulated in Table 6. With the same initial SOC, the MAE and RMSE of three algorithms were arranged from large to small by EKF, UKF, and STF sequence, thereby suggesting that STF was the most robust.

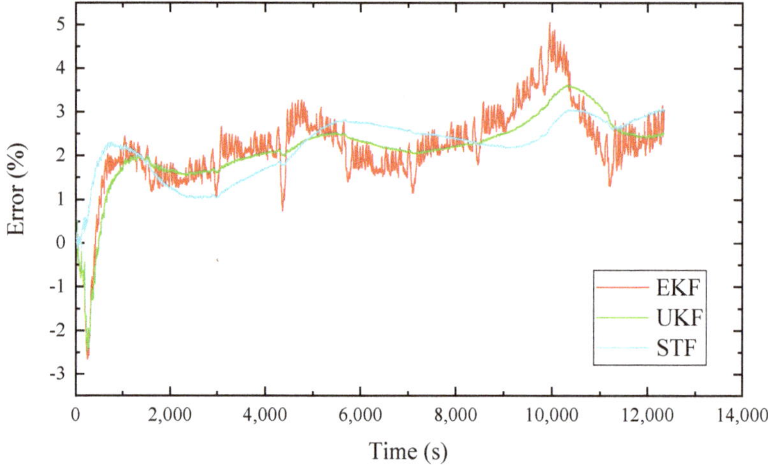

Figure 16. The estimated SOC error of the algorithms with -5 mV voltage offset at 30 °C.

Table 6. The MAE and RMSE of three algorithms when with voltage measurement offset.

Algorithm	MAE (%)	RMSE (%)
EKF	2.262	2.385
UKF	2.222	2.305
STF	2.186	2.280

6. Conclusions

We presented a modified ECM that took hysteresis and temperature effect into account for the first time. The proposed model aimed to improve the voltage tracking accuracy of the existing ECM whose applicability is often limited to steady current profiles and certain temperatures. With the offline hysteresis-SOC-T table correcting the OCV and the rest of the model parameters were estimated online, the improved model has the adaptability to ambient temperature rather than origin model. Then the improved model was applied to three model-based recursive estimators including EKF, UKF, and STF for SOC estimation. The procedures to execute each step for parameter identification and the algorithms were explained in detail. Lastly, the model and the estimators were validated by experimental test with commercial 18650 type $Li(Ni_{1/3}Co_{1/3}Mn_{1/3})O_2$ cells. The validation test consisted of several high dynamic FUDS microcycles, rest period, and a discharge process. This test was carried out under different temperatures ranging from 0 °C to 50 °C. The MAE and RMSE of the proposed model were smaller than the ordinary model in voltage estimation. In the same vein, the performances of EKF, UKF, and STF were also evaluated by MAE and RMSE. The STF algorithm outperformed UKF and EKF in estimation accuracy, robustness, and convergence behavior, thereby demonstrating a better solution for BMS application.

Author Contributions: C.Z. designed the experiments and wrote the paper; F.Y. and C.D. dominated and supervised the project; G.R. help to review and edit the paper.

Funding: This work was funded by the National Natural Science Foundation of China, grant number 51775393; the platform construction of Hubei province renewable energy intelligent vehicle, grant number 2016BEC116; the 111 Project, grant number B17034 and Research and Development of Products Test Cycle for New Energy Vehicles in China and the Fundamental Research Funds for the Central Universities, grant number WUT:2017-IVA-034.

Conflicts of Interest: The authors declare no conflicts of interest.

Nomenclature

C_N	Battery nominal capacity
U_L	terminal voltage
U_1	voltage over RC network
U_{oc}	open circle voltage
R_1	polarization resistance
C_1	polarization capacitance
I^+	battery charge current
I^-	battery discharge current
k	time step index
η	coulomb efficiency
h_k	hysteresis voltage
γ	positive constant to tune the rate of decay
$S(k)$	experimental data (battery voltage or SOC) at step k
$\hat{S}(k)$	estimated value (battery voltage or SOC) at step k
$k\,|\,k$-1	prior estimation state
ω_k	system process noise
v_k	measurement noise
ξ	identification parameters
χ	a n-dimension vector for state space
λ	scaling parameter
W_i^m	weights of mean
W_i^c	weights of covariance
α	small positive number
β	state distribution parameter
κ	scaling factor
Q_k	covariance matrix of the state noise
R_k	covariance matrix of the measurement noise
K_k	Kalman gain
ε_k	residual error
V_k	residual error sequence covariance matrix
μ_k	fading factor

List of abbreviations

ANN	Artificial Neural Network
BMS	Battery Management System
BSF	Battery Size Factor
CDKF	Central-Difference Kalman filter
ECM	Equivalent Circuit Model
EKF	Extended Kalman Filter
EMF	Electro-Motive Force
EV	Electric Vehicles
FL	Fuzzy Logic
FUDS	Federal Urban Driving Schedule
HPPC	Hybrid Pulse Power Characterization
KF	Kalman filter
MAE	Mean Absolute Error
OCV	Open Circuit Voltage
RLS	Least Square Method
RMSE	Root Mean Squared Error
SOC	State of Charge
STF	Strong Tracking Filter
SVM	Support Vector Machine
UKF	Unscented Klaman Filter

References

1. Li, Z.; Huang, J.; Liaw, B.Y.; Zhang, J. On state-of-charge determination for lithium-ion batteries. *J. Power Sources* **2017**, *348*, 281–301. [CrossRef]
2. Waag, W.; Fleischer, C.; Sauer, D.U. Critical review of the methods for monitoring of lithium-ion batteries in electric and hybrid vehicles. *J. Power Sources* **2014**, *258*, 321–339. [CrossRef]
3. Ismail, M.M.; Hassan, M.A.M. The state of charge estimation for rechargeable batteries based on artificial neural network techniques. In Proceedings of the IEEE Control Decision and Information Technologies (CoDIT), Hammamet, Tunisia, 6–8 May 2013; pp. 733–739.
4. Ismail, M.; Dlyma, R.; Elrakaybi, A. Battery state of charge estimation using an Artificial Neural Network. In Proceedings of the 2017 IEEE Transportation Electrification Conference and Expo (ITEC), Chicago, IL, USA, 22–24 June 2017.
5. Burgos, C.; Sáez, D.; Orchard, M.E.; Cárdenas, R. Fuzzy modelling for the state-of-charge estimation of lead-acid batteries. *J. Power Sources* **2015**, *274*, 355–366. [CrossRef]
6. Jiani, D.; Zhitao, L.; Youyi, W. A fuzzy logic-based model for Li-ion battery with SOC and temperature effect. In Proceedings of the 11th IEEE Control & Automation (ICCA), Taichung, Taiwan, 18–20 June 2014; pp. 1333–1338.
7. Álvarez Antón, J.C.; García Nieto, P.J.; de Cos Juez, F.J.; Sánchez Lasheras, F.; González Vega, M.; Roqueñí Gutiérrez, M.N. Battery state-of-charge estimator using the SVM technique. *Appl. Math. Modell.* **2013**, *37*, 6244–6253. [CrossRef]
8. Hansen, T.; Wang, C.-J. Support vector based battery state of charge estimator. *J. Power Sources* **2005**, *141*, 351–358. [CrossRef]
9. Sheng, H.; Xiao, J. Electric vehicle state of charge estimation: Nonlinear correlation and fuzzy support vector machine. *J. Power Sources* **2015**, *281*, 131–137. [CrossRef]
10. Surendar, V.; Mohankumar, V.; Anand, S.; Prasanna, V.D. Estimation of State of Charge of a Lead Acid Battery Using Support Vector Regression. *Procedia Technol.* **2015**, *21*, 264–270. [CrossRef]
11. Alvarez Anton, J.C.; Garcia Nieto, P.J.; Blanco Viejo, C.; Vilan Vilan, J.A. Support Vector Machines Used to Estimate the Battery State of Charge. *IEEE Trans. Power Electron.* **2013**, *28*, 5919–5926. [CrossRef]
12. Campestrini, C.; Horsche, M.F.; Zilberman, I.; Heil, T.; Zimmermann, T.; Jossen, A. Validation and benchmark methods for battery management system functionalities: State of charge estimation algorithms. *J. Energy Storage* **2016**, *7*, 38–51. [CrossRef]
13. Plett, G.L. Extended Kalman filtering for battery management systems of LiPB-based HEV battery packs Part 1. Background. *J. Power Sources* **2004**, *134*, 252–261. [CrossRef]
14. Plett, G.L. Extended Kalman filtering for battery management systems of LiPB-based HEV battery packs Part 2. Modeling and identification. *J. Power Sources* **2004**, *134*, 262–276. [CrossRef]
15. Plett, G.L. Extended Kalman filtering for battery management systems of LiPB-based HEV battery packs Part 3. State and parameter estimation. *J. Power Sources* **2004**, *134*, 277–292. [CrossRef]
16. Plett, G.L. High-Performance Battery-Pack Power Estimation Using a Dynamic Cell Model. *IEEE Trans. Veh. Technol.* **2004**, *53*, 1586–1593. [CrossRef]
17. Dai, H.; Sun, Z.; Wei, X. Online SOC Estimation of High-power Lithium-ion Batteries Used on HEVs. *IEEE Veh. Electron. Saf.* **2006**, 342–347.
18. Mastali, M.; Vazquez-Arenas, J.; Fraser, R.; Fowler, M.; Afshar, S.; Stevens, M. Battery state of the charge estimation using Kalman filtering. *J. Power Sources* **2013**, *239*, 294–307. [CrossRef]
19. Sepasi, S.; Ghorbani, R.; Liaw, B.Y. Improved extended Kalman filter for state of charge estimation of battery pack. *J. Power Sources* **2014**, *255*, 368–376. [CrossRef]
20. Pérez, G.; Garmendia, M.; Reynaud, J.F.; Crego, J.; Viscarret, U. Enhanced closed loop State of Charge estimator for lithium-ion batteries based on Extended Kalman Filter. *Appl. Energy* **2015**, *155*, 834–845. [CrossRef]
21. Pan, H.; Lü, Z.; Lin, W.; Li, J.; Chen, L. State of charge estimation of lithium-ion batteries using a grey extended Kalman filter and a novel open-circuit voltage model. *Energy* **2017**, *138*, 764–775. [CrossRef]
22. Li, Z.; Zhang, P.; Wang, Z.; Song, Q.; Rong, Y. State of Charge Estimation for Li-ion Battery Based on Extended Kalman Filter. *Energy Procedia* **2017**, *105*, 3515–3520.

23. Sun, F.; Hu, X.; Zou, Y.; Li, S. Adaptive unscented Kalman filtering for state of charge estimation of a lithium-ion battery for electric vehicles. *Energy* **2011**, *36*, 3531–3540. [CrossRef]
24. Xiong, R.; Gong, X.; Mi, C.C.; Sun, F. A robust state-of-charge estimator for multiple types of lithium-ion batteries using adaptive extended Kalman filter. *J. Power Sources* **2013**, *243*, 805–816. [CrossRef]
25. Roscher, M.A.; Sauer, D.U. Dynamic electric behavior and open-circuit-voltage modeling of LiFePO4-based lithium ion secondary batteries. *J. Power Sources* **2011**, *196*, 331–336. [CrossRef]
26. García-Plaza, M.; Eloy-García Carrasco, J.; Peña-Asensio, A.; Alonso-Martínez, J.; Arnaltes Gómez, S. Hysteresis effect influence on electrochemical battery modeling. *Electr. Power Syst. Res.* **2017**, *152*, 27–35. [CrossRef]
27. Dong, G.; Wei, J.; Zhang, C.; Chen, Z. Online state of charge estimation and open circuit voltage hysteresis modeling of LiFePO 4 battery using invariant imbedding method. *Appl. Energy* **2016**, *162*, 163–171. [CrossRef]
28. Li, H.; Song, Y.; Lu, B.; Zhang, J. Effects of stress dependent electrochemical reaction on voltage hysteresis of lithium ion batteries. *Appl. Math. Mech.* **2018**, *39*, 1453–1464. [CrossRef]
29. Christopherson, J.P. *Battery Test Manual For Electric Vehicles*; Idaho National Laboratory: Idaho Falls, ID, USA, 2015.
30. Li, Y.; Wang, C.; Gong, J. A combination Kalman filter approach for State of Charge estimation of lithium-ion battery considering model uncertainty. *Energy* **2016**, *109*, 933–946. [CrossRef]
31. Xiong, R.; Sun, F.; Gong, X.; Gao, C. A data-driven based adaptive state of charge estimator of lithium-ion polymer battery used in electric vehicles. *Appl. Energy* **2014**, *113*, 1421–1433. [CrossRef]
32. Li, D.; Ouyang, J.; Li, H.; Wan, J. State of charge estimation for LiMn2O4 power battery based on strong tracking sigma point Kalman filter. *J. Power Sources* **2015**, *279*, 439–449. [CrossRef]

Article

Extending Battery Lifetime by Avoiding High SOC

Evelina Wikner * and Torbjörn Thiringer

Department of Electrical Engineering, Chalmers University of Technology, 41296 Gothenburg, Sweden;
torbjorn.thiringer@chalmers.se
* Correspondence: evelina.wikner@chalmers.se; Tel.: +46-32-772-16-50

Received: 4 September 2018; Accepted: 1 October 2018; Published: 4 October 2018

Abstract: The impact of ageing when using various State of Charge (SOC) levels for an electrified vehicle is investigated in this article. An extensive test series is conducted on Li-ion cells, based on graphite and NMC/LMO electrode materials. Lifetime cycling tests are conducted during a period of three years in various 10% SOC intervals, during which the degradation as function of number of cycles is established. An empirical battery model is designed from the degradation trajectories of the test result. An electric vehicle model is used to derive the load profiles for the ageing model. The result showed that, when only considering ageing from different types of driving in small Depth of Discharges (DODs), using a reduced charge level of 50% SOC increased the lifetime expectancy of the vehicle battery by 44–130%. When accounting for the calendar ageing as well, this proved to be a large part of the total ageing. By keeping the battery at 15% SOC during parking and limiting the time at high SOC, the contribution from the calendar ageing could be substantially reduced.

Keywords: electric vehicle; Plugin Hybrid electric vehicle; Li-ion battery; modelling; measurements

1. Introduction

Electrified vehicles (EVs) are today becoming more and more common. Despite the significant decreases in the cost of the battery, the battery itself is still the dominating cost in the electric propulsion system. It is thus of very high importance that the battery in the vehicle degrades as little as possible during the life of the vehicle.

It is well known that high temperature has a negative impact on battery ageing [1–5]. Another factor that affects the ageing is the Depth of Discharge (DOD), where a larger DOD increases the ageing [6,7]. High charge and discharge rates accelerate the ageing [6–8]. For the investigated test cell, current rates over 3C are considered as high and rapidly deteriorates the cell if used for longer than a few seconds.

In vehicle applications, the discussion has been very focused on the driving range of the vehicle. A study by S. Karlsson [9] on drive patterns showed that the most common trip is 30–50 km long and that the majority of trips were even shorter.

The driving range of EVs today are from 100 km [10] and up to 499 km [11], meaning that the battery in most occasions only will be used in a small DOD before it is charged again. It is therefore of high interest to investigate the effects this user pattern can have on the vehicle battery.

Only few studies concerning small DODs can be found in the literature, and even fewer are considering the ageing effect of the placement of the DOD in the State of Charge (SOC) window. The ageing as a function of SOC levels for LiFePO$_4$ (LFP) battery cells was investigated in [1,2] and it was found that high and low SOC levels are substantially degrading the cell. Similar results were presented for an NMC cell [3]. However, extensive testing on large commercial cells in small DODs at different SOC levels have not been documented to this date.

The purpose of this article is to investigate the ageing dependency on SOC, using an extensive test series on large commercial cells as foundation. An empirical model is used to quantify this impact in synthetic drive cycles as well as more realistic drive cycles and user cases. Furthermore, a final objective is to establish the impact of calendar ageing, i.e., to plan the charging so that a more favourable SOC level can be used for the battery while the vehicle is parked.

2. Methods

2.1. Battery Cell Test Set-Up

A 26 Ah commercial pouch battery cell is used as reference object. The SOC window is defined, based on the recommendations from the cell manufacturer, by a lower and upper voltage limit of 2.8 and 4.15 V.

The cells are put in a test set-up as presented in Figure 1. The set-up comprises of the test cell, current connections, voltage and temperature sensors, a bakelite bottom and two aluminum plates to represent the cell placed in a battery pack. The equipments used for the lifetime testing were a MACCOR Series 4000, a PEC SBT0550, and a Digatron MCT 100-05-08 ME.

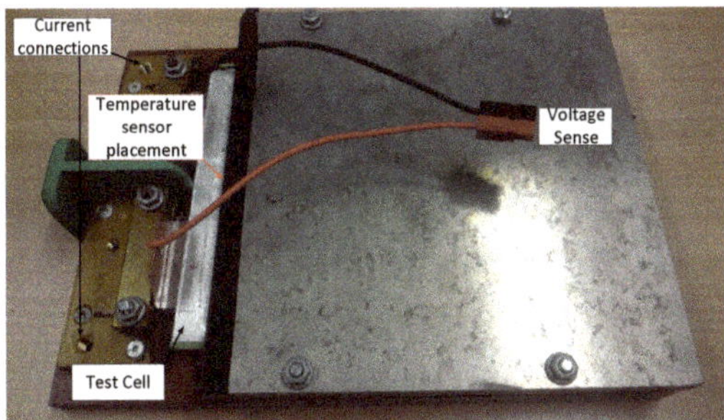

Figure 1. Test cell with holder.

Apart from registering the voltage and current, the temperature of the cell was also monitored. The ambient temperature was controlled by climate chambers Vötsch VT3 7034, VT3 4060, VC 4033, and Climate Temperature System (CTS) T-40/350.

The cells were tested using symmetric synthetic constant current (CC) cycles in different SOC levels, C-rates, and temperatures. The charge and discharge currents were always kept the same. A test cycle with a 1C charge and 1C discharge is here represented by the shorter notation 1C. In Table 1, the full test matrix is provided.

Table 1. Cycling test matrix for developing and verifying the empirical model for various State of Charge (SOC) intervals, temperatures, and C-rates.

SOC Interval (%)	25 °C			35 °C		
	1C	2C	4C	1C	2C	4C
0–10	-	x	-	-	x	-
10–20	x	x	-	x	x	x
20–30	-	x	-	-	-	-
40–50	-	x	-	-	x	-
60–70	x	x	x	x	x	x
70–80	-	x	-	-	-	-
80–90	-	x	-	-	-	-
0–30	-	-	-	-	x	-

2.2. Test Procedure

The capacity as well as the impedance of the battery cells were investigated on a regular basis during the lifetime, using an RPT (Reference Performance Test) procedure, at approximately each

200 Full Cycle Equivalent (FCE). The RPT procedure contains a 1C capacity measurement, followed by discharge and charge pulses at 5C and 1C at every 10% SOC level.

2.3. Measurement Results

To be able to compare the tested cells, regardless of DOD used during the tests, all test cycles were normalised to FCEs. In Figure 2, the capacity retention with various SOC levels are given as function of FCE. The lifetime tests were performed over a three-year period. Tests in higher SOC levels and higher C-rates did not require as much test time as tests conducted in lower SOC levels and using lower C-rates.

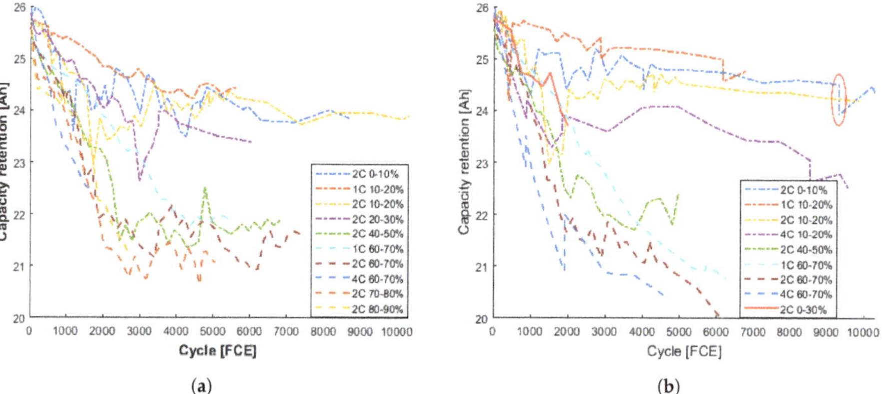

Figure 2. The ageing data used to parameterise the empirical ageing model in (a) 25 °C and (b) 35 °C. In (b), the data is also used to verify the model included. The vertical changes in capacity indicates stops in the testing due to maintenance, i.e., a rest period.

During this long test period, maintenance of the test equipment had to be performed at several occasions. An RPT test was performed before and after a planned maintenance stop, thus showing two results for the same cycle number where some of the results showed rather big differences. One of this events is marked with a read ellipse in Figure 2b where a vertical drop can be seen in the capacity curve. During these rest periods, the cells are in most cases losing capacity, though in the case for the aggressive test using a 4C rate, the cell instead recovered capacity. The cells cycled at low SOC levels had an average of three of these planned stops, while the shorter tests conducted at high SOC levels only had two in average. Furthermore, the data have higher fluctuations than expected, compared to cycle tests conducted using a large DOD (see [6]). Too few reports have been published on testing in small DODs to explain this. Similar fluctuations, though not as large, can be seen in the data presented by Schmalstieg et al. [3], supporting that tests in small DODs introduce a behaviour not seen when cycled in a large DOD.

The data show two distinct ageing patterns, one for the lower SOC levels and one for the higher SOC levels. This has also been observed when performing post mortem on the investigated cells where two cells in 60–70% SOC and 10–20% SOC were studied. The study showed that cycling in 60–70% SOC generated greater loss of cathode material, thicker SEI (Solid Electrolyte Interface) layer, and higher resistance on the anode compared to the cell cycled in 10–20% SOC [12]. Surprisingly, in the lower SOC levels, even the 4C test is performing well. All the 10% DOD data were used to develop the empirical ageing model. The 0–30% SOC data was used as a verification case for the model.

Calendar ageing in 25 °C and 35 °C at two SOC levels, 15 and 90% SOC, were performed for almost 700 days. The results, shown in Figure 3, clearly shows the difference between the low and high SOC levels, as well as the expected behavior for increased temperature.

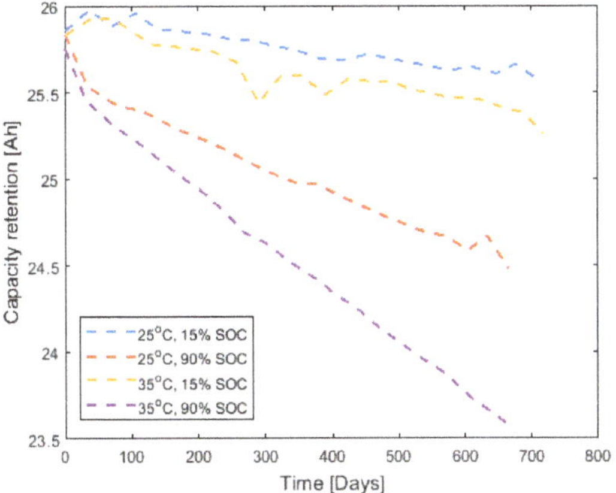

Figure 3. Calendar ageing in 25 °C and 35 °C at 15% and 90% State of Charge (SOC).

3. Model Development

By studying the shape of the capacity degradation as a function of FCE for the tested cells, it was found that a second order exponential function could represent the behaviour in an adequate way. The following equation was thus used

$$Cap(SOC, C\text{-}rate, T) = a \cdot e^{(b \cdot FCE)} + c \cdot e^{(d \cdot FCE)}. \tag{1}$$

The equation and its parameters are mathematical adaptations and can not be linked to a physical background. Since the initial capacity is known, the expression was simplified using $c = 26 - a$.

Equation (1) was fitted to all the data tested in 2C for both temperatures and the trend in the a, b and d parameters were studied. From this, a linear relation for parameter a was established as a function of SOC

$$a(SOC) = 6.2\frac{SOC}{90} + 0.093. \tag{2}$$

The fitting procedure was then repeated, now using a from Equation (2). Parameter b was found to be depending on SOC and C-rate as

$$b(SOC, C\text{-}rate) = \left(0.98\left(\frac{C\text{-}rate}{2}\right)^{3.3} + 0.01741\left(\frac{SOC}{20}\right)\right) \cdot$$
$$\left(\frac{-0.6045}{SOC^{2.4}} - 5.512 \cdot 10^{-4}\right) \cdot \left(\frac{SOC}{20}\right)^{(0.05C\text{-}rate^3 - 0.35C\text{-}rate^2 + 1)} \tag{3}$$

and the value for the b parameter as function of SOC for different C-rates is presented in Figure 4.

Now, using the values calculated from Equations (2) and (3) for a and b in Equation (1), the data were fitted again. The behaviour for parameter d with its SOC and temperature dependency was more difficult to capture. A decent fit to the data could only be achieved with a 4th order polynomial for the SOC dependence, not including the temperature dependency. With an R^2 value of only 0.8 for the SOC dependency, it was decided to use a look-up table for the d parameter (see Table 2). For the SOC intervals missing in the test data, interpolation was used to estimate the d parameter. To calculate the d parameter at temperatures between 25 and 35 °C, a linear relationship was used.

The parameter d showed also a dependency on C-rate. The difference between the $d(2C)$, $d(1C)$ and $d(4C)$ values in 35 °C was used to estimate the d value according to:

$$d(C\text{-}rate < 1, SOC < 50) = d(2C, SOC) + (C\text{-}rate - 1)\left(d(2C, 10\text{--}20) - d(1C, 10\text{--}20)\right),$$
$$d(C\text{-}rate < 1, SOC < 50) = d(2C, SOC) + (C\text{-}rate - 1)\left(d(2C, 60\text{--}70) - d(1C, 60\text{--}70)\right),$$
$$d(C\text{-}rate < 2, SOC < 50) = d(2C, SOC) + (2 - C\text{-}rate)\left(d(2C, 10\text{--}20) - d(1C, 10\text{--}20)\right),$$
$$d(C\text{-}rate < 2, SOC < 50) = d(2C, SOC) + (2 - C\text{-}rate)\left(d(2C, 60\text{--}70) - d(1C, 60\text{--}70)\right), \tag{4}$$
$$d(C\text{-}rate > 2, SOC \geq 50) = d(2C, SOC) + \frac{C\text{-}rate - 2}{2}\left(d(2C, 10\text{--}20) - d(4C, 10\text{--}20)\right),$$
$$d(C\text{-}rate > 2, SOC \geq 50) = d(2C, SOC) + \frac{C\text{-}rate - 2}{2}\left(d(2C, 60\text{--}70) - d(4C, 60\text{--}70)\right).$$

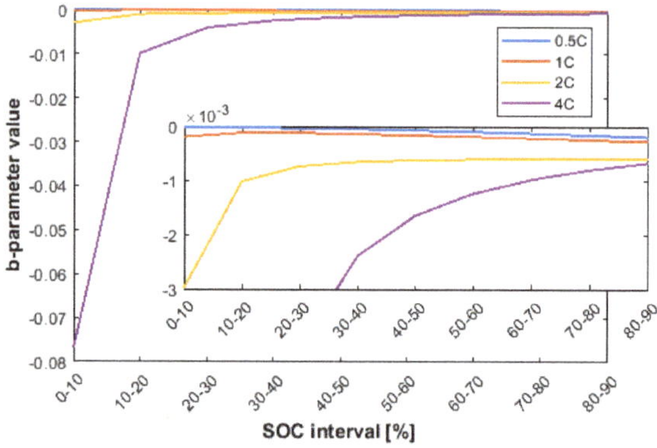

Figure 4. SOC and C-rate dependence for the b parameter in the ageing function.

Table 2. Look-up table for SOC and temperature dependency for parameter d.

SOC Interval (%)	0–10	10–20	20–30	30–40	40–50	50–60	60–70	70–80	80–90
$d(25\,^\circ C, 2C) =$	$[-6.620$	-3.210	-2.410	-3.700	-5.000	-2.550	-0.100	-0.010	$-0.001] \cdot 10^{-6}$
$d(35\,^\circ C, 2C) =$	$[-3.042$	-1.000	-0.400	-4.730	-9.000	-7.67	-6.331	-7.000	$-0.7] \cdot 10^{-6}$

The resulting performance of the degradation function after its parametrisation is shown in Figure 5a for the SOC behavior towards the experimental data for 25 °C, and in Figure 5b for 35 °C. The parametrised functions represent the measured results well. Furthermore, it can be clearly noted that the battery will last much longer if used in the lower SOC intervals even when using higher C-rates.

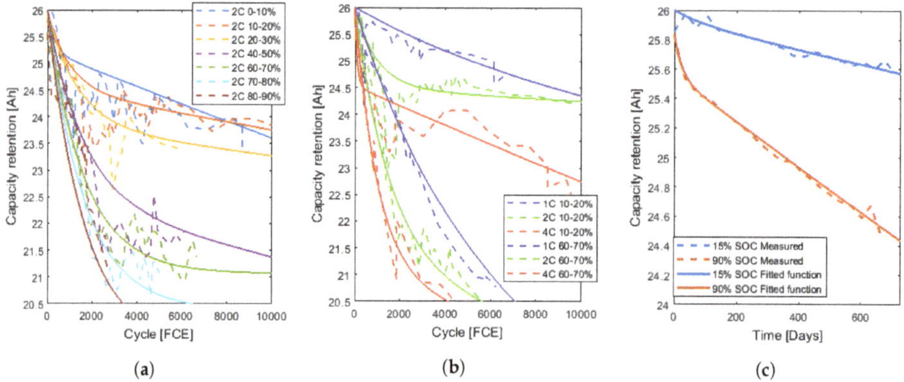

Figure 5. Parameterised ageing function together with input data, (**a**) 2C data in 25 °C; (**b**) different C-rates in 35 °C; and (**c**) calendar ageing in 25 °C at 15% and 90% SOC and the corresponding degradation equations.

For the calendar data, (1) was fitted to the data in 25 °C, though now as a function of days. The results are shown in Figure 5c, using the corresponding equations

$$Cap_{cal}(15\%, Days, 25\,°C) = 0.07433 \cdot e^{(-0.009545 \cdot Days)} + (26 - 0.07433) \cdot e^{(-1.900 \cdot 10^{-5} \cdot Days)}, \qquad (5)$$

$$Cap_{cal}(90\%, Days, 25\,°C) = 0.2900 \cdot e^{(-0.04173 \cdot Days)} + (25.843 - 0.2900) \cdot e^{(-6.153 \cdot 10^{-5} \cdot Days)}. \qquad (6)$$

3.1. Cycling Ageing Model

To translate the ageing function of the cell model into a battery pack ageing model, the capacity of the cells were normalized and from here on all the capacity degradation will therefore be presented in %.

In vehicles, the full capacity of a battery pack is normally not utilised, in order to extend the lifetime of the battery. When 100% SOC is displayed in the vehicle, i.e., fully charged, this could typically correspond to the single cells being charged to about 90% of the upper SOC limit given by the manufacturer. In the same way, the 0% SOC indication in the vehicle are normally not the 0% rated SOC of the battery cells. Typically, this will be around 10–15% of the battery cell SOC.

For this model, we will disregard the lower limit and allow the use until 0% SOC for the battery cell. The model validity is limited to 90% SOC as an upper limit. With these limits, the model can be run in intervals up to 90% DOD.

To calculate the capacity degradation in intervals larger than 10% DOD, the contributions from each 10% DOD were added together. For example, a 30% DOD in 0–30% SOC adds the contribution for 0–10%, 10–20% and 20–30% SOC. Figure 6 shows experimental data and the model results for this interval using a 2C rate in 35 °C. As can be noted, the ageing function predicts the result well also for the 0–30% SOC case.

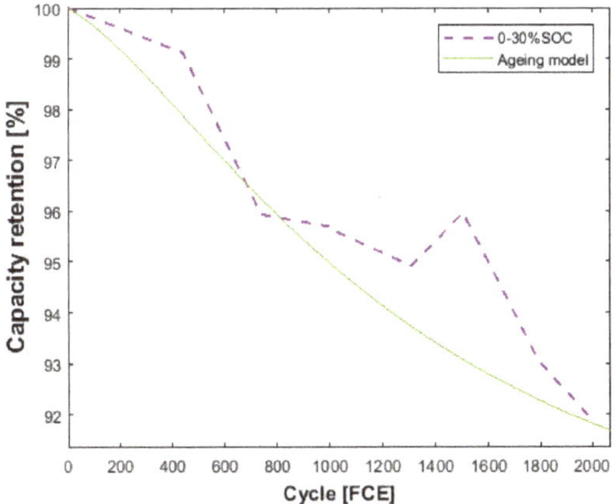

Figure 6. Capacity degradation in 0–30% SOC with 2C charge and discharge for the ageing function together with experimental data in 35 °C.

3.2. Vehicle Model and Vehicle Parameters

In order to investigate the impact in a vehicle application, three vehicles are designed and an energy consumption model is set up, based on the force balance of the vehicle according to

$$F_{acc} = m_v \frac{d}{dt} v(t) = F_{powertrain} - |F_{brake}| - |F_a| - |F_r| - F_g, \tag{7}$$

where the different forces are described as

$$\text{Air drag:} \quad F_a(t) = \frac{1}{2}\rho_a A_f c_d (v_{car} - v_{wind})^2, \tag{8}$$

$$\text{Rolling resistance:} \quad F_r(t) = c_r m_v g cos(\alpha), \quad v_{car} > 0, \tag{9}$$

$$\text{Grading:} \quad F_g(t) = m_v g sin(\alpha), \tag{10}$$

where ρ_A is the air density, c_d the coefficient of aerodynamic resistance, A the frontal area of the vehicle, v_{car} vehicle speed, v_{wind} wind speed, c_r the rolling resistance coefficient, m the vehicle mass, g gravitation constant on earth, and α road inclination. Both the wind speed and road inclination are set to 0, since this will only introduce small errors [10], and for this study these discrepancies will not affect the results.

The air drag force and rolling resistance force are always working against the direction of the movement. Many EVs have a regenerative braking system, though, for this study, the brake force is only considered to be a friction based brake system.

To account for the losses in the different components in the powertrain, an efficiency factor was used. The efficiency factor for an electric powertrain depends on all the efficiencies of the separate components in the powertrain, which, in turn, depend on their different working ranges, i.e., the vehicle speed and torque. The efficiency has been measured to be 0.6–0.9 [10], and for this work an efficiency factor of 0.8 has been used for the whole drivetrain.

Three vehicles are used representing a small Battery Electric Vehicle (BEV), Vehicle 1, a medium-sized BEV, Vehicle 2, and a Plugin Hybrid Electric Vehicle (PHEV), Vehicle 3. The data of the vehicles are presented in Table 3.

Table 3. Vehicle data for the three different vehicles used in the different case studies [10].

Parameter	ρ_{air} (kg/m³)	A_f (m²)	c_d (-)	c_r (-)	m_v (kg)	g (m/s²)	r (-)	α (deg)	W_{batt} (kWh)
Vehicle 1	1.225	2.13	0.35	0.015	1100	9.81	0.29	0	17
Vehicle 2	1.225	2.32	0.35	0.015	1700	9.81	0.33	0	24
Vehicle 3	1.225	2.32	0.35	0.015	1700	9.81	0.33	0	10

4. Drive Cycle and Drive Cases

4.1. Drive Cycle

The drive cycle used to estimate the energy consumption for the vehicles is the Worldwide Harmonized Light Vehicles Test Cycle (WLTC) for class 3b vehicles. For mixed driving, the full WLTC is used, for city driving the medium WLTC, and the Extra-High for highway driving. Figure 7 shows the WLTC with its corresponding parts.

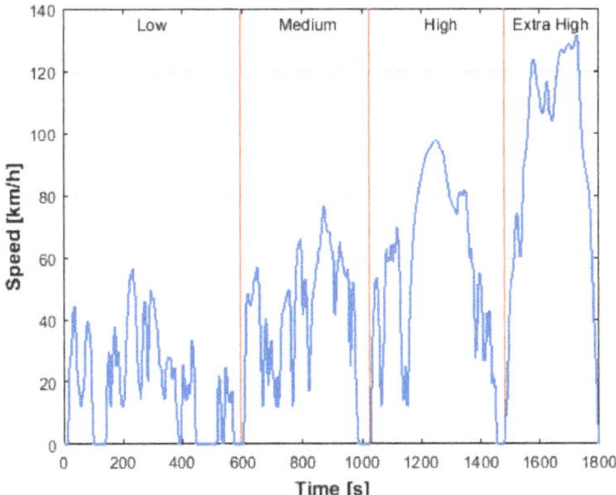

Figure 7. The Worldwide Harmonized Light Vehicles Test Cycle (WLTC) drive cycle for Class 3b vehicles used to estimate the energy consumption [13].

To achieve the distance in the two user cases, described in Section 4.4, the different drive cycles are repeated until the targeted distance is reached.

The C-rate required for the different drive cases is estimated by dividing the DOD required for the drive cycle with the operation time for the drive cycle according to

$$C\text{-}rate = \frac{SOC_{init} - SOC_{end}}{100 \cdot t_{drivecycle}[h]}. \tag{11}$$

4.2. Synthetic Drive Cases

To easier separate the effect of driving in different SOC intervals, some synthetic drive cases were investigated. Table 4 shows the different synthetic cases studied.

The battery temperature was set to 30 °C, since the temperature in the battery increases during usage and the battery cooling system often is designed to keep the temperature at or slightly below this temperature [14].

Table 4. Synthetic drive cases at 30 °C in various SOC, Depth of Discharge (DOD) and C-rates.

Parameters Case	DOD (%)	C-Rate	T_{batt} (°C)	SOC_{init} (%)
S1	10	0.5, 1, 2	30	30
S2	10	0.5, 1, 2	30	50
S3	10	0.5, 1, 2	30	90
S4	20	1	30	30
S5	20	1	30	50
S6	20	1	30	90
S7	50	1	30	50
S8	50	1	30	90

The benefit of using the synthetic drive cases is that they can provide very distinct and clear results. However, they are of course not fully representative towards real driving and in order to also provide some more realistic driving patterns, such cases were therefore also designed and investigated.

4.3. Drive Cycle Study

Table 5 displays the results from the vehicle model while running the different drive cycles with the different vehicles.

Table 5. Results from the vehicle model for the three vehicles running the Worldwide Harmonized Light Vehicles Test Cycle (WLTC) drive cycles.

	Drive Cycle	WLTC	WLTC Medium	WLTC Extra High
	Distance (km) [13]	23.3	4.8	7.2
	Time (s) [13]	1800	433	323
Vehicle 1	**C-rate**	0.39	0.23	1.09
	DOD (%)	19	3	10
Vehicle 2	**C-rate**	0.35	0.22	0.95
	DOD (%)	18	3	9
Vehicle 3	**C-rate**	0.85	0.53	2.27
	DOD (%)	42	6	20

Based on these results, the three vehicles were tested to see how long lifetime they would achieve if only running full discharges, i.e., 90% DOD, for the three different drive cycles. Five full discharges/charges during one week (4.5 FCE/week) and a battery temperature of 30 °C were assumed.

The vehicles will reach different distances where Vehicle 3 only has a short fully electric drive range. It should be mentioned that, in a real case, the hybrid mode would probably be used in the highway driving, unless a choice for pure electric mode exists. Still, for the sake of presenting the impact on battery ageing, we will assume only fully electric of a fully internal combustion engine (ICE) mode for this vehicle.

4.4. User Case Study

Going over to more realistic drive cases, two different cases for the three vehicles are investigated. Now, the drive distance is the same for all vehicles in order to show what lifetime the different vehicle batteries would achieve. In the case where the driving distance is too far for Vehicle 3, as mentioned before, fully electric drive is assumed until the battery is depleted, first after that is the ICE started.

User 1 has a driving distance of 25 km to work and works five days a week. During weekends, the user drives 75 km twice. All the trips are on highway and the WLTC Extra High is used to estimate the C-rate. The car is charged after each trip to 50% SOC (62% SOC for Vehicle 3) during weekdays and fully charged on weekends. The reference case is that the driver always charges fully whenever possible, and then leaves the vehicle with fully charged battery during parking.

User 2 has a driving distance of 25 km to work and works five days a week. During weekends, the user drives 75 km twice. The trips to work consist of mixed driving and the full WLTC is used to estimate the C-rate. For the longer trips in the weekend, the WLTC Extra High is used to estimate the C-rate. The car is charged after each trip to 50% SOC during weekdays and fully charged on weekends. The reference case is again that the driver always charges to full whenever possible.

The different cases with the parameters used as input to the ageing model for the three vehicles are shown in Table 6.

Table 6. Drive cases translated into DOD, inital State of Charge (SOC_{init}), C-rate, and Full Cycle Equivialent (FCE) per week to be used as ageing model input values.

Parameters Case: User, Vehicle	DOD1 (%) (SOC_{init} (%))	Trips Week	C-Rate	T_{batt} (°C)	DOD2 (%) (SOC_{init} (%))	Trips Week	C-Rate	T_{batt} (°C)	FCE Week
1: User 1, Vehicle 1	30 (50)	10	1.09	30	89 (89)	2	1.09	30	4.8
ref: User 1, Vehicle 1	30 (90)	10	1.09	30	89 (90)	2	1.09	30	4.8
2: User 1, Vehicle 2	26 (50)	10	0.95	30	77 (77)	2	0.95	30	4.1
ref: User 1, Vehicle 2	26 (90)	10	0.95	30	77 (90)	2	0.95	30	4.1
3: User 1, Vehicle 3	62 (62)	10	2.27	30	90 + ICE (90)	2	2.27	30	8
ref: User 1, Vehicle 3	62 (90)	10	2.27	30	90 + ICE (90)	2	2.27	30	8
4: User 2, Vehicle 1	21 (50)	10	0.39	30	89 (89)	2	1.09	30	3.9
ref: User 2, Vehicle 1	21 (90)	10	0.39	30	89 (90)	2	1.09	30	3.9
5: User 2, Vehicle 2	19 (50)	10	0.35	30	77 (77)	2	0.95	30	3.4
ref: User 2, Vehicle 2	19 (90)	10	0.35	30	77 (90)	2	0.95	30	3.4
6: User 2, Vehicle 3	46 (50)	10	0.85	30	90 + ICE (90)	2	2.27	30	6.4
ref: User 2, Vehicle 3	46 (90)	10	0.85	30	90 + ICE (90)	2	2.27	30	6.4

5. Results

The ageing model are in several cases extrapolated to C-rates lower than 1C. Although this means going outside the range where experiments could be used to calibrate the model, it is assumed that the trends regarding SOC- and temperature dependence are similar also for lower C-rates. By studying the calendar ageing data in Figure 3, it can be observed that, for 0 C-rate, the trend is very similar to the trend seen for the 1C results.

5.1. Synthetic Drive Cases

The result from the pure synthetic cases which were presented in Table 4 are given in Figure 8 as function of FCEs. The well known fact that increasing C-rate and increasing DOD are increasing the ageing can be seen. It also shows that the SOC level has a huge impact on the ageing, where higher C-rates in the lower SOC levels even generate less ageing than the lower C-rates in the higher SOC levels. As mentioned before, extrapolation was used for the lower C-rates seen in Figure 8a. The results produced are reasonable, although it would have been desirable to have experimental 0.5C cycling data.

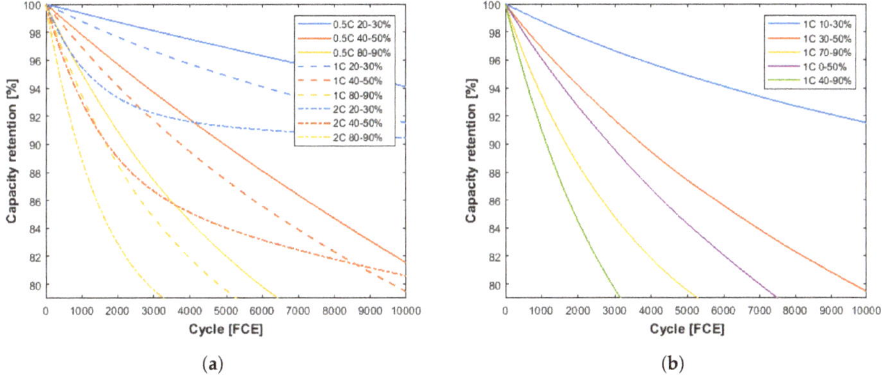

Figure 8. Model result at 30 °C starting at different SOC levels in (**a**) 10% Depth of Discharge (DOD) with 0.5C, 1C and 2C; and (**b**) 20% and 50% DOD with 1C.

5.2. Drive Cycle Study

The degradation of the battery as function of years for the three vehicles while running 90% DOD for the WLTC drive cycles are shown in Figure 9. Important to note is that these results only represent the impact of ageing due to driving and no consideration has been taken to the calendar ageing. The actual battery lifetime will therefore be shorter than the shown results. However, here the focus is on the comparison of the impact from different types of driving, calendar ageing will be investigated in Section 5.4.

The results show, as expected, that Vehicle 3 with the highest C-rate has the shortest lifetime in all three cycles. As the C-rate is reduced, this difference in lifetime is also reduced.

For the city driving, WLTC medium, see Figure 9a, all vehicles reach close to the same lifetime. Vehicles 1 and 2 show almost the same ageing trajectory. As expected, this is also the drive cycle that provides the longest lifetime.

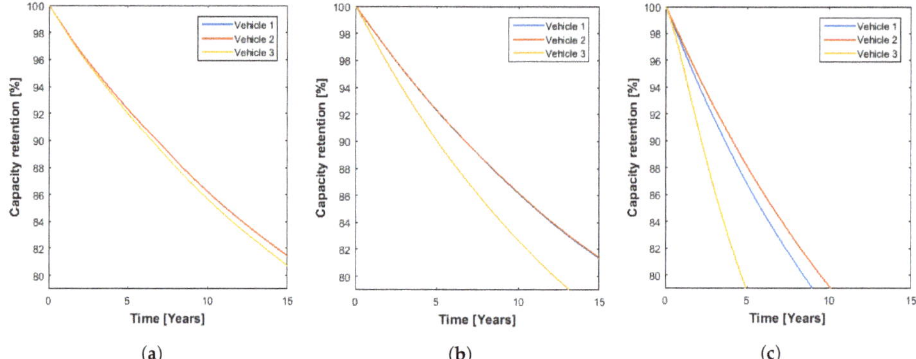

Figure 9. Ageing for the three vehicles while assuming driving according to, (**a**) WLTC medium; (**b**) WLTC; and (**c**) WLTC Extra High for five full charge/discharges per week all year around with a battery temperate of 30 °C.

The highway driving, Figure 9c, is the case showing the most aggressive ageing. Vehicles 1 and 2 manage to meet the common set warranty level of eight years [10], though not Vehicle 3.

5.3. User Case Study

It is obvious that using a larger battery in relation to the vehicle size will increase the lifetime. The result of the lifetime from the two user cases, for the three different vehicles, is presented in Figure 10. As expected, the longest lifetime is achieved for Vehicle 2, which has the largest battery in relation to the vehicle size. Vehicle 3, which has the smallest battery, in relation to the vehicle size, ageing.

In Figure 10a, the spread in lifetime for the vehicles is very large. The difference is mainly related to the C-rate, though also to the DOD and the SOC placement of the DOD. Vehicle 3 has the highest C-rate, however by not fully charging the battery (only to 62% SOC) when driving the shorter trips, the lifetime can be improved with more that 1.4 years, 59%, compared to Case 3 ref. For Vehicles 1 and 2, this difference is even larger in number of years, showing that the placement of the DOD highly affects the lifetime, increasing the lifetime with 44% and 56%, respectively.

In the second user case, Figure 10b, the C-rates are lower as mixed driving is assumed for the trips during the weekdays. Here, the DOD is also smaller and all vehicles can start at 50% SOC. The improvement in lifetime by placing the DOD at a lower SOC level is even more prominent here, especially for Vehicle 3 with an improved lifetime of 130%. Vehicles 1 and 2 each have a 70% lifetime increase.

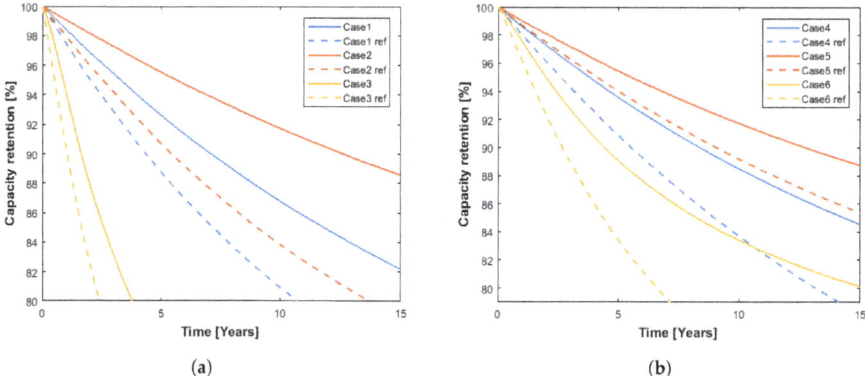

Figure 10. Model result for the three different vehicles at 30 °C while used according to (**a**) User 1; and (**b**) User 2.

5.4. Cycling and Calendar Ageing

Until now, only cycling ageing has been considered. The next step is to add calendar ageing. The cycling and calendar ageing were added together by the number of FCE driven each day and the percentage of hours in storage per day, $time_{cal}$, according to

$$Cap_{case1}(Cycling, Calendar) = Cap(FCE/day) + time_{cal} \cdot Cap_{cal}(SOC, Days, 25\,^{\circ}C). \qquad (12)$$

As can be seen in Figure 3, the temperature highly affects the ageing. The average temperature during a day is highly dependent on the location in the world and season; here, an average temperature of 25 °C has been used.

From (5) and (6), the lifetime, until reaching 80% capacity retention, when only considering calender ageing, is 10 and 32 years, for 15 and 90%, respectively. According to this, storing the battery at 15% SOC instead of 90% SOC will triple the lifetime.

The calendar ageing combined with driving was studied in the WLTC and WLTC Extra High drive cycle case for Vehicles 1 and 3. It is assumed that a full charge of the vehicle battery takes 8 h and the vehicle is fully charged five times a week, i.e., 40 h/week. The driving time differs, depending on cycle and vehicle, between 2–12 h/week. This results in 69–75% calendar ageing time each day. For a simpler comparison the same calendar time, 72%, was used in the following.

In Table 7, the contribution from calendar ageing are presented as percent reduced lifetime compared to the pure cycle ageing. For Vehicle 1 driven according to WLTC, only a low C-rate is required and the cycling ageing is very moderate, resulting in an approximate lifetime of 16 years at 80% capacity retention. Including calendar ageing at 90% SOC, the lifetime is reduced to just above six years, a 61% lifetime reduction. Storing the battery in 15% SOC results in a lifetime of 11 years, a 35% lifetime reduction. For Vehicle 3, the lifetime reduction when including the calendar ageing for the WLTC drive case at 90% SOC is 55%, and only 29% at 15% SOC.

Table 7. Ageing reduction when adding calendar ageing at two different SOC levels to the cycling ageing for six different cases. The values in brackets are only included for comparison since the cases are physically not possible.

Case	Calendar Time (%)	Cycling and Charging Time (%)	@15% SOC (%)	@90% SOC (%)	5/7 Days @90% and 2/7 Days @15% SOC (%)
WLTC, Vehicle 1	72	28	35	61	56
WLTC, Vehicle 3	72	28	29	55	49
WLTC Extra High, Vehicle 1	72	28	21	46	40
WLTC Extra High, Vehicle 3	72	28	13	32	27
Case 1	72	28	37	(63)	(58)
Case 1 ref	72	28	(27)	52	47

For the WLTC Extra High, the C-rates required are higher resulting in a shorter cycle lifetime. For Vehicle 3, the cycle lifetime is approximately 5 years and combining the calendar ageing reduces it to four and three years for 15% and 90% SOC, respectively. In this case, the contribution from the calendar ageing is much smaller, only reducing the lifetime with 13% and 32%. The corresponding values for Vehicle 1 is a reduction of 21 and 46%, respectively.

Now, considering one of the user cases, Case 1, where the vehicle is only used for 5 h/week and charged during 42.7 h/week, resulting in the vehicle battery being in storage 72% of the time.

In Figure 11, the results from Case 1 and Case 1 ref is presented, and three calendar ageing examples are shown. In the first example, it is assumed that the battery is used in accordance with Case 1, stored at 15% SOC and that the charging is planned to be finished just before usage. In the second example, Case 1 reference, the battery is charged as soon as the driving is finished and then stored fully charged, i.e., 90% SOC. In the third example, a mixed charge behaviour is assumed. In the reference case, the cell is only discharged to 60% SOC during weekdays and the battery is therefore assumed to be charged to 90% after each drive, i.e., storing the battery at 90% SOC 5 days a week.

During the two remaining days, the battery is assumed to be stored at low SOC, i.e., 15% SOC, and charged before the upcoming drive.

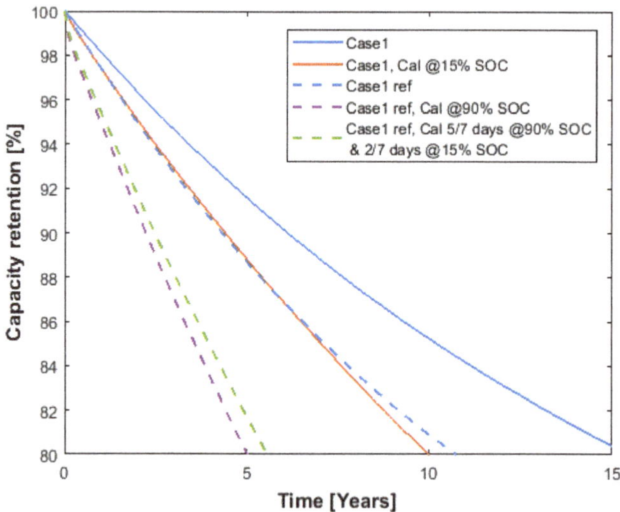

Figure 11. Resulting capacity degradation when including calendar ageing in 25 °C for user case 1 at 15%, case 1 reference at 90% SOC and a combination of 15% and 90%.

The effect of the calendar ageing is indeed important; in Case 1, keeping the battery at low SOC levels for as long as possible reduces the lifetime significantly, with 37%. In Case 1 ref, this impact is even larger. Storing the battery fully charged leads to a 52% reduction in lifetime. The small change by storing the battery at 15% SOC two days a week can prolong the lifetime with more than half a year, only reducing the lifetime due to calendar ageing with 47%.

All cases presented in Table 7 are not physically possible for Case 1 and Case 1 ref. However, by running the model according to the above stated examples for both cases, a clear trend can be seen; avoiding high SOC can prolong the lifetime by limiting the effects from the calender ageing.

6. Discussion

The lifetime tests on this cell showed, contradicting previous studies on LFP [1] and NMC cells [3], that the low SOC levels is less detrimental compared to higher SOC levels. Additionally, even higher C-rates show less ageing compared to the tests conducted in higher SOC levels. An important factor that needs to be pointed out is the definition of the SOC window. Schmalstieg et al. [3] defined the SOC window from 2.5–4.2 V, compared to the 2.8–4.15 V used for the here studied cells. If we consider the corresponding voltage level for the SOC intervals instead, the results produced by Schmalstieg [3] then prove to be similar to the ones presented in this study. From this, we can conclude that how we define the SOC window is important since the ageing is strongly linked to the SOC level, i.e. the voltage level of the cell.

Even though the impact of high SOC on ageing is a well known fact, few studies have focused on the impact of ageing when not using the full battery capacity. The costumers wishing for a longer driving range of EVs have controlled the research so far. As shown in the *Swedish car movement data project* [9], the most common trips are 30–50 km, and today the driving range of the EVs are longer than this, opening up the field for updated battery management strategies and, with this, the possibility of prolonging the battery lifetime.

The synthetic cases studied in Figure 8 show how the C-rate, DOD, and SOC level impact the ageing. The fastest ageing is achieved for S8, 1C 40–90%, while the same DOD, though in 0–50%, S7, manages more than double the number of FCE until reaching 80% capacity retention. Other cases showing faster ageing are S6, only using 20% DOD in 70–90% SOC, and all three C-rates for S3, 10% DOD in 80–90% SOC. These results indicates that the SOC level is more important than the C-rate.

The S1, S2 and S3 cases show even more clearly that the impact of SOC level is higher than the impact of C-rate. Important to point out is that, as the SOC level is increased, the impact of C-rate also seems to increase, S3 using 2C shows similar ageing as S8 though only using 10% DOD.

In S1-3, the model was extrapolated outside the available measured data, in order to estimate the ageing in 0.5C. Even if the trend seems fairly straightforward, the results for C-rates lower than 1C have to be treated with prudence.

For vehicle applications, the type of driving is very important for the ageing. Higher speeds require substantially higher C-rates, and the losses for the air drag grows with the square of the vehicle speed. If the battery is small, as for Vehicle 3, the required C-rate will be high, resulting in faster ageing.

For the three vehicles studied in this work, driving the distance of 25km on highway used 30, 26 and 62% DOD, respectively. When using mixed driving, only 21, 19, and 46% DOD were required to complete the trip. The highway driving required higher C-rate compared to the mixed driving, resulting in faster ageing. For all three vehicles, the choice to charge only up to 50% SOC highly improved the lifetime, for Vehicle 3 with 1.4 years. In the mixed driving case, the extension of the lifetime was even higher.

Including calendar ageing to the model showed that the lifetime could be reduced with more than 60% compared to the case only considering cycling ageing. This contribution could be reduced to 35% lifetime reduction by planning the charging, thus maintaining a low SOC level during parking.

Intuitively, the importance of calendar ageing is larger when the cycling ageing is moderate and the expected lifetime long. When the cycling ageing is rapid, the calendar ageing is less influential. For the most aggressive drive case, Vehicle 3 in WLTC Extra High, the lifetime reduction was as low as 13%, assuming storage of the battery at 15% SOC.

The consideration of DOD and the SOC placement needs to be studied further now that the BEV and PHEV becomes more common. Understanding the effects of this holds great potential for prolonging the lifetime of the battery in a vehicle application, as this study has shown.

In this study, it has been shown that much can be gained by taking a deeper look at the user pattern. If this is better understood, it can be used to reduce the costs of the vehicle through optimisation of the battery, and it can also give guidance to the user of how to better utilise the battery technology.

7. Conclusions

The final conclusion from this study is that there is a huge potential for prolonging the battery lifetime by avoiding high SOC values. Additional prolonging of the lifetime can also be reached by only charging the battery with the needed energy, using a small DOD, and to do this just before the driving. This strategic planning of the charging will limit the impact from the calendar ageing.

Author Contributions: Conceptualization, E.W. and T.T.; Methodology, E.W.; Software, E.W.; Validation, E.W.; Formal Analysis, E.W.; Investigation, E.W.; Resources, T.T.; Data Curation, E.W.; Writing—Original Draft Preparation, E.W.; Writing—Review and Editing, E.W. and T.T.; Visualization, E.W.; Supervision, T.T.; Project Administration, T.T.; Funding Acquisition, T.T.

Funding: This research was funded by the Swedish Energy Agency through the Battery Foundation, ABB Corporate Research, Västerås, Sweden, and Volvo Car Group, Göteborg, Sweden.

Acknowledgments: The authors would like to thank the Swedish Energy Agency, Volvo Car Group, Göteborg, Sweden, and ABB Corporate Research, Västerås, Sweden for the financing of this work.

Abbreviations

The following abbreviations are used in this manuscript:

BEV	Battery Electric Vehicle
CC	Constant Current
C-rate	Current required to discharge the battery in 1 h
DOD	Depth of Discharge
EV	Electric Vehicle
FCE	Full Cycle Equivalent
ICE	Internal Combustion Engine
LFP	Lithium Iron Phosphate
LMO	Lithium Manganese Oxide
NMC	Lithium Nickel Manganese Cobalt Oxide
PHEV	Plugin Hybrid Electric Vehicle
RPT	Reference Performance Test
SEI	Solid Electrolyte Interface
SOC	State of Charge
WLTC	Worldwide Harmonized Light Vehicles Test Cycle

References

1. Groot, J. State-of-Health Estimation of Li-Ion Batteries: Cycle Life Test Methods. Licentiate Thesis, Chalmers University of Technology, Gothenburg, Sweden, 2012.
2. Groot, J. State-of-Health Estimation of Li-Ion Batteries: Ageing Models. Ph.D. Thesis, Chalmers University of Technology, Gothenburg, Sweden, 2014.
3. Schmalstieg, J.; Käbitz, S.; Ecker, M.; Sauer, D.U. From accelerated aging tests to a lifetime prediction model: Analyzing lithium-ion batteries. In Proceedings of the World Electric Vehicle Symposium and Exhibition (EVS27), Barcelona, Spain, 17–20 November 2013; pp. 1–12. [CrossRef]
4. Ramadass, P.; Haran, B.; White, R.; Popov, B.N. Capacity fade of Sony 18650 cells cycled at elevated temperatures Part I. Cycling performance. *J. Power Sources* **2002**, *112*, 606–613. [CrossRef]
5. Wright, R.B.; Motloch, C.G.; Belt, J.R.; Christophersen, J.P.; Ho, C.D.; Richardson, R.A.; Bloom, I.; Jones, S.A.; Battaglia, V.S.; Henriksen, G.L.; et al. Calendar-and cycle-life studies of advanced technology development program generation 1 lithium-ion batteries. *J. Power Sources* **2002**, *110*, 445–470. [CrossRef]
6. Wikner, E. Lithium Ion Battery Aging: Battery Lifetime Testing and Physics-Based Modeling for Electric Vehicle Applications. Licentiate Thesis, Chalmers University of Technology, Gothenburg, Sweden, 2017.
7. Wang, J.; Liu, P.; Hicks-Garner, J.; Sherman, E.; Soukiazian, S.; Verbrugge, M.; Tataria, H.; Musser, J.; Finamore, P. Cycle-life model for graphite-LiFePO4 cells. *J. Power Sources* **2011**, *196*, 3942–3948. [CrossRef]
8. Ning, G.; Haran, B.; Popov, B.N. Capacity fade study of lithium-ion batteries cycled at high discharge rates. *J. Power Sources* **2003**, *117*, 160–169. [CrossRef]
9. Karlsson, S. *The Swedish Car Movement Data Project*; Final Report; Chalmers University of Technology: Gothenburg, Sweden, 2013.
10. Grunditz, E. Design and Assessment of Battery Electric Vehicle Powertrain, with Respect to Performance, Energy Consumption and Electric Motor Thermal Capability. Ph.D. Thesis, Chalmers University of Technology, Gothenburg, Sweden, 2016.
11. Tesla. Model 3. Available online: https://www.tesla.com/sv_SE/model3 (accessed on 4 September 2018).
12. Björklund, E.; Wikner, E.; Younesi, R.; Brandell, D.; Edström, K. Influence of state-of-charge in commercial $LiNi_{0.33}Mn_{0.33}Co_{0.33}O_2/LiMn_2O_4$-graphite cells analyzed by synchrotron-based photoelectron spectroscopy. *J. Energy Storage* **2017**, *15*, 172–180. [CrossRef]

13. DiselNet, Emission Test Cycles. Worldwide Harmonized Light Vehicles Test Cycle (WLTC). Available online: https://dieselnet.com/standards/cycles/wltp.php (accessed on 3 September 2018).
14. Damköhler, F. (Ed.). Thermal management in vehicles with electric drive system. *Porsche Eng. Mag.* **2011**, *1*, 34–36.

Article

Experimental Evaluation and Prediction Algorithm Suggestion for Determining SOC of Lithium Polymer Battery in a Parallel Hybrid Electric Vehicle

Insu Cho [1], Jongwon Bae [1], Junha Park [1] and Jinwook Lee [2,*]

[1] Department of Mechanical Engineering, Graduate School, Soongsil University, Seoul 06978, Korea; chois@soongsil.ac.kr (I.C.); rainmaker@soongsil.ac.kr (J.B.); junha6316@ssu.ac.kr (J.P.)

[2] School of Mechanical Engineering, Soongsil University, Seoul 06978, Korea

* Correspondence: immanuel@ssu.ac.kr; Tel.: +82-2-820-0929

Received: 3 September 2018; Accepted: 10 September 2018; Published: 13 September 2018

Abstract: The necessity of hybrid vehicles and electric vehicles is widely known for reasons such as fossil fuel depletion, climate change, emission norms mandated by regulations, and so on. Expansion of the hybrid vehicle market is a realistic way to respond to fuel efficiency regulations. Hybrid electric vehicles are continuously challenged to meet cross-attribute performance while minimizing energy usage and component cost in a highly competitive automotive market. Current optimization strategy for a parallel hybrid requires much computational time and relies heavily on the drive cycle to accurately represent driving conditions in the future. With increasing application of the lithium-ion battery technology in the automotive industry, development processes and validation methods for the battery management system (BMS) have attracted attention. The purpose of this study is to propose an algorithm to analyze charging characteristics and improve accuracy for determining state of charge (SOC), the equivalent of a fuel gauge for the battery pack, during the regenerative braking period of a TMED type parallel hybrid electric vehicle.

Keywords: parallel hybrid electric vehicle; regenerative braking; fuel consumption characteristics; energy efficiency; state of charge; lithium polymer battery

---

1. Introduction

The necessity of hybrid vehicles and electric vehicles is widely known for reasons such as fossil fuel depletion, climate change, emission norms mandated by regulations, and so on. With the addition of the electric motor, battery, and associated power electronics, the cost of powertrain and hence vehicles rise, often a hindrance for original equipment manufacturers (OEMs) and end-customers [1].

The hybrid electric vehicle (HEV) is an alternative to reduce the high dependence on petroleum products, because HEVs retain characteristics attributed to conventional vehicles such as performance, safety and trustworthiness, and reduced fuel consumption. Some modifications are necessary in the vehicle longitudinal dynamics equation to provide a specific power management control system because of the electrical power source addition that complements conventional engine and powertrain systems [2].

In 2017, the U.S. Environmental Protection Agency (EPA) approved fuel efficiency standards revealing high fuel efficiency levels. Expansion of the hybrid electric vehicle market is a realistic way to respond to fuel efficiency regulations. Recently, various technologies have been developed to improve efficiency of hybrid systems and reduce prices including configuration of a parallel mild hybrid system [3].

The automotive vehicle market has seen an increase in the number of hybrid electric vehicles, and forecasts predict additional growth [4].

Hybrid electric vehicles are continuously challenged to meet cross-attribute performance while minimizing energy usage and component cost in a highly competitive automotive market. As electrified vehicles become mainstream in the marketplace, hybrid customers are expecting more attribute refinements combined with the enhanced fuel economy benefits [5].

Energy management of HEVs is a difficult task due to the complexity of the total system in terms of electrical, mechanical, and thermal behavior [6]. HEVs are complex hardware systems often controlled by software that is complex to maintain, time-consuming to calibrate, and not always guaranteed to deliver optimal fuel economy. Hence, coordinated, systematic control of different subsystems of HEV is an attractive proposition [7].

Current optimization strategy for a parallel hybrid requires much computational time and relies heavily on the drive cycle to accurately represent the driving conditions in the future [8].

The 'State of Charge' of a battery is an estimation of remaining energy (percent). It is like a fuel gauge. Namely, SOC is the equivalent of a fuel gauge for the battery pack in a battery-driven electric vehicle. Measuring and knowing the SOC of a battery or battery bank is beneficial when applied towards alternative energy, or in any situation wherein we need to know its condition. It is critical to estimate and know the SOC of the secondary battery cells, defined as the available capacity (in Ah) and expressed as a percentage of its rated capacity.

With increasing application of lithium ion battery technology in the automotive industry, development processes and validation methods for the battery management system have attracted more attention. One fundamental function of the BMS is to continuously estimate the battery's SOC and state of health (SOH) to guarantee a safe and efficient operation of the battery system. For SOC as well as SOH estimations of a BMS, there are certain nonideal situations in a real vehicle environment, such as measurement inaccuracies, variation of cell characteristics over time, and so forth, that will influence the outcome of battery state estimation in a negative way. Quantifying such influence factors demands extensive measurements [9].

Dheenadhayalan, et al. presented a new approach to accurately estimate the state of charge of a lithium-ion battery based on an extended Kalman filter. This method uses equivalent circuits of a lithium-ion battery to develop state and observer equations of the extended Kalman filter [10].

Zhang, et al. proposed a big data-based algorithm to build a battery pack dynamic model and a probabilistic model for energy consumption prediction [11].

Arasaratnam, et al. revealed that to estimate the SOC of Li-ion batteries, we derive a normalized state-space model based on Li-ion electrochemistry and apply a Bayesian algorithm. Simulation results reveal that the PST can estimate SOC with accuracy higher than 95% without experiencing divergence. Bayesian algorithm is obtained by modifying Potter's square root filter and naming the Potter SOC tracker (PST) in this paper [12].

As a research on energy storage device for electric vehicles, Rukan Genc et al. expressed that super capacitor (SC) is suffering plenty of limitation factors: high leakage current, thermal aging, high equivalent series resistance (ESR), low voltage window, and so forth. To overcome such drawbacks, one should consider the variety of materials in a smart way. There are materials providing higher capacity values such as graphite. However, their performance in SC is limited and saturated. We believe new intelligent materials with alternative sources may have high impact in engineering to produce next-generation SC while there is plenty of space on the materials selection compared to Li-ion batteries. So, such hybrid SC can be used as a standard high-power source in electrical vehicles in the near future [13].

Waiard Saikong et al. studied comparison of energy consumption of electric vehicles in three different energy storage systems consisting of lead-acid battery, lithium-ion battery, and hybrid energy storage system (HESS) incorporating lithium-ion battery and ultra-capacitor. As a result, lowest specific energy and power cause battery oversizing of the lead-acid type to use with an intermittent load. For city traffic under frequent stop-and-drive conditions, HESS is an appropriate solution. However, lithium-ion battery and HESS do not have a significant difference on the extra urban driving cycle route [14].

In this study, the research target is to evaluate charging characteristics and propose a new algorithm for improving determination accuracy of SOC during the regenerative braking period of TMED type parallel hybrid electric vehicles.

2. Experimental Setup

In this study, vehicle dynamometer (EC Type, Jastec Co., Seongnam, Korea) and the hybrid electric vehicle (2011 YF Sonata HEV, Hyundai Motor Co., Seoul, Korea) were used for vehicle experiment, and data were acquired by the chassis dynamometer, current measurement device (PicoScope, Pico Technology, Cambridgeshire, UK), and OBD-II monitoring system (Cantalker, D&K Information Communication Technology, Anyang, Korea).

On-board diagnostics (OBD) is an automotive term referring to a vehicle's self-diagnostic and reporting capability. OBD systems provide the vehicle owner or service technician access to the status of various vehicle subsystems. The amount of diagnostic information available via OBD has varied widely since its introduction in the early 1980s versions of on-board vehicle computers.

Modern OBD implementations use a standardized digital communications port to provide real-time data in addition to a standardized series of diagnostic trouble codes, or DTCs, that allow one to rapidly identify and remedy malfunctions within the vehicle.

2.1. Chassis Dynamometer and Experimental Vehicle

The test vehicle was operated on an EC system dynamometer as shown in Figure 1. This method is a device for measuring driving force generated from the driving wheels of the vehicle by using the EC method. When the roller is rotated by the car wheel, the flywheel of the belt-connected brake (PAU) is rotated. When a current is applied to the coil of the brake PAU, it is magnetized and becomes an electromagnet, and a magnetic field is formed. This magnetic field generates an eddy current in the flywheel and forms an electromotive force in the direction opposite to rotational direction, and acts as a brake to measure power of the experimental vehicle [15].

The chassis power-measuring device used in this study conforms to the US BAR97 (Bureau of Automotive Repair) certification standard and supports IM 240 (Inspection & Maintenance Driving Cycle), ASM 2525 (Acceleration Simulation Mode), and CVS 75 (Constant Volume Sampler) modes, which are test modes of gasoline and LPG (Liquefied Petroleum Gas) vehicles according to current South Korean regulations. It also supports diesel vehicle test modes KD147, LugDown3, and NEDC (New European Driving Cycle).

The TMED type hybrid vehicle is a domestic vehicle and uses a clutch between the engine and motor to control the engine. The engine uses only the motor when starting from stop state and low speed, and only the power of the engine with high efficiency at the constant speed with low load. In a state requiring a large output such as acceleration and gradeability, the engine and motor are used simultaneously to improve fuel economy. Table 1 shows specifications of the hybrid electric vehicle used in this study.

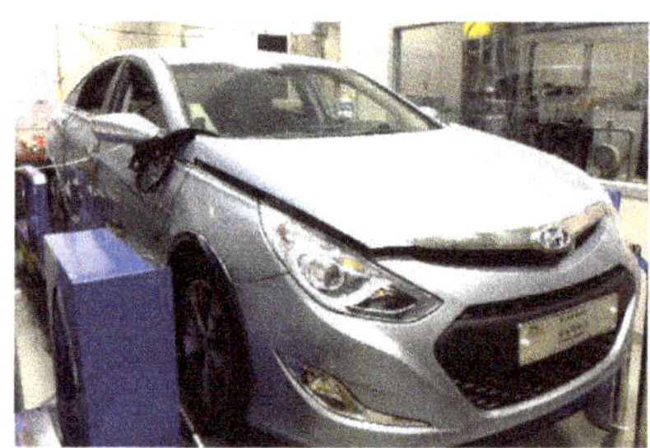

Figure 1. TMED type parallel hybrid vehicle and chassis dynamometer.

Table 1. Specifications of HEV used in this study.

HEV Specification	
Engine	2.0 L Atkinson cycle Double overhead cam
Max. Power	150 PS
Max. Torque	18.3 kg m
Motor	30 kW (41 PS)/20.9 kg m
HSG	8.5 kW
Battery	270 V Lithium polymer 5.3 Ah 1.5 kWh

2.2. Data Acquisition Device

To measure actual current change in this experiment, a precise current measurement was performed using a high-battery battery ammeter clamp (9287 Universal Clamp on CT, Hioki Co., Nagano, Japan) as shown in Figure 2a. An ammeter clamp was installed on high-voltage wire connected to the high-voltage battery and drive motor. Measured current data was stored through the PicoScope shown in Figure 2b and charge and discharge states were analyzed by analyzing measurement data. Figure 2c shows the real-time estimation of SOC, Motor RPM, Engine RPM, HSG RPM, Injection Duration, and Voltage. Specifications of the ammeter clamp are shown in Table 2. Figure 3 shows an overall schematic diagram of the experimental setup used in this study.

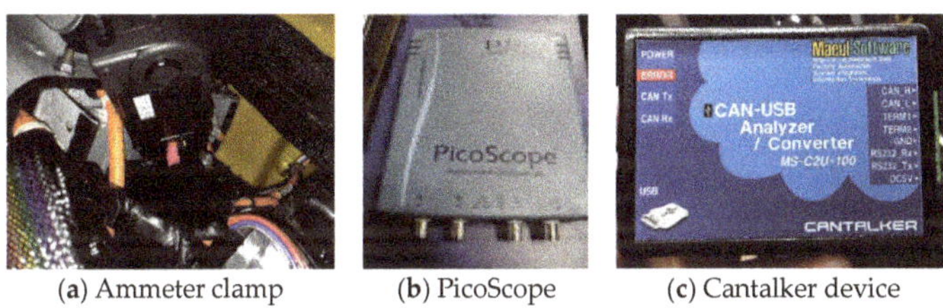

(**a**) Ammeter clamp (**b**) PicoScope (**c**) Cantalker device

Figure 2. Data acquisition devices for measuring power signal of HEV.

Table 2. Specifications of Clamp Current Sensor.

9278 UNIVERSAL CLAMP ON CT (HIOKI)	
Rate current	200 A AC/DC (continuous 350 A)
Frequency band width	DC to 100 kHz (±f.s)
Accuracy	±rdg. ±f.s
(DC and 45 to 66 Hz)	phase±
Max. circuit voltage	600 V peak (insulated wire)
Core jaw dia.	20 mm (0.79 in)

Figure 3. Overall schematic diagram of experimental apparatus.

2.3. Vehicle Driving Cycle and Experimental Condition

IM240 mode was performed for the operating mode for analyzing charging characteristics in regenerative braking operation according to the initial SOC state as shown in Figure 4. IM240 cycle is a chassis dynamometer schedule developed and recommended by the U.S. EPA for emission testing of in-use, light-duty vehicles in inspection and maintenance (I&M) programs implemented in a number of states. The EPA has also developed a guidance document specifying IM240 emission standards for use in I&M testing programs.

If the hybrid vehicle does not operate the accelerator pedal according to SOC management strategy, there is a situation wherein the battery is charged at the time of deceleration during inertia running. Since the deceleration section in the middle of the mode as shown in Figure 4 is the deceleration section due to inertia running, in this study, it is not regenerative braking during braking, and is excluded from the target.

This is because regenerative braking due to brake actuation is a clear reference to characterizing battery charging separately from hybrid operating modes.

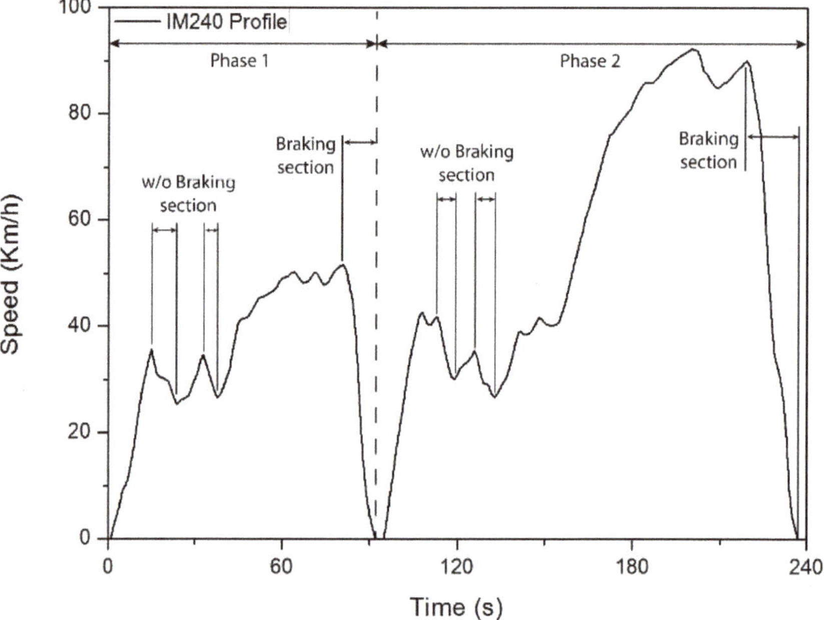

Figure 4. Driving profile of IM240 mode.

The energy of the hybrid vehicle used in this study is calculated by calculating J (Joule, kWh) to calculate the actual charged electric energy E (Energy, kWh), obtained by the following formula:

$$1\,J = 2.778 \times 10^{-4}\,Wh$$

$$1431\,Wh = 1431 \times \frac{1}{2.778 \times 10^{-4}}\,J$$

$$= 5{,}151{,}188\,J = 5151.188\,kJ$$

$$Gain\,E : \frac{Regeneration\,J}{5151.188} \times 100$$

In this study, regenerative braking period is an energy recovery mechanism that slows a vehicle into a form that can be stored in a secondary battery, and is ranged to the point wherein the vehicle speed is 0 after the brake is applied. Data interval of SOC and current (OBD, PicoScope) are extracted and analyzed in 350 ms units.

Experimental conditions are shown in Table 3, and initial SOC state was 35% and 65% (tolerance of IM240 mode is within ± 3 km/h). IM240 is a 240 s test representing a 3.1 km route with an average speed of 47.3 km/h and a maximum speed of 91.2 km/h.

Table 3. Experimental Conditions.

Parameter	Specification
Driving cycle	IM 240 Driving mode
Ambient temp.	24 ± 3 (°C)
Fan speed	Fan speed
Eco-mode	On
Test procedure	Confirm RBS → Vehicle running state (speed, SOC) in IM240 → GDS data processing → Analysis and comparison

3. Results and Discussion

Figure 5 shows the section wherein vehicle speed is zero from the moment when braking signal is input as the result of the IM240 test under SOC 35% and 65% conditions. Characteristic of the hybrid vehicle that optimizes SOC is that the engine operates from the start, operating in low SOC state, and begins to charge and maintains constant charging state even after the vehicle stops. On the contrary, when SOC is 65%, only the output main motor is driven, and the engine is in the resting state in the regenerative braking period.

As the vehicle used in this study is applied to the algorithm that maintains the SOC standard at approximately 50%, it is judged that variation rate is low at the condition where SOC is relatively high.

A total of three experiments were performed with respect to two regenerative brakes (Sections A and B) under SOC 35% and 65% conditions. As shown in Figure 5, A is a low-speed section for regenerative braking at approximately 50 km and B is a high-speed section for regenerative braking at approximately 90 km. In this study, all data in OBD and PicoScope were matched by Python's spline interpolation model at 0.35 s intervals. Spline interpolation is a form of interpolation wherein the interpolant is a special type of piecewise polynomial called a spline. Spline interpolation is often preferred over polynomial interpolation because the interpolation error can be made small even when using low-degree polynomials for the spline. Spline interpolation prevents the problem of Runge's phenomenon, when oscillation can occur between points interpolating using high-degree polynomials.

In the case of the OBD signal, SOC evaluation value is output based on calculated current after passing through the algorithm of the manufacturer's own, but it can be inferred that the PicoScope is caused by directly calculating current charged in the battery.

3.1. Comparison of Estimated Battery Charge

Table 4 summarizes changes in charge level due to the SOC signal from the OBD of the vehicle and current measured directly from the battery for hybrid to the PicoScope. Error value, the difference between the previous two values, is shown, and charge amount is also calculated for the SOC 65% condition in the same method.

From error results shown in Tables 4 and 5, current error measured in the PicoScope is larger than SOC fluctuation obtained through the OBD.

As a result of performing the experiment three times, error occurs in the result obtained with OBD and the PicoScope. After displaying measured charge in the repeated experiment according to each SOC condition, the difference between the OBD and the PicoScope was indicated. Although there is no consistent error in the specific measurement method, the difference according to the SOC reveals a larger error in the state of low charge (SOC 35%).

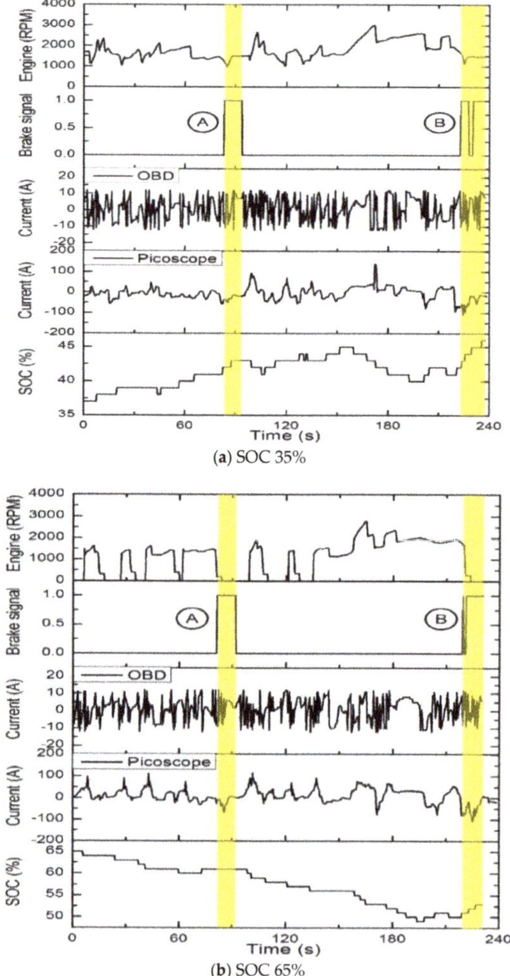

(**a**) SOC 35%

(**b**) SOC 65%

Figure 5. Charge–discharge current and regenerative braking characteristics.

Table 4. Comparing Electric Charging Variation between SOC and PicoScope at SOC 35%.

SOC 35% Condition	1st Test		2nd Test		3rd Test	
	SOC (OBD)	SOC (PicoScope)	SOC (OBD)	SOC (PicoScope)	SOC (OBD)	SOC (PicoScope)
Section A	1.62	2.00	1.46	1.00	1.80	2.00
Section B	3.29	3.24	3.49	4.00	3.26	3.97
Error (SOC 35%)	1st test		2nd test		3rd test	
Section A	0.38		0.46		0.20	
Section B	0.05		0.50		0.71	

Table 5. Comparing Electric Charging Variation between SOC and PicoScope at SOC 65%.

SOC 65% Condition	1st Test		2nd Test		3rd Test	
	SOC (OBD)	SOC (PicoScope)	SOC (OBD)	SOC (PicoScope)	SOC (OBD)	SOC (PicoScope)
Section A	0.65	0.00	0.84	1.00	0.81	0.00
Section B	2.28	2.00	1.65	2.00	2.61	1.89
Error (SOC 65%)	1st test		2nd test		3rd test	
Section A	0.65		0.16		0.81	
Section B	0.28		0.35		0.72	

Figure 6 shows the error of each condition by linear fit, and absolute value of the slope was analyzed. Each case represents a total of three experiments with the hybrid vehicle under the same conditions. The slope of error between zone A and zone B is approximately 3.67 times under SOC 35% condition and 2.75 times under 65% condition. SOC error increased because charging is performed more frequently in low-SOC condition. The amount of recovered electric energy in the high-speed Section B is large, influencing increase of the charge amount error. So, algorithms and applications for SOC estimation with higher accuracy are needed.

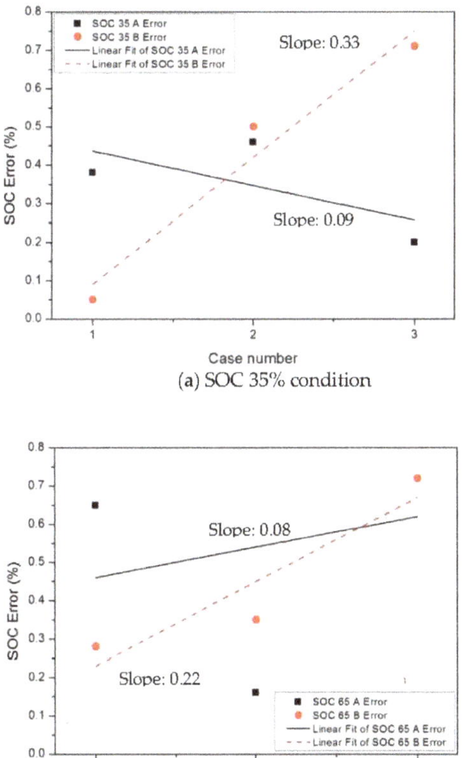

Figure 6. Linear fitting of error values between OBD data and PicoScope data.

3.2. Suggestion of SOC Calculation Algorithm

To improve accuracy for determining SOC, a complex experimental method is needed. To solve this problem, we propose an algorithm as shown in Figure 7 that reduces error due to current integration by measuring and integrating charge current of the secondary battery by the SOC at each braking deceleration section (low speed, medium speed, high speed).

So, two charging states are defined for the secondary battery, that is, the charging lower limit value and the charging upper limit value. When charge state is between upper limit value and lower limit value, the secondary battery is charged with increased voltage during braking, and the secondary battery is no longer charged in the driving phase.

First, during braking, the battery SOC is estimated at corresponding deceleration per deceleration section, and battery current is measured at predetermined time intervals during braking period.

Thereafter, the SOC value is calculated based on accumulated current value measured for a predetermined time, and compared with the SOC of the immediately preceding same condition, it is determined if the SOC value is equal to or smaller than a particular value (a).

If it is not less than the particular value (a), it is judged as an error. If it is less than the particular value (a), this value is compared with the SOC value at the start of braking.

If it is not less than the specified value (a), it is judged an error. If it is less than the specified value (a), this value is compared with the SOC value at the start of braking. If the value is less than the specific value (b), the actual SOC calculation result is transmitted to determine the final SOC value. If it is more than the specific value (b), it is judged an error.

When it is determined as an error, the arithmetic mean value of the two values is calculated and stored as the final SOC value.

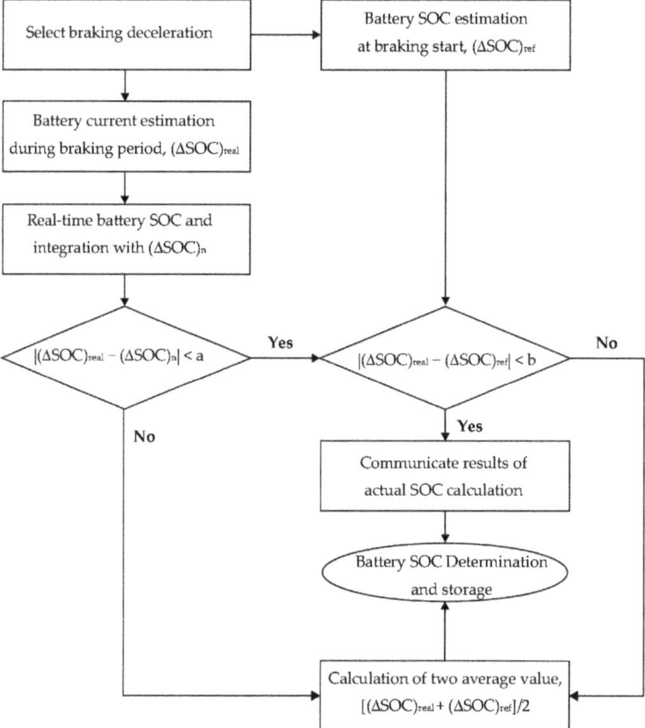

Figure 7. Algorithm flowchart for SOC determination.

4. Conclusions

This study analyzes charging characteristics of the TMED parallel hybrid electric vehicle during the regenerative braking period and proposes an algorithm for improving accuracy of SOC determination. Conclusions obtained by this study can be summarized as follows:

(1) In the low-speed Section A, the error of the charge amount due to the regenerative braking was less influenced by initial SOC. The amount of recovered electric energy in the high-speed Section B is large, influencing increase of the charge amount error.

(2) The error slope between Zone A and Zone B is approximately 3.67 times at SOC 35% and 2.75 times at 65%. SOC error increased because charging is performed more frequently in low-SOC condition.

(3) To reduce error due to current integration, we proposed an SOC estimation algorithm that measures and integrates the amount of charge current of the secondary battery by SOC at each braking deceleration section (low speed, medium speed, high speed).

(4) Also, it is expected that power performance of an electric vehicle through an energy supply device such as an ultracapacitor can be optimized, and improvement of power prediction accuracy of a more complicated system will be required.

Author Contributions: I.C. and J.P. conducted an HEV vehicle experiment to acquire data; J.B. supported the setting of the experimental equipment of the HEV vehicle and conducted preliminary experiments; I.C. and J.P. arranged and analyzed the acquired data; I.C. and J.L. analyzed the data and wrote the paper; J.L. supervised and advised all parts in this paper.

Funding: This research was funded by **MOTIE** (Ministry of Trade, Industry and Energy) in Republic of Korea.

Acknowledgments: This research was supported by the R&D project on Industrial Core Technology (2018) of **MOTIE** (Ministry of Trade, Industry and Energy) in Republic of Korea.

Conflicts of Interest: The authors declare no conflicts of interest.

Nomenclature

ASM 2525	Acceleration simulation mode 2525
BMS	Battery management system
CVS 75	Constant volume sampler mode 75
DTC	Diagnostic trouble codes
ESR	Equivalent series resistance
HESS	Hybrid energy storage system
IM 240	Inspection & maintenance driving cycle 240
NEDC	New European driving cycle
OBD	On-board diagnostics
OEM	Original equipment manufacturer
PST	Potter SOC tracker
SC	Super capacitor
SOC	State of charge
TMED	Transmission-mounted electrical device

Greek Symbols

Δ	Delta

Subscripts

real	Calculated current value
ref	Reference, calculated SOC value
n	Numerate value and integration of SOC from calculated current value

Superscripts

°   Degree

Abbreviation

f.s   Full scale
rdg   Percentage error relative to the reading

References

1. Prajapati, G.; Chachra, D.; Koona, R.; Warule, P.; Baidya, K. *Development of a P3 5-Speed Hybrid AMT*; SAE Technical Paper 2017-26-0090; SAE International: Warrendale, PA, USA, 2017. [CrossRef]
2. Eckert, J.J.; Corrêa, F.C.; Santiciolli, F.M.; Costa, E.D.; Dionísio, H.J.; Dedini, F.G. *Parallel Hybrid Vehicle Power Management Co-Simulation*; SAE Technical Paper 2014-36-0384; SAE International: Warrendale, PA, USA, 2014. [CrossRef]
3. Jung, D.B.; Cho, S.W.; Park, S.J.; Min, K.D. Application of a Modified Thermostatic control Strategy to Parallel Mild HEV for Improving Fuel Economy in Urban Driving Conditions. *Int. J. Automot. Technol.* **2016**, *17*, 339–346. [CrossRef]
4. Gahagan, M. *Lubricant Technology for Hybrid Electric Automatic Transmission*; SAE Technical Paper 2017-01-2358; SAE International: Warrendale, PA, USA, 2017. [CrossRef]
5. De Hesselle, E.; Grozde, M.; Adamski, R.; Rolewicz, T.; Erazo, M. *Hybrid Powertrain Operation Optimization Considering Cross Attribute Performance Metrics*; SAE Technical Paper 2017-01-1145; SAE International: Warrendale, PA, USA, 2017. [CrossRef]
6. Mustafa, R.; Schulze, M.; Eilts, P.; Küçükay, F. *Intelligent Energy Management Strategy for a Parallel Hybrid Vehicle*; SAE Technical Paper 2014-01-1909; SAE International: Warrendale, PA, USA, 2014. [CrossRef]
7. Sengupta, S.; Gururaja, C.; Hingane, S.; Prajwal, A.K.; Maniar, M.; Mikuláš, O.; Pekar, J. *Evaluation of Model Predictive and Conventional Method Based Hybrid Electric Vehicle Supervisory Controllers*; SAE Technical Paper 2017-01-1253; SAE International: Warrendale, PA, USA, 2017. [CrossRef]
8. Meng, Y.; Currier, P. *A System Efficiency Approach to Parallel Hybrid Control Strategies*; SAE Technical Paper 2016-01-1156; SAE International: Warrendale, PA, USA, 2016. [CrossRef]
9. Li, S.; Stapelbroek, M.; Pfluger, J. *Model-In-The-Loop Testing of SOC and SOH Estimation Algorithms in Battery Management Systems*; SAE Technical Paper 2017-26-0094; SAE International: Warrendale, PA, USA, 2017. [CrossRef]
10. Dheenadhayalan, P.; Nair, A.; Manalikandy, M.; Reghu, A.; John, J.; Rani, V.S. *A Novel Method for Estimation of State of Charge of Lithium-Ion Battery Using Extended Kalman Filter*; SAE Technical Paper 2015-01-1183; SAE International: Warrendale, PA, USA, 2015. [CrossRef]
11. Zhang, Z.; Huang, M.; Chen, Y.; Gao, D. *Big-Data Based Online State of Charge Estimation and Energy Consumption Prediction for Electric Vehicles*; SAE Technical Paper 2016-01-1200; SAE International: Warrendale, PA, USA, 2016. [CrossRef]
12. Arasaratnam, I.; Ahmed, R.; El-Sayed, M.; Tjong, J.; Habibi, S. *Li-Ion Battery SoC Estimation Using a Bayesian Tracker*; SAE Technical Paper 2013-01-1530; SAE International: Warrendale, PA, USA, 2013. [CrossRef]
13. Genc, R.; Alas, M.O.; Harputlu, E.; Repp, S.; Kremer, N.; Castellano, M.; Colak, S.G.; Ocakoglu, K.; Erdem, E. High-Capacitance Hybrid Supercapacitor Based on Multi-Colored Fluorescent Carbon-Dots. *Sci. Rep.* **2017**, *7*, 11222. [CrossRef] [PubMed]
14. Saikong, W.; Kulworawanichpong, T. Comparative Study of Energy Consumption for Electric Vehicles with Various On-board Energy Storage Systems. *Energy Procedia* **2017**, *138*, 81–86. [CrossRef]
15. Chung, J.H.; Kim, J.S.; Kim, J.W.; Lee, J.W. Study of Fuel Consumption Characteristics and Regenerative Braking Recovery Rate in a TMED Type Parallel Hybrid Electric Vehicle. *Trans. Korean Soc. Mech. Eng. B* **2016**, *40*, 485–494. [CrossRef]

 *applied sciences*

Article

Optimized Multiport DC/DC Converter for Vehicle Drivetrains: Topology and Design Optimization

Duong Tran [1,2], Sajib Chakraborty [1,2], Yuanfeng Lan [1,2], Joeri Van Mierlo [1,2] and Omar Hegazy [1,2,*]

1   Department of Electrical Machines and Energy Technology (ETEC) & MOBI Research Group, Vrije Universiteit Brussel (VUB), Pleinlaan 2, Brussels 1050, Belgium; dai-duong.tran@vub.be (D.T.); sajib.chakraborty@vub.be (S.C.); Yuanfeng.Lan@vub.be (Y.L.); joeri.van.mierlo@vub.be (J.V.M.); omar.hegazy@vub.be (O.H.)
2   Flanders Make, Heverlee 3001, Belgium
*   Correspondence: omar.hegazy@vub.be; Tel.: +32-2629-2992

Received: 20 June 2018; Accepted: 9 August 2018; Published: 11 August 2018

Abstract: DC/DC Multiport Converters (MPC) are gaining interest in the hybrid electric drivetrains (i.e., vehicles or machines), where multiple sources are combined to enhance their capabilities and performances in terms of efficiency, integrated design and reliability. This hybridization will lead to more complexity and high development/design time. Therefore, a proper design approach is needed to optimize the design of the MPC as well as its performance and to reduce development time. In this research article, a new design methodology based on a Multi-Objective Genetic Algorithm (MOGA) for non-isolated interleaved MPCs is developed to minimize the weight, losses and input current ripples that have a significant impact on the lifetime of the energy sources. The inductor parameters obtained from the optimization framework is verified by the Finite Element Method (FEM) COMSOL software, which shows that inductor weight of optimized design is lower than that of the conventional design. The comparison of input current ripples and losses distribution between optimized and conventional designs are also analyzed in detailed, which validates the perspective of the proposed optimization method, taking into account emerging technologies such as wide bandgap semiconductors (SiC, GaN).

Keywords: interleaved multiport converter; multi-objective genetic algorithm; hybrid electric vehicles; losses model; wide bandgap (WBG) technologies; Energy Storage systems

---

1. Introduction

The recent technological developments in the fields of batteries, electric motors and power electronics interface (PEI) support electro-mobility transition. These advances introduce several possibilities, generating a broad variety of powertrain architectures as presented in [1]. Multiport converters (MPCs) are increasingly attracting research interest. By employing MPC, it is possible to diversify the energy sources so that power system availability can be increased in hybrid electric powertrain systems. MPCs can provide a unique solution to combine multiple energy sources (i.e., battery, supercapacitor, fuel Cell), which have different voltage-current (V-I) characteristics and energy density versus power density performances. Figure 1 illustrates the power distribution role of MPC in the Electric Variable Transmission (EVT)-based powertrain, which has been recognized as a promising and emerging technology for vehicles.

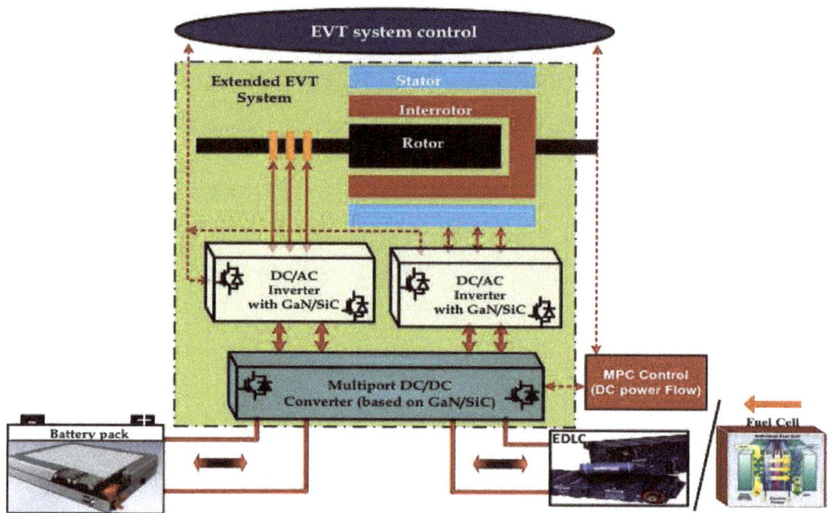

Figure 1. Multiport Converter integrated into the EVT-based powertrain.

A family of MPCs is classified as non-isolated and isolated topology. In an isolated topology, the sources are usually connected to a half bridge converter to achieve DC-AC conversion, which allows the use of a high frequency transformer for high voltage ratios. In addition, the transformer enables the galvanic isolation between the inputs and outputs. Furthermore, with a transformer, it is easier to connect several outputs at different voltage level by properly selecting the number of turns of the secondary winding. However, for high power applications, the transformer is a bulky component. Thus, in vehicular applications, non-isolated topologies are preferred. Non-isolated MPC can be divided into parallel ports topologies and shared components topologies. The advantage of a shared component topology is that less switches are needed and thus the price is expected to be lower; however some topologies as presented in [2] are unable to deliver energy simultaneously. Parallel ports instead inherently increase the system reliability as the ports can be driven either simultaneously or independently, relying on different active components [3]. The advantage of paralleling the ports in a single converter is the gain in flexibility on the energy management techniques, compared to shared component. In fact, the ports can be controlled separately. In addition, better packaging and thermal management can be achieved compared to standard DC/DC converters. Despite being lighter compared to isolated converters, weight and cost is the main drawback in confront of shared components MPCs. Therefore, the interleaving technique can be applied to reduce the global converter weight and cost. Several MPCs have been developed based on [3] as in [4,5], proving a high efficient and compact solution for vehicle applications with a centralized control. Figure 2 shows a typical configuration of non-isolated bidirectional interleaved MPC in the vehicle powertrain.

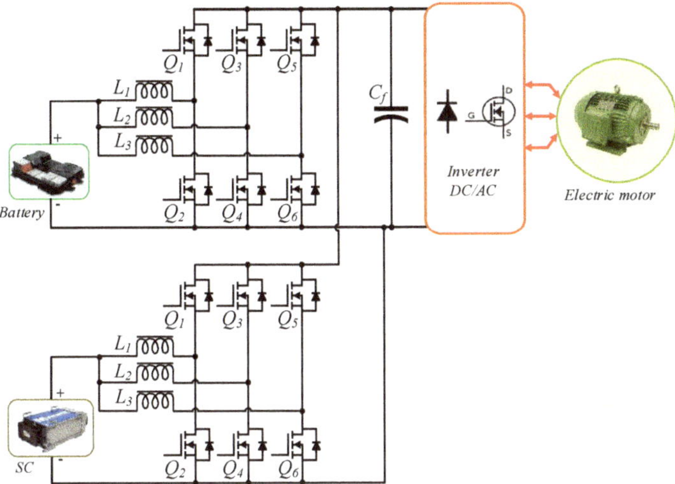

Figure 2. Non-isolated MPC using an interleaved bidirectional converter for each port in the Full Electric Bus powertrain.

The design of MPC power electronics system requires multidisciplinary knowledge and a large number of design variables in different engineering fields (electrical, magnetic, thermal, mechanical). The ability and expertise of the designer may end up with a good, but not optimal design. It may require more effort for further iterations through hardware prototype testing to obtain better performance in term of efficiency and weight. Therefore, mathematical optimization techniques and computer-aided software have been developed to tackle the design problem. In the literature, optimization for the power converter design can be classified into two main techniques: the gradient-based techniques using the derivative information and the metaheuristic-based techniques using the stochastic search. Several gradient-based methods have been employed for the optimization problem of power converter. Seeman et al. [6] used the Nonlinear Programming (NP) based on Lagrangian functions to optimize a switched-capacitor converter. Wu et al. [7] used the Augmented Lagrange Penalty Function (ALPF) technique to optimize a half-bridge dc-dc converter. Sergio et al. [8] utilized the Sequential Quadratic Programming (SQP) for a boost power-factor-correction converter optimization. However, the main drawback of gradient-based algorithms is that if the design space contains several local minima, there is a possibility that a gradient-based optimizer may be trapped by a local minimum, and the result depends on the selection of the initial design point. So far in the literature, no existing gradient-based algorithms are able to find the global optimization solution [9]. Furthermore, the gradient-based methods are mathematically guided algorithms, which require stringent mathematical formulations, causing a complexity of the system when variables increase. The metaheuristic-based optimization method was thus developed to solve the derivative-free and multi-objective problem with a large number of variables. Metaheuristic methods imitate the best features in nature, based on natural selection and social adaption. Among numerous metaheuristic methods, Genetic Algorithm (GA) [10] and Particle Swarm Optimization (PSO) [11] have been widely utilized to design the circuity of a power converter. The GA can be applied to optimize the medium-frequency transformer [12] of isolated converter, heatsink and bus capacitor volumes [13] of a three-phase inverter to archive minimum weight, losses and cost, with respect to constraints of design specification and physical limitation of components. The PSO, combined with Differential Evolution (DE), helps find an optimal transformer design for the Dual-Active-Bridge converter [14], the resonant tank of isolate bidirectional series resonant converter [15], and the inductor using EE core geometry [16]. So far, almost all researches have formulated a single objective formulation (efficiency, or weight, or cost [8]) or aggregated multiple conflicting objectives (weight, and loss, and cost) into one single objective. The multi-objective optimization of transformer design was solved by the Non-dominated Sorted Genetic Algorithm (NSGA-II) [12]; however, the final design selected from Pareto-solutions was not explained clearly.

In this paper, a new optimization methodology as shown in Figure 3 is proposed for the non-isolated interleaved MPC. The main characteristics of the interleaved converter are analyzed by predefined specifications such as maximum power P_{max}, input voltage V_{in}, output voltage V_{out} and required input current ripples I_{ripple} for battery and Supercapacitor (SC) ports, to derive objective functions that can be used for optimization problem formulation. A multi-objective genetic algorithm and Average Ranking technique are then employed to find three design variables (the number of phases N_{ph}, switching frequency f_{sw}, and core index representing geometry parameters of the core) to simultaneously minimize three trade-off objectives: weight of inductors, converter losses and input current ripples. To closely attain a practical design, a database was developed, which included commercial available inductor cores (23 cores) and Insulated Gate Bipolar Transistor (IGBT) modules (8 IGBT modules) for the optimization process. A hypothesis is that an optimal solution can be found in the database. The SOLIDWORKS software (Solidworks Premium 2018, Dassault Systèmes SolidWorks Corporation, Waltham, MA, USA, 2018) is then used to visualize the physical structure of optimal inductors that are imported into the COMSOL Multiphysics (Version 5.3a, COMSOL, Inc., Burlington, MA, USA, 2018), a Finite Element Method (FEM)-based software, to simulate the electromagnetic field of the designed inductor. The curve fitting Matlab function is also used to plot the inductance value in the function of air-gap and number of turns. The simulation results show reduction of weight in the optimized design compared to a conventional design.

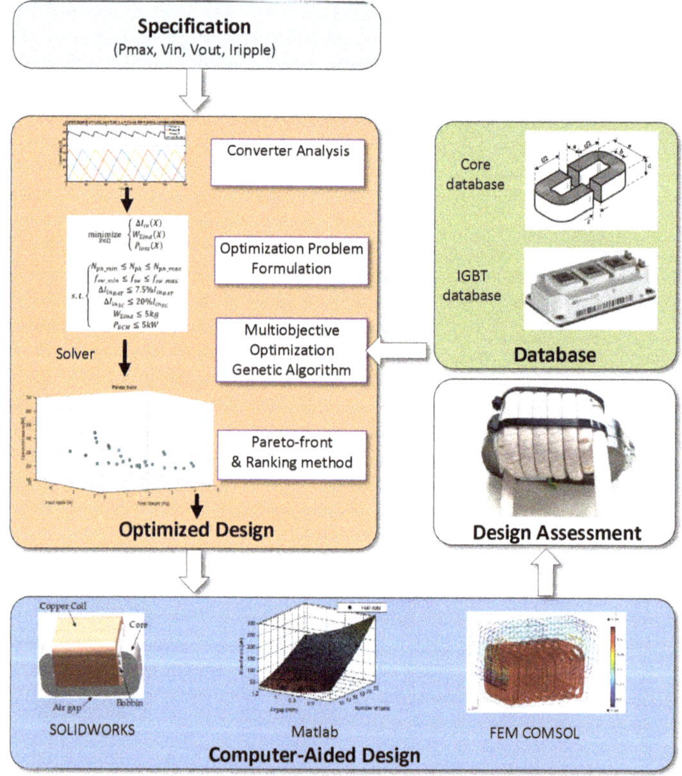

Figure 3. Proposed design optimization methodology.

The organization of this paper includes six sections. Section 2 presents an analysis of input current ripple, weight of inductor, and converter losses. Section 3 formulates the multiple objectives optimization problem. Section 4 explains about the proposed design framework based on NSGA-II and Average Ranking method and Section 5 discusses improvement in the optimized design compared to the conventional design. The conclusions are given in Section 6.

2. Analysis of Non-Isolated Interleaved DC-DC Converter

As analyzed in the Introduction, isolated MPCs are usually used for low-power systems due to the limitation of magnetic designs for transformers. Non-isolated MPC topologies are more suitable for high-power powertrain system of vehicles. Thus, in this paper, the topology of MPC in Figure 2 has been selected for optimization. The MPC consists of two Interleaved Bidirectional Converters (IBC) interfacing with a battery port and SC port, respectively. The objective of design optimization is to minimize input current ripple, converter losses, and inductor weight of IBC for each port. Some key parameters are foreseen intuitively to have an impact on optimization objectives. Firstly, if the switching frequency f_{sw} increases, the size of the inductor core can be reduced; however, switching loss is increased. Secondly, the more number of phases added, the more current flowing in each phase can be reduced, leading to less semiconductor losses and reduction in inductor sizing. However, this adds more weight to the power electronics system. Finally, a bulky inductor can reduce the input current ripple that is important for battery lifespan; however, it introduces more weight and core losses. Therefore, the relationship of optimization objectives and design variables needs to be thoroughly analyzed.

2.1. Input Current Ripple

In the IBC, the phase interleaving technique enables one to decrease the input current ripple by shifting each interleaved phase by $360°/N_{ph}$ such that the current is cancelled out, as shown in Figure 4a. More phases are added in the interleaved converter; the ΔI_{in} peak is further reduced for each additional phase added. However, even though the amplitude of the ripples is reduced, the frequency of the ripples increases with increase in the number of phases.

The input current ripple cancellation effect of an interleaved converter in the Continuous Conduction Mode (CCM) has been analyzed and quantified in [17–19]. However, their derived equations are complicated to use in formulating the optimization problem. For the sake of convenience in the optimization process, we rewrite the function of input current ripple in terms of the duty ratio.

According to [17–19], the function of input current ripple ΔI_{in} with regard to the duty ratio D can be recognized as a parabolic equation, $\Delta I_{in} = aD^2 + bD + c$, as shown in Figure 4b. As can be seen, if the IBC has N_{ph} phases, the peak of current ripple occurs separately in N_{ph} regions of duty ratio. Each region is associated with an integer number $k \in [0, N_{ph}-1]$.

The vertex of the parabola and the points where ΔI_{in} is zero (dashed red circles in Figure 4b) are considered to determine the coefficients a, b and c. It is noted that the peak of the inductor current ripple $\Delta \hat{I}_L$ in one single phase is calculated as Equation (1), therefore, the peak of the input current ripple $\Delta \hat{I}_{in}$ becomes Equation (2):

$$\Delta \hat{I}_L = \frac{V_o}{4 f_{sw} L} \tag{1}$$

$$\Delta \hat{I}_{in} = \frac{\Delta \hat{I}_L}{N_{ph}} \tag{2}$$

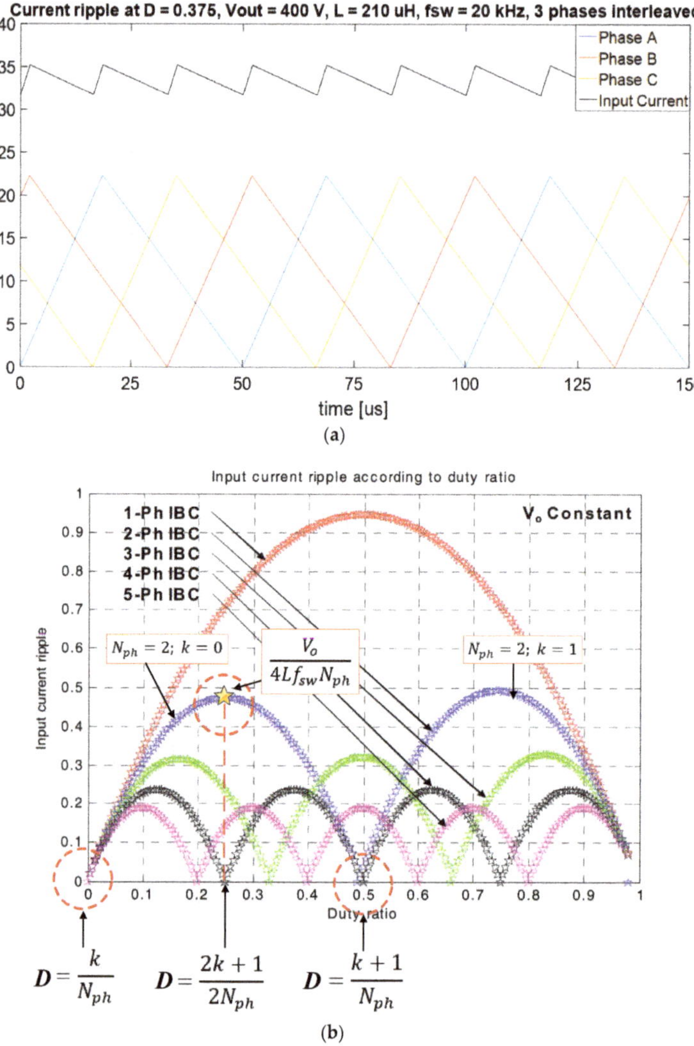

Figure 4. (**a**) Reduction of input current ripple through phase interleaving. (**b**) Normalized peak-to-peak input current ripple as a function of the duty ratio.

As a result, the input current ripple analytical equation can be derived by the following system in Equation (3):

$$\begin{cases} a\left(\dfrac{k}{N_{ph}}\right)^2 + b\dfrac{k}{N_{ph}} + c = 0 \\[2ex] a\left(\dfrac{k+1}{N_{ph}}\right)^2 + b\dfrac{k+1}{N_{ph}} + c = 0 \\[2ex] a\left(\dfrac{2k+1}{2N_{ph}}\right)^2 + b\dfrac{2k+1}{2N_{ph}} + c = \dfrac{V_o}{4Lf_{sw}N_{ph}} \end{cases} \tag{3}$$

By solving the system, the analytical expression of ΔI_{in} is derived as Equations (4) and (5):

$$\Delta I_{in} = \frac{V_o\left(D - \frac{k}{N_{ph}}\right)}{L f_{sw}}\left[1 - N_{ph}\left(D - \frac{k}{N_{ph}}\right)\right] \tag{4}$$

$$1 - \frac{V_{max}}{V_o} < D < 1 - \frac{V_{min}}{V_o} k \in [0, N_{ph} - 1] \tag{5}$$

2.2. Weight of Inductors.

The weight of an inductor mainly consists of the weight of copper coil (or winding) and the weight of core. An inductor design is illustrated in Figure 5. The weight of a coil W_{coil} is dependent on the physical structure of winding (i.e., length, diameter, number of turns, and number of layers) while the weight of the core W_{core} is dependent on the type of core (i.e., material, shape). The air gap g is added to prevent saturation in the inductor.

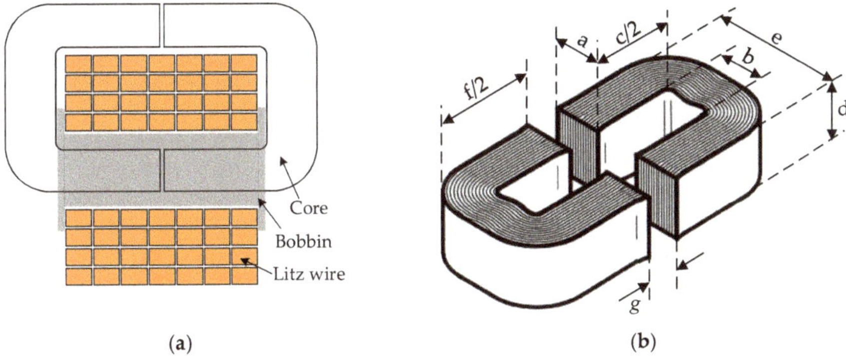

Figure 5. (**a**) Structure of an inductor; (**b**) Geometry parameters of a core.

To determine the weight of a coil W_{coil} (kg), it is first necessary to calculate the number of turns N_t (turn) as in Equation (6); $K_u(-)$ is utilization factor, W_a (mm^2) is window area, A_{cu} (mm^2) is cross section of the wire:

$$N_t \leq \frac{K_u W_a}{A_{cu}} \tag{6}$$

The diameter of the wire conductor D_{wire} (mm) is calculated as Equation (7), where J_{rms} $\left(\frac{A}{mm^2}\right)$ is current density:

$$D_{wire} = \sqrt{\frac{4}{\pi}\frac{\hat{I}_L}{J_{rms}}} \tag{7}$$

which gives the number of layers N_{layer} as in Equation (8):

$$N_{layer} = floor\left(\frac{D_{wire} N_t}{c}\right) + 1 \tag{8}$$

where the function $floor\ (X)$ returns the nearest integer less than or equal to X. And c (mm) is the height window of the core as shown in Figure 5b.

The number of turns on each layer $N_{t,\ layer}$ (turn) is then calculated as Equation (9):

$$N_{t,\ layer} = round\left(\frac{N_t}{N_{layer}}\right) \tag{9}$$

where the function $round(X)$ returns to the nearest integer with X.

TThe length of the coil l_{wire} (mm) as in Equation (10) is then based on the number of turns N_t, the number of layers N_{layer} and the geometric parameters of the core (a and d in Figure 5b).

$$l_{wire} = \sum_{m=1}^{N_{layer}} 2N_{t,layer}[a + d + D_{wire}(2m - 1)] \tag{10}$$

The weight of the coil W_{coil} (kg) is finally calculated as in Equation (11), where ρ_{cu} ($8940 \frac{kg}{m^3}$) is the mass density of copper:

$$W_{coil} = \rho_{cu} l_{wire} A_{cu} \tag{11}$$

For one port, the total weight of inductors $W_{\Sigma ind}$ (kg) is based on the number of inductors according to the number of phases N_{ph}, the weight of core W_{core}, the weight of coil W_{coil}. (assumed that the weight of bobbin W_{bobbin} is constant):

$$W_{\Sigma ind} = N_{ph}(W_{coil} + W_{core} + W_{bobbin}) \tag{12}$$

The inductor design is not straightforward. In practical design, the core is selected from available commercial products and the wire is decided by the amplitude of the inductor current. For the sake of optimization, it is necessary to derive the dependence of the inductance value on the design specifications (i.e., input current, switching frequency, output voltage). The main idea is that the designed inductor should guarantee the pre-defined current ripple which is also dependent on the duty cycle and the switching frequency f_{sw}. In addition, the core of the inductor is not saturated. The inductance value L (μH) can be derived from a second-degree polynomial equation as Equation (13). The detailed derivation steps are explained in the Appendix A.

$$L^2 I_L^2 + L\left[\frac{I_L(1 - D_{max})D_{max}}{f_{sw}} - K_u W_a J_w A_c B_{max}\right] + \left[\frac{V_o(1 - D_{max})D_{max}}{2f_{sw}}\right]^2 = 0 \tag{13}$$

From Equation (13), it is possible to calculate the maximum inductance that can be achieved by a given core. It is obvious that out of the two roots of (13), only the real root has a physical meaning.

2.3. Losses of Converter

Losses of one phase consist of IGBT losses (conduction loss and switching loss), inductor losses (conduction loss and core loss) and air-gap loss. The loss caused by the skin effect can be neglected.

2.3.1. IGBT Losses and Diode Losses

The losses of IGBTs (P_{loss_IGBT}) and diodes (P_{loss_D}) are due to the conduction losses and switching losses, which are evaluated based on [20], but neglecting the effect of the temperature variation.

$$P_{loss\_IGBT} = I_{Srms}^2 r_{CE} + V_{CE}I_s + \left(\frac{V_o}{V_{cc}}\right)^{1.2}\left[E_{off}\left(\frac{I_{s,rms}}{I_c}\right) + E_{on}\left(\frac{I_{s,rms}}{I_c}\right)\right]f_{sw} \tag{14}$$

$$P_{loss\_D} = I_{Drms}^2 r_f + V_{F0}I_D + \left(\frac{V_o}{V_{cc}}\right)^{0.6} E_{rr}\left(\frac{I_{d,rms}}{I_c}\right)^{0.6} f_{sw} \tag{15}$$

where the IGBT and diode characteristics (r_{CE}, V_{CE}, V_{CC}, I_c, E_{off}, E_{on}, r_f, V_{F0}, I_F and E_{rr}) are given by the IGBT and diode datasheets; in addition, the effect of temperature variation is neglected. I_S, $I_{S,rms}$, I_D and $I_{D,rms}$ are the switch and diode current.

2.3.2. Inductor Losses

The inductor losses consist of conduction loss P_{cond_L}, core loss P_{core_L}, and air-gap loss P_{gap_L}. As shown in Equation (17), the conduction loss P_{cond_L} known as ohmic loss is dependent on the internal resistance of winding R_L. The core loss P_{core_L} as in Equation (18) are produced from the flux density ripple B_{ac}, which is proportional to the inductor current ripple ΔI_L. The core losses are estimated based on the charts given by the manufacturer (METGLAS, Inc., CC core) [21]. In addition, high-frequency gap loss P_{gap_L} in nanocrystalline cores [22] can be computed as in Equation (19).

$$P_{loss\_L} = P_{cond\_L} + P_{core\_L} + P_{gap\_L} \tag{16}$$

$$P_{cond\_L} = R_L I_{L,rms}^2 \tag{17}$$

$$P_{core\_L} = W_t \left(6.5 f_{sw}^{1.51} B_{ac}^{1.74} \right)$$
$$B_{ac} = \frac{0.4 \pi N_t \Delta I_L 10^{-4}}{g} \tag{18}$$

$$P_{gap\_L} = k_g g c^{1.65} f_{sw}^{1.72} B_{ac}^2 \tag{19}$$

where $k_g = 1.68 \times 10^{-3}$ a numerical constant, c (mm): the depth of the iron core.

The total losses of interleaved converter for one port is calculated in Equation (20).

$$P_{loss} = N_{ph} \left(P_{loss\_IGBT} + P_{loss\_D} + P_{loss\_L} \right) \tag{20}$$

3. Optimization Problem Formulation

The optimization process aims at the optimal set of the inductor, the number of phases, and switching frequency to minimize three objective functions: the input current ripple (ΔI_{in} as shown in Equation (4)), the total losses of the converter (P_{loss} as shown in Equation (20)), and the weight of all inductors ($W_{\Sigma ind}$ as shown in Equation (12)) that contributes critically to the whole weight of the converter. It is assumed that the weight of other components such as heat sink, IGBT modules, bus bar, and filter capacitor is fixed during the optimization process. The multi-objective optimization problem is mathematically presented in Equation (21).

$$
\begin{aligned}
&\underset{X \in \Omega}{\text{minimize}} \left\{
\begin{array}{l}
\Delta I_{in}(X) \\
W_{\Sigma ind}(X) \\
P_{loss}(X)
\end{array}
\right. \\
&\quad s.t. \left\{
\begin{array}{l}
N_{ph\_min} \leq N_{ph} \leq N_{ph\_max} \\
f_{sw\_min} \leq f_{sw} \leq f_{sw\_max} \\
\Delta I_{in_{BAT}} \leq 7.5\% I_{in_{BAT}} \\
\Delta I_{in_{SC}} \leq 20\% I_{in_{SC}} \\
W_{\Sigma ind} \leq 5\,\text{kg} \\
P_{BCM} \leq 5\,\text{kW}
\end{array}
\right.
\end{aligned} \tag{21}
$$

The design vector X containing three variables (continuous f_{sw}, discrete N_{ph}, and discontinuous Core index) must be found in the feasible solution space Ω and subject to several constraints according to design specifications, physical limitation, and component safe operating areas. The minimum CCM power is added in the optimization routine to consider the negative effect of interleaving to ensure the Boundary Condition Mode (BCM) is at high power. In fact, as the current is split into several phases, the power at which the converter work in BCM is given by Equation (22):

$$P_{BCM} = N_{ph} V_{in} \frac{\Delta I_L}{2} \tag{22}$$

4. NSGA-II Optimizer for the Proposed Optimization Design Framework

Figure 3 shows the flowchart of the proposed optimization framework based on Non-dominated Sorting Genetic Algorithm-II (NSGA-II) for the converter of individual ports in the MPC. The NSGA-II is the second version of the famous "Non-dominated Sorting Genetic Algorithm" based on the work of Prof. Kalyanmoy Deb et al., which solves non-convex and non-smooth single and multi-objective optimization problems. The detailed working principle of the NSGA-II can be found in [23]. The NSGA-II can achieve good performance while solving a type of multidimensional problem defined in Equation (21) with discontinuous variable, providing highly accurate results with a reduced number of evaluations. The design methodology uses the large database of commercial standard core and IGBT modules to guarantee a hypothesis that the optimal core can be found in the database.

The main principle of the NSGA-II is that each design solution is represented by its chromosome made of the different genes, where the genes represent the integer associated with the design variable; the new design solutions are then produced by reproduction of the parents (design choices). The Blend Crossover (BLX) during the reproduction is based on the arithmetical average of each gene and a random variable α. In addition, polynomial mutation is also considered to introduce diversity in the design choices.

Based on the principle of NSGA-II [23], in this research an existing Multi-Objective Genetic Algorithm (MOGA) has been modified. A built-in Matlab function called "gamultiobj" is used to create modified NSGA-II. In the Matlab setting "gaoptimset", three primary functions are adapted: crossover operators ('CrossoverFcn'), mutation operators ('MutationFcn') and the population selection of the next generation ('CreationFcn').

The presence of multiple objectives in a problem results in a set of Pareto-optimal solutions known as Pareto-front instead of a single optimal solution. A solution is called a Pareto-optimal solution if none of the objective functions can be improved in value without degrading some of the other objective values. Without any further information, one of these Pareto-optimal solutions cannot be said to be better than the other one, which demands a designer to find as many Pareto-optimal solutions as possible. Thus, the Average Ranking (AR) [24] is employed to underpin the final solution from the Pareto-front. In the AR method, the tensor A is formulated as a three-dimensional matrix in Equation (23):

$$a_{ijk} \in A = \begin{cases} 1, & when\ f_k(s_i) < f_k(s_j) \\ 0, & when\ f_k(s_i) = f_k(s_j) \\ -1, & when\ f_k(s_i) > f_k(s_j) \end{cases} \tag{23}$$

where f is the objective function and s is the design solution. a_{ijk} records -1, 0 or 1 depending on whether the choice s_i is better, equal to, or worse than s_j on objective k. The AR method calculates a score for each choice s_i by summing the ranks of s_i for each objective. For example, with 3 objectives, if s_i is 2nd best on two of those objectives and 5th best on the other, its AR score will be $2 + 2 + 5 = 9$. Therefore, for $s_i \in P$, where P is the Pareto set as Equation (24):

$$AR(s_i) = \sum_k \sum_{j \neq i} (|P| + 1) - a_{ijk} \tag{24}$$

The inner sum calculates a score for s_i for a given objective, and this will be 1 if s_i is the best on that objective. Generally, $z + 1$ if z members of P are better at that objective.

5. Optimization Results Assessment

To validate the proposed optimization framework, the optimized design and a conventional design are needed to satisfy the same specification as summarized in Table 1.

Table 1. Design specification for MPC.

Notation	Description	Unit	Battery Port	SC Port
P_0	Rating power	kW		30
N_{ph}	Number of phases	[−]		3
V_o	Output voltage	V		400
V_{in_max}	Maximum input voltage	V	250	400
V_{in_min}	Minimum input voltage	V	200	200
I_{in}	Input current	A	150	166.67
ΔI_{in}	Input ripple current	A	15	16.67
$\hat{I}_{in}$	Input peak current	A	157.5	175
I_{o_min}	Minimum output current	A		5

In the proposed optimization design, the upper bound and lower bound of design variables are defined in advance, as shown in Table 2. To be more detailed, the setting of NSGA-II parameters is shown in Table 3.

Table 2. Design variable bounds.

Design Variables	Symbol	Lower Bound	Upper Bound	Unit
Number of phases	N_{ph}	2	4	—
Switching frequency	f_{sw}	20	100	kHz
Core database			23	

Table 3. Parameter settings used in GA.

Parameters	Value
Generation number	200
Population size	50
Crossover probability	0.85
Mutation probability	0.1

After executing the NSGA-II optimizer for the problem in Equation (21), the Pareto optimal solutions are sketched in Figure 6. The final optimal solution found by AR method shows that the optimal number of phases is three and optimal switching frequency is 60 kHz. The optimal core is AMCC50 from Metglas®Inc, a subsidiary of Hitachi Metals America, Ltd, Conway, SC, USA.

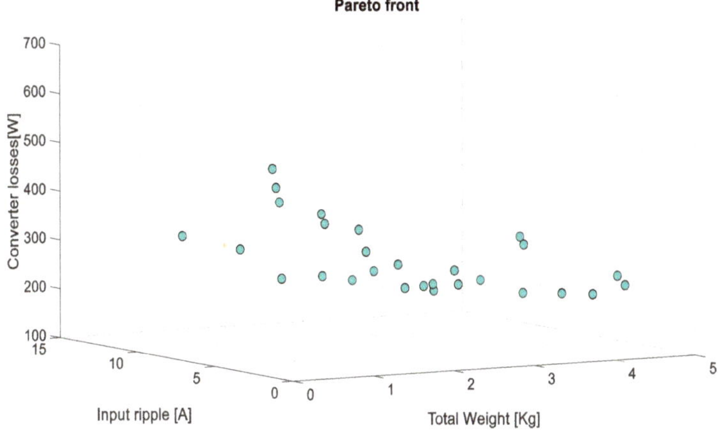

Figure 6. A three-dimensional Pareto optimal solution.

For the sake of a fair comparison, the switching frequency and number of phases, both found from the proposed methodology, are kept unchanged to design inductors in the conventional design that are based on well-established equations of a boost converter.

The inductance value of conventional design, as shown in Equation (25), should be sufficient to ensure the predefined input current ripple under the worst case (minimum input voltage V_{in_min}).

$$L_{conv} = \frac{V_{in\_min} \times (1 - D_{max}) \times D_{max}}{f_{sw} \times I_{o\_min}} \tag{25}$$

The selected core is AMCC50 with cross-section area $A_C = 400$ mm^2, the maximum flux density $B_{max} = 1.2$ T. Without air gap, number of turns in the conventional design can be calculated by Equation (26).

$$n_{conv} = \frac{L_{init} \times I_{L\_peak}}{A_C \times B} \tag{26}$$

The conventional and optimal design of inductors for battery and SC port are shown in Table 4. As can be seen, the inductance values increase from 166 µH (conventional design) to 177 µH (optimized design) for battery port, and from 148 µH (conventional design) to 150 µH (optimized design) for SC port even though the number of turns for both are reduced. The number of turns are reduced from 19 turns (conventional design) to 17 turns (optimized design) for the battery port, and from 18 turns (conventional design) to 15 turns (optimized design) for the SC port. It is understandable since the air-gaps 0.55 mm are added into the optimized inductor designs for two ports. As the results, the values of three objective functions are decreased considerably.

Table 4. Comparison between the conventional and optimized design for Battery and SC ports.

Notation & Description		Unit	Battery Port			SC Port		
			Conventional	Optimized	Δ_{BAT}	Conventional	Optimized	Δ_{SC}
L	Inductance	µH	166	177	-	148	150	-
n	Number of turns for inductor	turns	19	17	-	18	15	-
g	Air gap for inductor core	mm	0	0.55	-	0	0.55	-
$W_{\Sigma ind}$	Weight of inductors for 3 phases	kg	3.41	2.73	20%	3.17	2.61	17.6%
W_{loss}	Power losses of converter at full load (30 kW)	kW	1.41	1.36	3.5%	0.74	0.72	2.02%
	Power losses of converter at low load (5 kW)	kW	0.55	0.51	7.2%	0.39	0.38	2.56%
$\Delta \hat{I}_{in}$	Maximum input current ripple	A	15	9.32	38%	16.67	11	34%

As shown before in Equation (12), the weight of the inductor consists of the weight of core, the weght of coil that is dependent on the length of wire, and the weight of the bobbin. The weight of AMCC50 core is 586 g and the weight of bobbin is 14 g. The Litz wire used for the wire of inductors is rectangular HF-LITZ WIRE covered with Polyester PET tape from Von Roll Isola France SA, Belfort, France, which has 0.2 kg/meter, 2600 strands, and the diameter of each strand is 0.1 mm. According to Equation (10), the length of wire for inductor coil can be calculated. Afterward, the weight of an inductor can be determined in a function of number of turns. For fast calculation, the relation between the weight of an inductor and the number of turns can be assumed as a 2nd-order polynomial curve that can be derived by using the fitting-curve function in Matlab as shown in Equation (27). Figure 7 illustrates the relation described in Equation (27). As can see from Table 4, the total weight of inductors is reduced by 20% and 17.6% for the battery port and the SC port, respectively.

$$W_{ind}(\text{N}) = 0.43 \times N^2 + 19 \times N + 596.8 \tag{27}$$

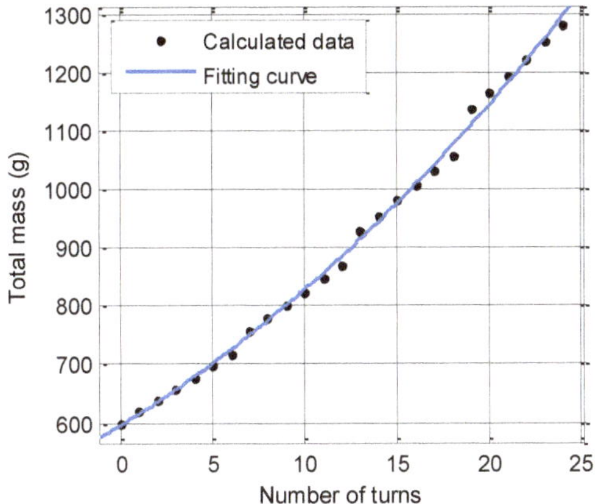

Figure 7. Weight of an inductor in function of number of turns.

Since the input current ripple and converter losses are highly dependent on the inductance value, it is important to ensure that the real inductor can obtain value as close as possible to that of theoretical inductance. To fulfill this purpose, the commercial COMSOL Multiphysics software is employed to compute the model of the inductor based on Finite Element Method. Figure 8 illustrates the inductor design in SOLIDWORKS that is imported to the COMSOL software.

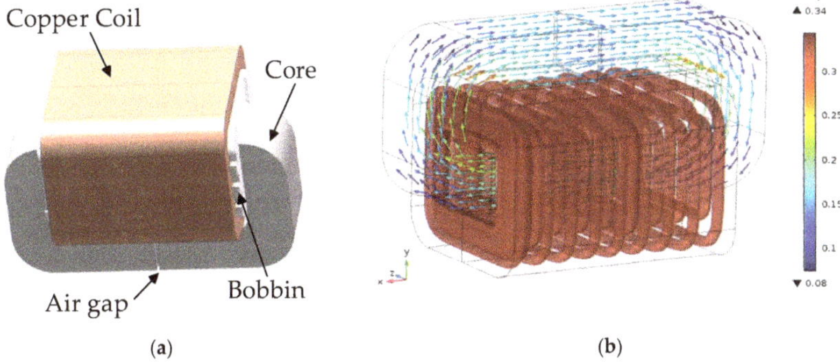

Figure 8. (a) SOLIDWORKS inductor design, (b) Magnetic flux distribution and current distribution of inductor in the COMSOL Multiphysics (Version 5.3a, COMSOL, Inc., Burlington, MA, USA, 2018).

The material name of the core is METGLAS Alloy 2605SA1 that has high saturation flux density (1.56T) and a low loss resulting from micro-thin Metglas ribbon (25 μm). To ease the computation time, the fitting technique is also used to find the inductance value in a function of air-gaps and number of turns. To do so, 16 inductance values are generated from COMSOL Multiphysics with air-gap range from 0–1.2 mm and number of turns from 10–22 turns. Using Matlab fitting function, a 2nd-order polynomial surface can be found as Equation (28).

$$L(N_t, g) = 12.36 \times N_t - 123.3 \times g + 0.36 \times N_t^2 - 12.22 \times N_t \times g + 122.3 \times g^2 - 4.7 \qquad (28)$$

where L is inductance (μH); g is airgap (mm). The impact of air gap and number of turns is simulated by FEM. The result is fitted by 2-degree polynomial function, mentioned as above. Figure 9 illustrates the impact of air-gap and the number of turns on the inductance value.

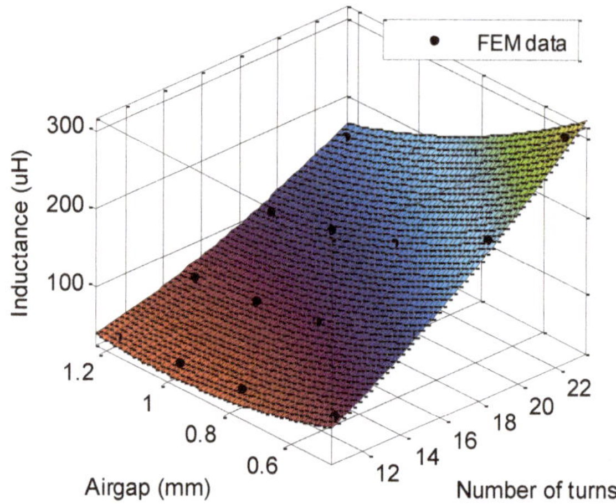

Figure 9. Inductance value in function of air-gap and number of turns.

To calculate the reduction of losses, the SiC-based semiconductor switch (2MBI150U2A-060) is used and switching frequency is kept as 60 kHz. The power losses of converters at full load 30 kW are reduced by 3.5% and 2.02% for battery port and SC port, respectively. More reduction of total loss, 7.2% and 2.56% for battery and SC port respectively, can be seen at low load 5 kW. The losses distribution at full load is shown in detail in Figure 10.

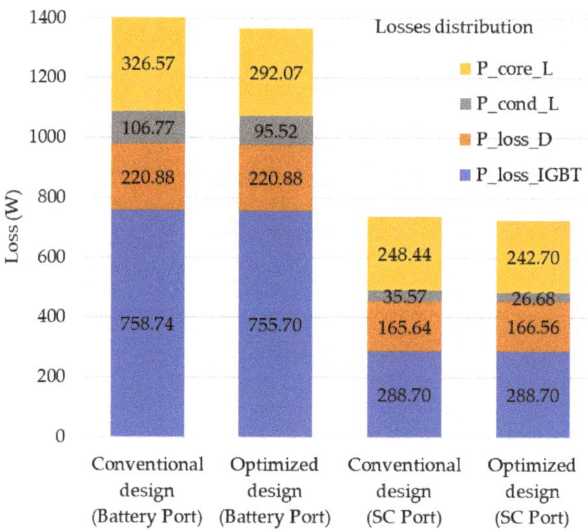

Figure 10. Losses distribution comparison at full load 30 kW.

The maximum input current ripple current, which is considered as a third objective function is also reduced by 38% and 34% for battery and SC converter ports, respectively. All diminution values that are shown in Table 4 validate the optimization methodology based on a multi-objective genetic algorithm and Average Ranking technique.

6. Conclusions

Several MPC topologies have been proposed in recent years with the aim to decrease the weight and component counts of DC/DC converters applied in hybrid drivetrain systems. Thus, the complex multidisciplinary design of these converters is a key challenge in the development phase of vehicle or machine drivetrains.

In this paper, a new optimization methodology based on Nondominated Sorting Genetic-Algorithm-II has been developed for MPC design to find optimization variables: a number of phases, inductor design, and switching frequency. The Average Ranking method is proposed to finalize the optimal solution among several Pareto-front solutions. Theoretically, the optimized design can archive better performance than the conventional design in terms of weight of inductors, input current ripple, and converter losses. The Finite Element Method such as COMSOL software is used to validate inductor designs, which is a crucial step for the future work. The proposed optimization process opens up new possible configurations in the optimization of MPC. Future research will involve the development of high-fidelity models of inductor design considering fringing effects, and fabrication of the prototype of MPC to validate design methodology compared to conventional design.

Author Contributions: D.T. has written the manuscript and presented the optimization methodology; S.C. analyzed multiport converter; Y.L. designed and analyzed the inductors by COMSOL and SOLIDWORK, J.V.M. and O.H. reviewed and edited the manuscript and they also provided supervision guidance to this research.

Funding: This research was funded by EMTECHNO project (Emerging Technologies in Multiport Systems for Energy Efficient Drivetrains), grant number IWT150513.

Acknowledgments: We acknowledge Flanders Make and VLAIO for the support to our research group.

Conflicts of Interest: The authors declare no conflict of interest.

Appendix A

This Appendix section explains how to derive the Equation (13) in which the inductance value can be found from a given core and design specification. Main steps are highlighted as following.

Appendix A.1. Maximum Flux Density

The reluctance of the core can be neglected compared to the one of the air gap results in Equation (A1):

$$N_t i = \frac{\phi g}{\mu_0 A_c} \tag{A1}$$

where N_t is the number of turns, ϕ the magnetic flux, g the air gap, μ_0 the vacuum magnetic permeability and A_c the net cross-sectional area of the core.

Given a peak winding current $\hat{I}_L$, it is desired to operate the core flux density at a peak value below the saturation flux density B_{max}. Therefore:

$$N_t \hat{I}_L = B_{max} \frac{g}{\mu_0} \tag{A2}$$

Appendix A.2. Inductance

The inductance is related to the number of turns N_t and the reluctance.

$$L = \frac{\mu_0 A_c N_t^2}{g} \tag{A3}$$

Appendix A.3. Winding Area

The wire must fit through the core window W_a. However, the wire does not pack perfectly which reduces the utilization factor K_u of the core window. Furthermore, insulation and the bobbin itself take some other place causing the utilization factor K_u to drop to values between 0.3 and 0.6. As a result, the number of turns in the core is limited by Equation (A4).

$$N_t A_{cu} \leq K_u W_a \tag{A4}$$

where A_{cu} is the conductor cross section.

Appendix A.4. Conductor Cross Section

The conductor needs to carry the peak current; therefore, another constraint is as Equation (A5):

$$A_{cu} \geq \frac{\hat{I}_L}{J_w} \tag{A5}$$

By substituting Equation (A5) into Equation (A4) and then into Equation (A2) an expression of the maximum air-gap for the maximum number of turns can be derived in Equation (A6):

$$g = \frac{K_u W_a J_w \mu_0}{B_{max}} \tag{A6}$$

It is clear that is a function of only the material and geometry of the core and it expresses the needed air gap to avoid that the material saturates when the core window is filled with conductors. In standard design techniques, L is calculated by the specifications on the ΔI_L. However, in this design algorithm, ΔI_L is a design variable, while the specification is set to ΔI_{in} which can be met with the phase interleaving. As a result, L becomes a design variable and it is related to the selected core. Moreover, it is worth expressing the peak current explicitly in terms of the inductance L. Therefore, as Equation (A7):

$$\hat{I}_L = I_L + \frac{1}{2}\frac{V_o(1-d)d}{f_{sw}L} \tag{A7}$$

where I_L in the interleaved case is as Equation (A8):

$$I_L = \frac{P_{max}}{N_{ph}V_o(1-d_{max})} \tag{A8}$$

By substituting (A7) and a version of Equation (A1) rearranged in N_t into Equation (A2) a second-degree polynomial expression of L can be derived.

$$L^2 I_L^2 + L\left[\frac{I_L(1-d)d}{f_{sw}} - K_u W_a J_w A_c B_{max}\right] + \left[\frac{V_o(1-d)d}{2f_{sw}}\right]^2 = 0$$

References

1. Chan, C.C.; Bouscayrol, A.; Chen, K. Electric, Hybrid, and Fuel-Cell Vehicles: Architectures and Modeling. *IEEE Trans. Veh. Technol.* **2010**, *59*, 589–598. [CrossRef]
2. Dobbs, B.G.; Chapman, P.L. A multiple-input DC-DC converter topology. *IEEE Power Electron. Lett.* **2003**, *1*, 6–9. [CrossRef]
3. Solero, L.; Lidozzi, A.; Pomilio, J.A. Design of Multiple-Input Power Converter for Hybrid Vehicles. *IEEE Trans. Power Electron.* **2005**, *20*, 1007–1016. [CrossRef]
4. Jiang, W.; Fahimi, B. Multiport Power Electronic Interface—Concept, Modeling, and Design. *IEEE Trans. Power Electron.* **2011**, *26*, 1890–1900. [CrossRef]
5. Hegazy, O.M.; Baghdadi, E.; Van Mierlo, J.; Lataire, P.; Coosemans, T. Analysis and modeling of a bidirectional multiport DC/DC power converter for battery electric vehicle applications. In Proceedings of the 2014 16th European Conference on Power Electronics and Applications, Lappeenranta, Finland, 26–28 August 2014.
6. Seeman, M.D.; Sanders, S.R. Analysis and Optimization of Switched-Capacitor DC–DC Converters. *IEEE Trans. Power Electron.* **2008**, *23*, 841–851. [CrossRef]

7. Wu, C.J.; Lee, F.C.; Balachandran, S.; Goin, H.L. Design Optimization for a Half-Bridge DC-DC Converter. *IEEE Trans. Aerosp. Electron. Syst.* **1982**, *4*, 497–508. [CrossRef]

8. Busquets-Monge, S.; Crebier, J.C.; Ragon, S.; Hertz, E.; Boroyevich, D.; Gurdal, Z.; Arpilliere, M.; Lindner, D.K. Design of a boost power factor correction converter using optimization techniques. *IEEE Trans. Power Electron.* **2004**, *19*, 1388–1396. [CrossRef]

9. De Leon-Aldaco, S.E.; Calleja, H.; Aguayo Alquicira, J. Metaheuristic Optimization Methods Applied to Power Converters: A Review. *IEEE Trans. Power Electron.* **2015**, *30*, 6791–6803. [CrossRef]

10. Man, K.F.; Tang, K.S.; Kwong, S. Genetic algorithms: Concepts and applications [in engineering design]. *IEEE Trans. Ind. Electron.* **1996**, *43*, 519–534. [CrossRef]

11. Kennedy, J.; Eberhart, R. Particle swarm optimization. In Proceedings of the ICNN'95-International Conference on Neural Networks, Perth, WA, Australia, 27 November–1 December 1995.

12. Garcia-Bediaga, A.; Villar, I.; Rujas, A.; Mir, L.; Rufer, A. Multiobjective Optimization of Medium-Frequency Transformers for Isolated Soft-Switching Converters Using a Genetic Algorithm. *IEEE Trans. Power Electron.* **2017**, *32*, 2995–3006. [CrossRef]

13. Ledoux, C.; Lefranc, P.; Larouci, C. Pre-sizing optimization of an inverter and the passive components. In Proceedings of the 2011 14th European Conference on Power Electronics and Applications, Birmingham, UK, 30 August–1 September 2011.

14. Qin, H.; Kimball, J.W.; Venayagamoorthy, G.K. Particle swarm optimization of high-frequency transformer. In Proceedings of the IECON 2010-36th Annual Conference on IEEE Industrial Electronics Society, Glendale, AZ, USA, 7–10 November 2010.

15. Kundu, U.; Sikder, S.; Kumar, A.; Sensarma, P. Frequency domain analysis and design of isolated bidirectional series resonant Dc-dc converter. In Proceedings of the 2016 IEEE International Conference on Power Electronics, Drives and Energy Systems (PEDES), Trivandrum, India, 14–17 December 2016.

16. Mirjafari, M.; Balog, R.S. Multi-objective design optimization of renewable energy system inverters using a Descriptive language for the components. In Proceedings of the 2011 Twenty-Sixth Annual IEEE Applied Power Electronics Conference and Exposition (APEC), Fort Worth, TX, USA, 6–11 March 2011.

17. Zhang, S. Analysis and minimization of the input current ripple of Interleaved Boost Converter. In Proceedings of the 2012 Twenty-Seventh Annual IEEE Applied Power Electronics Conference and Exposition (APEC), Fort Worth, TX, USA, 6–11 March, 5–9 February 2012.

18. Choe, G.-Y.; Kim, J.-S.; Kang, H.-S.; Lee, B.-K. An Optimal Design Methodology of an Interleaved Boost Converter for Fuel Cell Applications. *J. Electr. Eng. Technol.* **2010**, *5*, 319–328. [CrossRef]

19. High Efficiency, High Density, PolyPhase Converters for High Current Applications. Available online: http://bee.mif.pg.gda.pl/ciasteczkowypotwor/Noty%20aplikacyjne/Linear%20Technology/an77f.pdf (accessed on 26 July 2018).

20. Determining Switching Losses of SEMIKRON IGBT Modules. Available online: https://www.semikron.com/dl/service-support/downloads/download/semikron-application-note-switchinglosses-en-2014-08-19-rev-00/ (accessed on 26 July 2018).

21. Metglas, Inc. Inductor Cores, Powerlite Technical Bulletin. Available online: http://www.metglas.com (accessed on 26 July 2018).

22. Wang, Y.; Calderon-Lopez, G.; Forsyth, A.J. High-Frequency Gap Losses in Nanocrystalline Cores. *IEEE Trans. Power Electron.* **2017**, *32*, 4683–4690. [CrossRef]

23. Deb, K.; Pratap, A.; Agarwal, S.; Meyarivan, T. A fast and elitist multiobjective genetic algorithm: NSGA-II. *IEEE Trans. Evol. Comput.* **2002**, *6*, 182–197. [CrossRef]

24. Corne, D.W.; Knowles, J.D. Techniques for highly multiobjective optimisation. In Proceedings of the 9th annual Genetic and Evolutionary Computation Conference, London, UK, 7–11 July 2007.

Article

Battery Aging Prediction Using Input-Time-Delayed Based on an Adaptive Neuro-Fuzzy Inference System and a Group Method of Data Handling Techniques

Omid Rahbari [1,2,*], **Clément Mayet** [1,2], **Noshin Omar** [1,2] and **Joeri Van Mierlo** [1,2]

[1] ETEC Department & MOBI Research Group, Vrije Universiteit Brussel (VUB), Pleinlaan 2, 1050 Brussel, Belgium; Clement.Mayet@vub.ac.be (C.M.); noshomar@vub.ac.be (N.O.); joeri.van.mierlo@vub.ac.be (J.V.M.)
[2] Flanders Make, 3001 Heverlee, Belgium
* Correspondence: omid.rahbari@vub.ac.be

Received: 3 July 2018; Accepted: 2 August 2018; Published: 4 August 2018

Abstract: In this article, two techniques that are congruous with the principle of control theory are utilized to estimate the state of health (SOH) of real-life plug-in hybrid electric vehicles (PHEVs) accurately, which is of vital importance to battery management systems. The relation between the battery terminal voltage curve properties and the battery state of health is modelled via an adaptive neuron-fuzzy inference system and a group method of data handling. The comparison of the results demonstrates the capability of the proposed techniques for accurate SOH estimation. Moreover, the estimated results are compared with the direct actual measured SOH indicators using standard tests. The results indicate that the adaptive neuron-fuzzy inference system with fifteen rules based on a SOH estimator has better performances over the other technique, with a 1.5% maximum error in comparison to the experimental data.

Keywords: state of health estimation; adaptive neuron-fuzzy inference system (ANFIS); group method of data handling (GMDH); artificial neural network (ANN); electric vehicles (EVs); capacity degradation; lithium-ion battery; time-delay input

---

1. Introduction:

Notwithstanding the Paris Agreement, a technological transient from a hydrocarbon-based economy to the post-petroleum era, there is less tangible projective evidence of declining fossil-fueled based economies all over the world. For instance, recent investigation into the projection period, conducted in 2017 by the U.S. Energy Information Administration [1], indicates that the demand for liquid fuels will increase from 95 to 113 million barrels per day. The proportion of the transportation demand to the petroleum demand and other liquid fuels has been predicted to increase from 54% to 56%, leading this sector to be the main topic of electrification [2]. Nevertheless, the electrification of the transportation sector with existing electrical infrastructure leads the power system to collapse. However, it can be prevented if electric vehicles are coordinated and scheduled for a proper charging time-period and rate. In addition, recent progress in harnessing renewable energy sources (RESs), and improving battery characteristics shows that it is possible to completely mitigate the impact of connecting a large fleet of electric vehicles (EVs) on the power system. The majority of scientists have reached a consensus on viable alternatives for fossil fuels, mainly wind and solar energy, which have relatively low generation costs as well as high generation potential, respectively. However, their fluctuations in output are a serious problem [3]. To alleviate the oscillations of renewable generation sources, the following four possible approaches have been proposed:

(1) Coupling renewable energy systems with different generation characteristics in wider distribution via the transmission grids;

(2) Responding to the demand by adapting consumption patterns;
(3) Employing fossil-fueled utilities as a traditional back-up (either for meeting peak demand or providing spinning reserve); and
(4) Equipping the grid with storage devices such as compressed air storage, battery storage, and hydro pump storage.

Nevertheless, these approaches suffer from different drawbacks and limitations. For instance, dealing with the uncertainties of the renewable energy sources with different characteristics that are subjected to their inherent dependency on the weather conditions is a challenging task. Concerning the second approach, adapting consumers' patterns would require a new infrastructure to control the consumers' equipment. Regarding the main drawback of the third solution, fossil-fueled utilities would increase the environmental concern, which is contradictory to the objective of the Paris Agreement. Moreover, electrical vehicles and electrical energy storage systems equipped with lithium-ion batteries assume important roles as both back-up supply systems and primary energy sources. Indeed, energy storage systems (ESS) and electrical vehicles can be used to manage the demand in response to severe times (e.g., when RESs have fluctuations and load exceeds generation). Therefore, ESSs and EVs (in vehicle-to-grid [V2G] services) have been considered as great candidates to provide regulation services for frequency fluctuation, voltage deviation, and ancillary services.

However, EVs and ESSs whose V2G capability decreases because the battery performance degrades over time, decreasing both the energy and power capabilities as a result of the dynamic nonlinear nature of the electrochemical reactions, which are impacted by external states such as charge and discharge methods, usage, temperature, and the chemical makeup of the cell. In the meanwhile, battery technology is developing rapidly and battery cells with higher energy and power densities are becoming available. Hence, improving the performance of the battery management system (BMS) is an equally important task to make the battery reliable, safe, and cost-effective [4]. Indeed, the accurate estimator algorithms are essential for the smart battery management to estimate and measure the functional states of the battery, and it should contain state-of-the-art mechanisms to protect the battery from hazardous and inefficient operating conditions. In this regard, extensive research has been carried out for lithium-ion battery systems, investigating their high power density, energy efficiency, fast charging capability, light weight, steady-state float current, wide operating temperature range, low self-discharging rate, and the possible memory effect [5].

Furthermore, both the prognostications and engineering maintenance are key figures in various industry sectors such as aerospace, chemical, automotive, and so forth. Hence, the obvious formidable obstacles to wholesale EVs is a lack of confidence in the battery life-time and performance [6], leading the authors to look into two intelligent algorithms, which are capable to be implemented in the existing BMS hardware. The state of health can be estimated and classified into offline and online procedures, which have different advantages and drawbacks in terms of accuracy, time duration, and implementation. Based on the advantages and disadvantages, vehicle manufacturers select a suitable technique according to the application. Battery capacity estimation, referring to energy capability, poses tremendous challenges to researchers, whose attempts have turned to the relationship between capacity fade and an increase in battery resistance. Nevertheless, it has been observed that the changes in battery impedance cannot be exactly related to the capacity fade. Moreover, this approach needs extensive laboratory investigations to establish the correlation function [7].

Considerable research has been recently conducted on state of health (SOH) estimation models, which can be split into the following groups: electrochemical models (EMs), equivalent circuit models (ECMs), and data-driven or black-box models [8,9]. Electrochemical models are established to replicate the growth of a solid electrode interface (SEI) in lithium-ion and describe its influence on capacity degradation. Indeed, they are built based on concentrated solution and porous electrode theories. This means that the electrochemical models describe and elaborate the basic understanding of the electrochemical reaction inside the battery [10]. The EM includes mutually coupled non-linear partial differential equations (PDEs), increasing the numerical complexity and computational efforts, which poses difficulties in the real-time implementation phase, or large-scale simulation as a life-time prediction [11]. In this regard, desperate attempts to reduce the numerical complexity have been recently made through model-order reduction. In the literature [9], a dual SOH and state of charge (SOC) estimation technique has been proposed, by applying the sliding mode technique to the reduced version of PDE, namely a single partial model. The results showed that the proposed technique can track the SOH and SOC accurately. The advantage of the EM approaches is their independence from environmental conditions. On the other hand, as mentioned previously, the EM approaches require intensive computational efforts for system identification, because of a great quantity of parameters [12]. Moreover, the EM approaches are usually created for a particular type of battery consisting of specific anode and cathode materials [13].

The EC models are featured with ease of implementation and parameterization, as well as acceptable modeling accuracy [14]. The EC model completely depends on the environmental and operating conditions (e.g., SOH and SOC). This dependency on model parameters, derived from the operating conditions, can be addressed and captured via a look-up table, needing extensive experimental efforts to collect a sufficient dataset to describe a broad range of operating conditioning for batteries. The ECM's parameters can be estimated and updated via open-loop or close-loop methods. For the later method, an accurate EC model is required [15], and the battery parameters should be updated according to the aging state of the battery, which is a challenging task. Many techniques have been developed and some combined algorithms have been used to estimate SOC (directly or indirectly through the estimation of the open circuit voltage [OCV]), consequently estimating the SOH, such as the extended Kalman filter (EKF) and unscented Kalman filter (UKF). The EKF and UKF are effective techniques for SOH estimation. For instance, in the literature [16], a novel joint SOC and capacity estimator based on EKF has been introduced. The results showed that the proposed technique can capture the variation of the parameters in varying operating conditions and battery aging. Similarly, the authors of [17] proposed a new technique for SOH and SOC estimation, employed Coulomb counting method (CCM) to estimate SOC, taking the benefits of EKF to reduce accumulative errors of CCM, due to the current sensor noises. Moreover, the SOH was estimated based on the relationship between the dis/charge current and estimated SOC. The results demonstrated a reasonable estimation of SOH and SOC. These techniques are called joint estimation, and can estimate the SOH of the battery as accurately as the battery is modeled. This means that the accuracy is highly dependent on how the battery is modeled. Moreover, large matrix operation and inversions are required, leading to a high complexity. Furthermore, the joint estimation method may have poor numerical conditioning and suffer from instability [7]. Nonetheless, for this method, a dual estimation technique has been implemented, meaning that instead of one estimation algorithm, two adaptive filters are used. One of the filters estimates SOC and the other one is employed for the estimation of the model parameters. Sometimes, instead of the second filter used for model parameters identification, evolutionary algorithms are used [18]; a battery model was established and then a genetic algorithm was used to identify the model parameters and then estimate the SOH. In the literature [19], a multi-scale framework EKF was introduced to effectively estimate the state and parameters of the ECM, applied to a Li-ion battery for the capacity and SOC estimation. The results indicated that the proposed technique has a less than 3% error for the SOC estimation. In contrast to the joint estimation, the dual-technique consists of two adaptive filters. This technique demands a lower computational effort and the dimensions of the respective model matrices are lower than the joint estimation technique. In the literature [20], an effective joint SOH and SOC estimation technique was introduced. In this work, KF and UKF were combined to predict the state of the battery. The result regarding the SOC estimation is promising; nevertheless, the error of the SOH indicator is around 20%. In the literature [15], an adaptive sliding mode observer was employed to estimate the SOH and SOC of the Li-ion battery. The ECM consisted of two resistor and capacitor networks; furthermore, the results showed a high performance and robustness on the SOH and SOC estimations. However, similar to the joint technique, an accurate battery model is essential for the SOC and SOH estimations.

Indeed, observer techniques, known as a close-loop method, whose adaptability and effectiveness are utterly dependent on the credibility of the EC models and the robustness of the technique [10].

As stated previously, the techniques employed in ECM, suffer from inaccuracy owing to the lack of thorough understanding of the electrochemical dynamics and physics of the battery [21]. This drawback could be lessened via data-driven models, utilizing the information of the measurement ensemble. Consequently, prior knowledge of electro-chemistry is not required as a result of their capability to work with imprecise data and their self-learning ability [22]. Machine learning is categorized under data-driven method, which are widely employed for battery SOH estimation. In the literature [23], a recurrent neural network was used to monitor the SOH of a high-power lithium-ion battery. Lu et al. [24] proposed a group method of data handling, recognized as a polynomial neural network, in order to estimate the SOH of Li-ion batteries, and the results show a 5% error vs. the experimental data. The authors have concluded that the technique is universally valid for other types of battery chemistries. More recently, Chaoui et al. [5] employed an artificial neural network technique to estimate SOC and SOH directly and simultaneously. The technique used in the article is a useful tool for analyzing the system dynamics that are subjected to uncertainties [25]. In the literature [26], a naive Bayes model was introduced to predict the remaining useful life of a battery under different operating conditions. The comparative results showed the superiority of the proposed technique over the support vector machine. To reduce and avoid the need for computing power and a complex battery model, as well as considering the random driving cycle, researchers have been compelled to investigate the capacity degradation phenomena corresponding to SOH during charging or discharging processes, which could be more predictable than those methods mentioned previously [27]. Eddahech et al. [28] proposed a constant-voltage (CV) step as an indicator of capacity degradation. Then, four battery technologies were compared to show that the implemented method is very accurate by comparison with the classic discharged capacity measurements.

Motivation, Objective, and Innovation Contribution

Considering the limitations of the measurement devices in the present BMS, many external features of the battery are hard or even impossible to obtain in actual operation. Moreover, the applications of the above-mentioned methods are also limited by the computational capability of a real BMS. To address the above drawbacks of the methods described in the literature, this article proposes two states of health estimation techniques for Li-ion batteries, and then, another technique has been developed and compared to show the robustness of the proposed technique in this field. In this article, the proposed method requires only two external states (voltage and current), making the method suitable for EV applications. The key contributions of this article are summarized as follows:

- Employing an input time-delayed strategy to handle dynamic information of system.
- The Adaptive Neruo-fuzzy Inference System (ANFIS) and group method of data handling (GMDH) techniques are employed to analyze the relational grade between the SOH and selected features.
- Developing two data-driven frameworks to estimate the SOH. This article utilizes the fuzzy C-means clustering algorithm to tune and adjust the ANFIS parameter in advance, to create the initial rules.
- Accurate and effective validation of the framework in comparison to recently published articles and other methods.

The paper is organized as follows: in Section 2, a brief introduction is done regarding the group method of data handling and adaptive neuro-fuzzy inference system; in Section 3, both the discussion and comparisons between the proposed techniques are provided. The outcome of the article is summarized and concluded in Section 4.

2. Proposed Techniques

Based on the literature, modeling the relation between external and internal states is required for battery state estimation. Consequently, a battery model is needed for accurate estimation. Moreover, batteries are complicated electrochemical devices with non-linear behavior, affected by various internal and external states. This behavior can be described by a model whose formulation comprised of both uncertain and unknown parameters, but structurally known. In addition, describing the relationship between the battery terminal voltage property and battery SOH is an arduous task. As known from the literature, the charging process of an EV battery system includes two sub-processes, constant-voltage (CV) charge and constant-current (CC) charge. Charging or discharging of a certain amount of capacity (Ah) leads to a lower voltage change in fresh battery cells, while the same amount of Ah creates a higher voltage change in an aged cell for the same type of battery. This principle, the determination of the differential voltage responses to the ampere-hours discharged or charged from the battery before and after discharging or charging, is almost employed as the capacity estimation method. So, in this method, after a certain amount of energy throughput, the variation of voltage response is calculated and compared to the experimental data. This method is a practical solution for battery capacity monitoring [29–31]. The advantage of this method could reside in low inputs.

As can be seen in Figure 1, the terminal voltage curves are plotted at three different SOH levels while the batteries were charged using constant-current charging profile. The terminal voltage curves considerably vary from cycle to cycle. For instance, the terminal voltage curve of the battery at the beginning of life (BOL) has a lower slope than the voltage curves at 71% SOH. In addition, the initial, mean, and final voltages are not equal in the voltage property curves at different SOH levels. Hence, it can be concluded that the SOH can be reflected by the terminal voltage curve in a specific charging/discharging process. In other words, the battery's terminal voltage generally decreases and increases when being discharged and charged, respectively. The charging and discharging processes of a fixed number of ampere-hours lead to a lower voltage change for a battery with a higher SOH (fresh battery). On the other hand, a higher voltage change takes place when the battery's SOH is lower (aged battery). Figure 1 shows the battery charging profile based Lithium-ion battery (LIB) at different SOH from 97% to 71%, aged at 25 °C. For instance, the blue line represents 97% SOH, has a lower slope than the red line, and corresponded to 95% of the nominal capacity. In addition, the line with 71% SOH has a bigger slope than the line with 95% SOH.

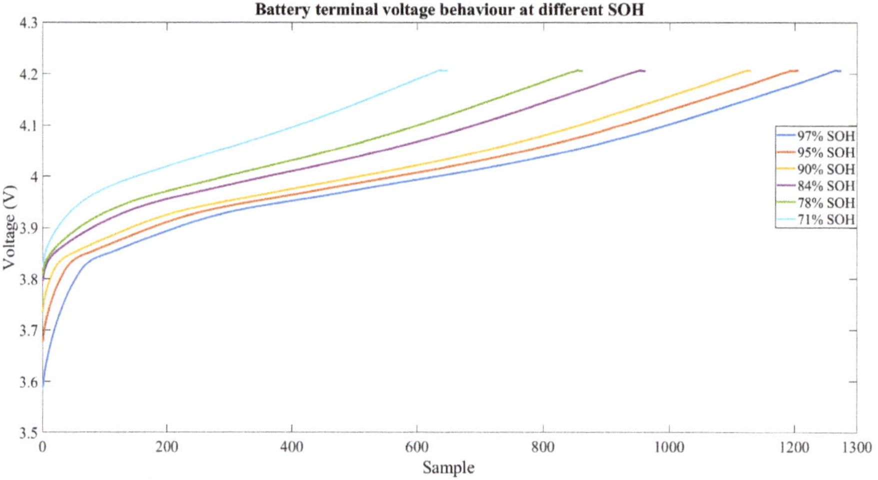

Figure 1. Terminal battery voltage at constant-current charging protocol (25 °C). SOH—state of health.

2.1. Group Method of Data Handling

The group method of data handling (GMDH) neural networks is a self-organized algorithm, meaning that the connections of the network (connections between neurons) are selected throughout the training phase to optimize the network [32]. In this approach, the neurons are completely not connected with the function nodes. Moreover, the number of layers, neurons in hidden layers, and active neurons are automatically configured, because of their self-organized capability. Furthermore, the network structure is modified until the best structure is accomplished, and thereafter, the optimized network defenses the dependency of the output values on the most notable input variables. It should be mentioned that GMDH can be employed in a wide range of fields, such as complex system modeling, forecasting, data mining, and knowledge discovery. The relation between inputs and outputs can be described as follows:

$$y = a_0 + \sum_{i=1}^{M} a_i x_i + \sum_{i=1}^{M}\sum_{j=1}^{M} a_{ij} x_i x_j + \sum_{i=1}^{M}\sum_{j=1}^{M}\sum_{k=1}^{M} a_{ijk} x_i x_j x_k + \dots \tag{1}$$

where $(x_1, x_2, \dots, x_M)$, $(a_1, a_2, \dots, a_M)$ and M are the input variables, the coefficient, and the number of input variables, respectively. By applying input data as a matrix, N point of observations of M variables are included. In the learning step, the network is tuned and estimates the coefficients of the polynomial, as described by Equation (2), and the remaining data samples are utilized to choose the optimal structure of the model, which can be realized by minimizing the error between the expected output (real value) and the estimated value. In this regard, Equation (3), known as a mean square error, is defined as a cost function of the algorithm.

$$y = a_0 + a_1 x_i + a_2 x_j + a_3 x_i x_j + a_4 x_i^2 + a_5 x_j^2 \tag{2}$$

$$\frac{1}{N}\sum_{n=1}^{N} (y_n - \hat{y}_n)^2 \tag{3}$$

where $\hat{y}_n$ and y_n are the estimated and expected values, respectively, and N is the length of the training dataset. The input variables are considered as pairs of (x_i, x_j), as can be seen by Equation (2). The regression polynomial is created and then iterations continue from two to three steps, until the mean square error of the test data converge to a constant value. The configuration of the group method of data handling is depicted in Figure 2.

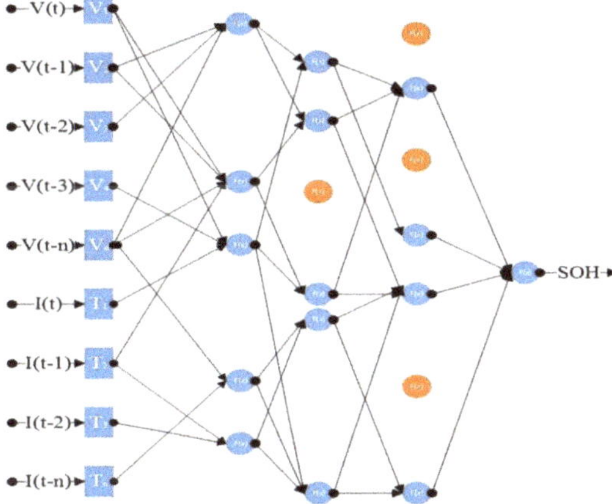

Figure 2. Group method of data handling (GMDH) structure.

Figure 2 illustrates the optimized structure, configured automatically by minimizing the cost function, as defined previously. Furthermore, some node functions were not connected to the network, as can be distinguished in Figure 2.

2.2. Adaptive Neuro-Fuzzy Inference system

Fuzzy logic (FL) is a robust system that transforms variables to mathematical language, which is consistent with the ability of human knowledge modeling. While, fuzzy logic tries to model either linear or non-linear systems, it is not possible to be trained by itself in a stochastic condition. Therefore, fuzzy logic systems are dependent on their operation rules, which should be defined by the experts who conclude, using their intuition, the parameters associated with membership functions. To overcome this problem, FL can be combined with artificial neural networks (ANNs), which have a remarkable ability to learn from imprecise data. Hence, combination of ANNs and FL procedures lead a better parameterization, which presents the fuzzy logic inference, known as the adaptive neuro-fuzzy inference system (ANFIS). Indeed, fuzzy logic and ANNs have both substantial benefits and drawbacks, which should be taken into consideration in terms of system modeling. In fuzzy logic language, called 'fuzzily', if–else statements are used to model the system by human knowledge. Although FLs are not capable of capturing measurement values, and use them to either adjust or modify the parameters like the Gaussian membership function variables, ANNs have the capability to be tuned and learnt by experimental data, leading a mathematical model not to be included in the system modeling, which can be possible by input–output mapping. Moreover, it has been demonstrated that the ANFIS is one of the techniques that can be utilized to any type of battery with various operating conditions (e.g., partial discharging, constant charge, and discharge processes) [33].

Two common fuzzy style inferences are Mamdani-style and Sugeno-style, which have been presented by Lotfi Zadeh and Takagi-Sugeno-kang, respectively [3]. To provide a better understanding, an ANFIS structure with two-input one-output is illustrated in Figure 3. The rule base considers two fuzzy 'if–then' rules of Takagi and Sugeno's type, which are as follows:

$$rule\ 1:\ If\ x \rightarrow A_1\ and\ y\ \rightarrow B_1,\ then\ Z_1 = p_1 x + q_1 y + r_1$$

$$rule\ 2:\ If\ x \rightarrow A_2\ and\ y\ \rightarrow B_2,\ then\ Z_2 = p_2 x + q_2 y + r_2$$

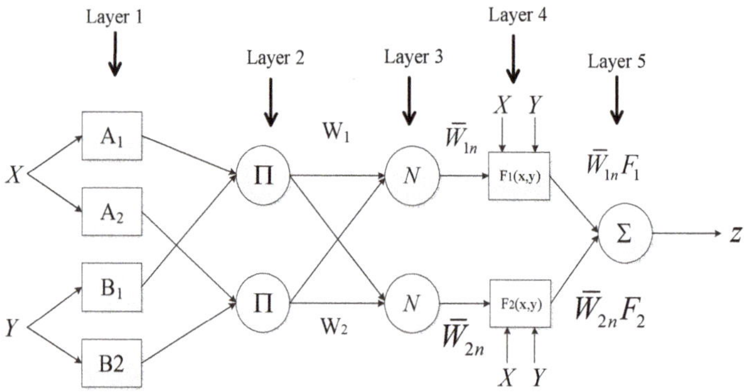

Figure 3. A general adaptive neural-fuzzy inference system [2].

The basic structure of ANFIS, considering as a fuzzy inference system, is a five-layered feedforward type, ANN, including different purpose-built types of nodes (e.g., non-weighted, adaptive, and non-adaptive connection links). The different layers can be classified into five-layers, which are as follows:

Layer 1   This layer is known as fuzzy-fication layer, which fuzzifies the input variables; every i node consists of a node function, which is $O_{1,i} = \mu_{A_i}(x)$, symbolized by A_i, x, O_{1i}, where A_i is the linguistic label according to the node function, x is the input to the node, and $O_{1,i}$ is the membership function of that, specifying the level for the assumed x. Hence, the membership function ascertains the membership level from the given input values. For a bell-shaped function, three parameters for each node should be defined, for which the maximum and minimum possible value are 1 and 0, respectively; where its generalized function can be mathematically described as follows:

$$\mu_{A_i}(x) = \frac{1}{1 + \left[((x - c_i)/(a_i))^2\right]^{b_i}} \tag{4}$$

where $\{a_i, b_i, c_i\}$ are the set parameters, called as premise parameters, μ is commonly chosen as bell-shaped or gauss-shaped, x is the first input variable, and the membership function variables are adjusted by changing the aforementioned parameters whenever the first input variable is fed to the ANFIS.

Layer 2   Is called 'fuzzy and', because in this layer, only 'AND' operators are allowed. This layer is utilized to compute the firing robustness of every rule. It means product operation (see Equation [5]) referred to the weighting factor of the corresponding rule, is used.

$$O_{2,i} = w_i = \mu_{A_i}(x_1) \times \mu_{B_i}(x_2) \; for \; i = 1,2 \tag{5}$$

Layer 3   Is known as 'normalization' term. The firing strength of each rule is normalized via computing the ration of each rule's firing strength to the total of each rules. In Equation (6), $\overline{w}_i$ is defined as the firing strength of each rule, as illustrated below:

$$O_{3,i} = \overline{w}_i f_i = \frac{w_i}{\sum w_i} = \frac{w_i}{w_1 + w_2}, \; for \; i = 1,2 \tag{6}$$

Layer 4   Is recognized as 'defuzzification'. This layer tries to compute the output of the previous layer, based on its node function; each node function is adaptive in accordance with the node function, as given by Equation (7).

$$O_{4,i} = \overline{w}_i \times f_i = \overline{w}_i(p_i x_i + q_i x_i + r_i), \; for \; i = 1,2 \tag{7}$$

where $\overline{w}_i$ is the output of the third layer and the parameters (p_i, q_i, r_i) are set parameters, which are being assumed by the conditions of the determined parameter. The parameters in the fuzzy inference layer are considered as consequent parameters.

Layer 5   Is called 'aggregation'. This layer is utilized to compute the total of the outputs of all of the rules to produce the overall ANFIS output, whose equation is represented as follows:

$$O_{5,i} = \sum_i \overline{w}_i f_i = \frac{\sum_i w_i f_i}{\sum_i w_i} = f_{out} \tag{8}$$

The aforementioned architecture is employed to adjust ANFIS model for SOH estimation, as discussed in the next section.

3. Result and Discussion

Many methods have been proposed in the literature to estimate SOH, whereby accurate battery parameters are needed to build the empirical model, which could be inefficient and expensive.

Nevertheless, the above developed techniques are capable of dealing with the complexity of the system modeling, insufficient data, and can still describe the system behavior.

3.1. Experimental Data

In this work, the experimental data from Prognostics Center of Excellence at National Aeronautics and Space Administration (NASA) Ames is employed to train and validate the proposed approaches [34]. This approach leads the comparison of the proposed techniques with that of recently published papers using the same dataset to be easier. The dataset consists of four batteries, aged through three different operational profiles conducting alternately in the dataset, namely impedance, charge, and discharge profiles. The impedance measurement process was performed by employing the electrochemical impedance spectroscopy (EIS) technique. Moreover, in the regular charge and discharge cycle, the batteries were charged and discharged at CC of 1.5 A and 2 A, respectively. In the charge step, 1.5 A is imposed to the batteries to reach the maximum voltage of 4.2 V, followed by the CV process, until the current decreased from 1.5 A to 20 mA. Nevertheless, in the discharge profile, the CC discharge step was conducted by reaching the voltage of 2.7 V, 2.5 V, 2.2 V, and 2.5 V for batteries, No. 05, 06, 07, and 18, respectively. As a consequence of reoccurring the above procedure, the capacity of the batteries reached 70% of the nominal capacity.

3.2. Short-Term State of Health Estimation

In this subsection, the performance of the short-term SOH estimation is presented by employing the proposed techniques. Both the GMDH and ANFIS are trained by the collected dataset. The inputs and the outputs of the system in the training phase are the battery terminal voltage and the SOH, respectively. The beginning-of-life (BoL), corresponding to a fresh battery, is defined as a 100% SOH, and the 167th cycle, when the capacity has reached the 1.4 Ah, is considered as the end-of-life. Moreover, the algorithm uses the unit-time-delays to consider the battery voltage at past time frames. The voltage is normalized, which is a standard procedure when such intelligent techniques are used. Thereafter, the normalized dataset after the computing and estimating procedures will be de-normalized. Owing to the capability of improvement in the read performance of the database, this technique is used. Indeed, each sample is divided by the maximum possible measurement. For instance, a measurement of 4.2 V constitutes as number 1, while 0 V is represented as number 0, and every other value is between 1 and 0. Furthermore, it should be noted that EVs are not always charged at a certain state of charge, which means that the technique should be able to estimate the SOH at different SOC levels, corresponding to different initial voltages. The proposed techniques, GMDH and ANFIS, were trained by the experimental results of battery No. 05. As mentioned previously, during the training phase, the structure and weights of GMDH and weights of ANFIS could be optimized and adjusted in terms of minimizing the error between the estimated SOH from the network, and the training targets from the experimental data. Then, the techniques are validated by employing the experimental data from battery No. 06. For the GMDH whose parameters are the maximum number of neurons in a layer, the maximum number of layers and selection pressure are set to 10, 5, and 0.6, respectively. It should be pointed out that the dataset for the training phase includes all of the voltage samples corresponding to 0% SOC to 100% SOC.

The GMDH parameters, maximum number of neurons in a layer, maximum number of layers, and selection pressure are set to 250, 10, and 0.6, respectively. For validation, battery No. 06 was used, whose experimental results were used to test the estimation accuracy of the GMDH technique. The actual and estimated SOH are depicted in Figure 4. The blue line shows the actual SOH and the red line indicates the estimated SOH at first and second cycles with 0.052 mean square error, and 0.23 root mean square error. It is observed that the relationship between the battery voltage and estimated SOH closely matches the actual test dataset. Moreover, the RMSE and MSE show that the GMDH has successfully discovered the effects of aging of the battery voltage behavior.

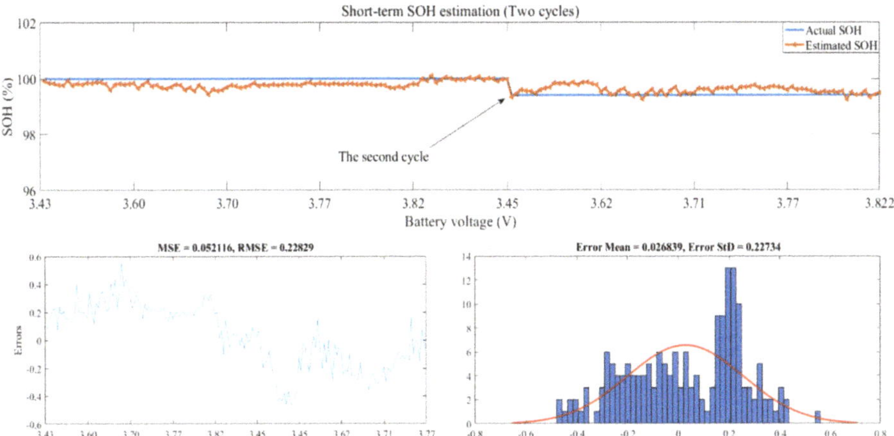

Figure 4. Experimental and estimated results of state of health (SOH) vs. battery voltage by employing GMDH (No. 06) for two cycles.

With regard to the second technique, as mentioned earlier, the combination of fuzzy logic and NNs leads to the ANFIS structure, which is classified under adaptive networks. Consequently, ANFIS has the ability to reach a conclusion from unclear and complex data, because of the fuzzy logic, with the capability to work from imprecise data [35]. In this regard, this technique is utilized to estimate the SOH from a set of curves whose shapes depend on the state of the system. Furthermore, the ANFIS cannot work without a training phase. Therefore, the battery terminal voltage during constant current charge profile at different SOH is prepared. Then, the membership functions should be adapted to the battery charge curves, which are diverse at different SOH levels. It should be pointed out that the constant-voltage sub-process is not included in the input dataset. The number of initial ANFIS rules for the first input was set to 15, these rules were generated using the fuzzy C-means (FCM) clustering method, and then the ANFIS was trained and tuned by the experimental results of battery No. 05. Moreover, the method used for optimization of the parameter of ANFIS, is a combination of back-propagation and least-square estimation. Note that the trained dataset consists of all of the voltage intervals, starting from 0% to 100% SOC. The dataset, related to the battery No. 06, is utilized to test the developed algorithm.

The errors between the experimental data (actual SOH) against the estimated SOH at different voltage levels are illustrated in Figure 5. The mean squared error (MSE) and root mean squared error (RMSE) are 0.009 and 0.094, respectively. As can be inferred from the results, the ANFIS has better performance compared with the GMDH. The results, shown in Figure 4, have a maximum error below 0.3. Moreover, the overestimation and underestimation is lower than that of the previous technique, which demonstrated the adaptive capability of the ANFIS technique.

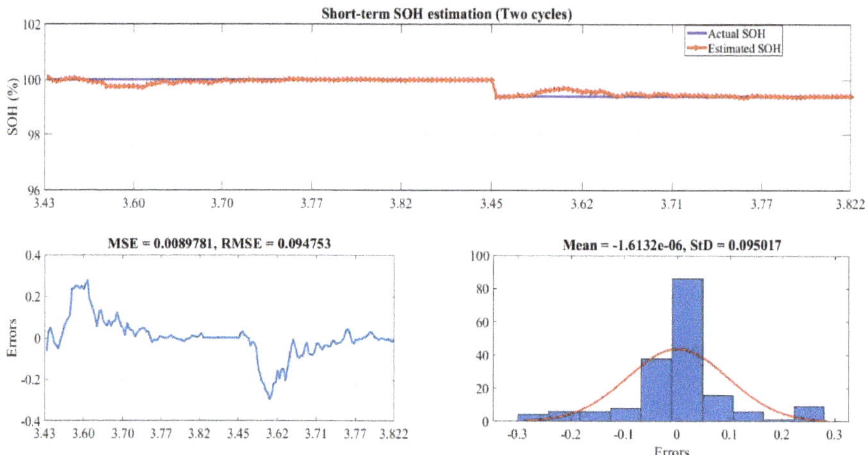

Figure 5. Experimental and estimated results of SOH vs. battery voltage by employing ANFIS (No.06) for two cycles.

3.3. Long-Term State of Health Estimation

In this subsection, the proposed techniques for the long-term battery state of health estimation are also evaluated. Note that in this procedure, all of the short-term SOH and voltage cycles are integrated to build one macro time scale concept. The charge data for 87 cycles of battery, No. 06, are employed to evaluate the proposed techniques for long-term estimation capability.

Figure 6 shows the long-term SOH estimation of battery No. 06. The obtained MSE and RMSE for the SOH estimation are 0.714, and 0.845, respectively. It can be seen that the GMDH, trained and tuned by battery No. 05, can be used to estimate the SOH for other batteries. Nevertheless, it is observed that, despite the better performance of GMDH for short-term estimation, in long-term SOH estimation, the fluctuation of GMDH is the most noticeable. According to Figure 6, the GMDH could not estimate the 1st, 21st, 54th, and 74th accurately. It can be concluded that the GMDH technique for long-term SOH estimation is instable.

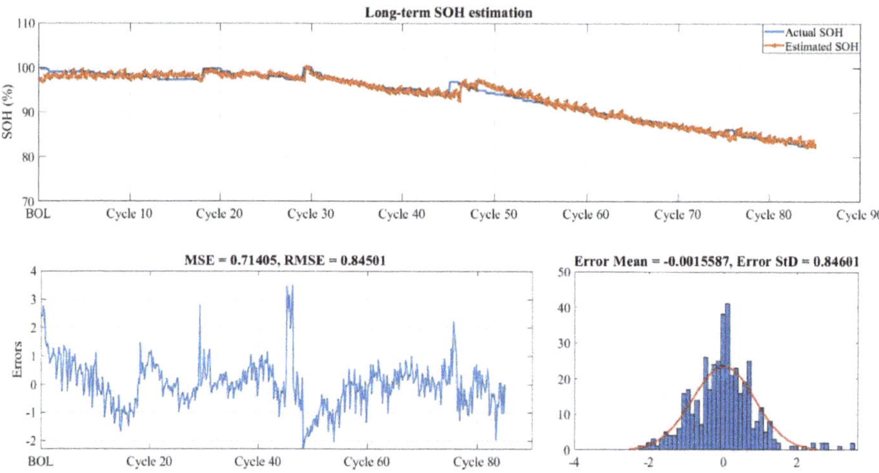

Figure 6. Long-term SOH estimation via GMDH for the 87 discharge cycles of battery No. 06.

The results of SOH estimation for battery No. 06 based on ANFIS, are plotted in Figure 7. As it is noted in the figure, the MSE and RMSE are 0.041 and 0.203, respectively, which shows a better stability from the GMDH for the long-term SOH estimation. It can be observed that the ANFIS has successfully learned the effect of capacity degradation on the battery terminal voltage. Therefore, overcharging and deep-discharging can be avoided, and also, the proposed techniques can be used for smart battery charging management, as ANFIS and GMDH have the capability to respond to an optimization algorithm as soon as they receive the inputs of the system.

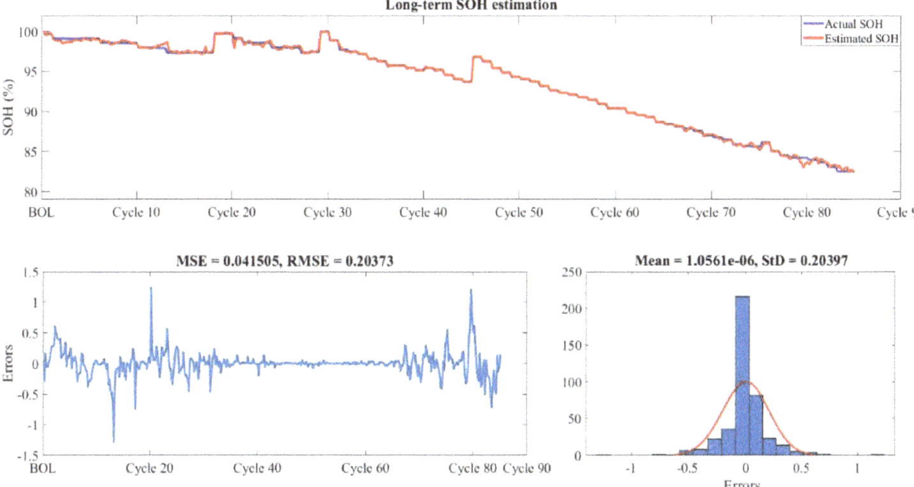

Figure 7. Long-term SOH estimation via ANFIS for the 87 discharge cycles of battery No. 06.

Table 1 presents the performance of the evaluation, comparing the proposed techniques with the recent published articles. As shown in the table, the ANFIS model obtains a much better performance over the GMDH model. For instance, the RMSE and MSE on battery No. 06 based on GMDH is 0.845 and 0.714, while the RMSE and MSE based on ANFIS is 0.203 and 0.041, respectively. Moreover, in terms of comparison, the present results and the resent published articles used same dataset from NASA, the performance of the models introduced in the literature [16,24,36] are compared in Table 1. As can be observed, the RMSE and MSE based on the ANFIS model are much better than the introduced models. Nevertheless, the following limitations need to be addressed in future studies:

1. While machine learning demonstrated an acceptable self-adaptation and high non-linearity modeling capability, a large amount of experimental data is required to obtain a high accuracy.
2. Although the introduced SOH method is more predictable and accurate under charging and discharging processes, it is not a usable method for plug-in hybrid electric vehicles (PHEVs)/PEVs when they are connected to smart charging infrastructure.

Table 1. Root mean square error (RMSE) results of long-term capacity estimations of adaptive neruo-fuzzy inference system (ANFIS), group method of data handling (GMDH), and a recent published article. MSE—mean square error; QGPER—quadratic polynomial mean function; DGA—geometry based approach.

Error	Ref. [36] QGPFR	Ref. [16] GPR-SE	Ref. [24] GMDH-DGA	Present Study ANFIS	Present Study GMDH
RMSE (battery No. 06)	5.12	1.7064	-	0.203	0.845
MSE (battery No. 06)	-	-	0.360	0.041	0.714

4. Conclusions

In this article, two data-driven techniques are developed for the state of health estimation. The developed techniques utilize an adaptive neuro-fuzzy inference system and group method of data handling to train the relation of the battery terminal voltage and state of health, enjoying the advantage over existing methods, as mentioned previously (e.g., lower inputs, described system behavior), with no need for computing power and a complex battery model. The comparative merit of the method and techniques implemented in this paper, compared to the existing ones in the literature, can be concluded in two main points. Firstly, the techniques are not dependent on any specific battery model, due to the fact that they are data-driven techniques, as can be inferred. The employed techniques can be applied to a great variety of battery technologies. Secondly, the battery operating dataset is applied to these techniques to analyze the internal structure, which is inaccessible. The comparison between the experimental and estimated results showed a robustness of the developed techniques, fast convergence performance, and outstanding accuracy for the battery health estimation.

Author Contributions: O.R. designed the study and mainly wrote the paper; N.O., C.M, and J.V.M revised and proofread the article.

Funding: This research received no external funding

Acknowledgments: We acknowledge the support of our research team from "Flanders Make".

Conflicts of Interest: The authors declare no conflicts of interest.

Abbreviations

ANN	artificial neural network
ANFIS	adaptive neruo-fuzzy inference system
BMS	battery management system
CC	constant current
CV	constant voltage
DG	distributed generation
DGA	geometry based approach
ESS	energy storage system
EV	electric vehicle
EKF	extended Kalman filter
G2V	grid-to-vehicle
GHG	greenhouse gas
GMDH	group method of data handling
GP	Gaussian process
HRES	hybrid renewable energy system
ITDNN	input time-delayed neural network
KF	Kalman filter
LS	least squares
NN	neural network
NEDC	new European driving cycle
MSE	mean squared error
PS	power system
PF	particle filter
QGPFR	quadratic polynomial mean function (GP)
RMSE	root mean square error
RBC	remaining battery capacity
SG	smart grid
SOC	state of charge
SOH	state of health

V2G	vehicle-to-grid
NPF	nonlinear predictive filter
MSE	mean square error
OCV	open circuit voltage
PHEV	plug-in hybrid electric vehicle

Nomenclature

$\hat{y}_n$	estimated values
y_n	expected values
M	number of input variables
x_i, x_j	pairs of input variables
$O_{1,i}$	membership function
A_i	linguistic label
(x_1, x_M)	Input variables
(a_1, a_M)	model coefficient

References

1. U.S. Energy Information Administration. *International Energy Outlook 2017*. 2017. Available online: https://www.csis.org/events/eias-international-energy-outlook-2017 (accessed on 1 August 2018).
2. Rahbari, O.; Omar, N.; Firouz, Y.; Rosen, M.A.; Goutam, S.; Van Den Bossche, P.; Van Mierlo, J. A novel state of charge and capacity estimation technique for electric vehicles connected to a smart grid based on inverse theory and a metaheuristic algorithm. *Energy* **2018**, *155*, 1047–1058. [CrossRef]
3. Rahbari, O.; Vafaeipour, M.; Omar, N.; Rosen, M.A.; Hegazy, O.; Timmermans, J.; Heibati, S.; Van DenBosschea, P. An optimal versatile control approach for plug-in electric vehicles to integrate renewable energy sources and smart grids. *Energy* **2017**, *134*, 1053–1067. [CrossRef]
4. Rahbari, O.; Omar, N.; Van Den Bossche, P.; Van Mierlo, J. A centralized state of charge estimation technique for electric vehicles equipped with lithium-ion batteries in smart grid environment. In Proceedings of the 2018 IEEE International Conference on Industrial Technology (ICIT), Lyon, France, 20–22 February 2018; pp. 1721–1725. [CrossRef]
5. Chaoui, H.; Ibe-Ekeocha, C.C.; Gualous, H. Aging prediction and state of charge estimation of a LiFePO$_4$ battery using input time-delayed neural networks. *Electr. Power Syst. Res.* **2017**, *146*, 189–197. [CrossRef]
6. Safari, M. Battery electric vehicles: Looking behind to move forward. *Energy Policy* **2018**, *115*, 54–65. [CrossRef]
7. Waag, W.; Fleischer, C.; Uwe, D. Critical review of the methods for monitoring of lithium-ion batteries in electric and hybrid vehicles. *J. Power Sources* **2014**, *258*, 321–339. [CrossRef]
8. Yang, D.; Wang, Y.; Pan, R.; Chen, R.; Chen, Z. State-of-health estimation for the lithium-ion battery based on support vector regression. *Appl. Energy* **2017**. [CrossRef]
9. Dey, S.; Ayalew, B.; Pisu, P. Combined Estimation of State-of-Charge and State-of-Health of Li-ion Battery Cells Using SMO on Electrochemical Model. In Proceedings of the IEEE Workshop on Variable Structure Systems, Nantes, France, 29 June–2 July 2014.
10. Pan, H.; Lü, Z.; Wang, H.; Wei, H.; Chen, L. Novel battery state-of-health online estimation method using multiple health indicators and an extreme learning machine. *Energy* **2018**, *160*, 466–477. [CrossRef]
11. Marcicki, J.; Canova, M.; Conlisk, A.T.; Rizzoni, G. Design and parametrization analysis of a reduced-order electrochemical model of graphite/LiFePO$_4$ cells for SOC/SOH estimation. *J. Power Sources* **2013**, *237*, 310–324. [CrossRef]
12. Tong, S.; Klein, M.P.; Park, J.W. On-line optimization of battery open circuit voltage for improved state-of-charge and state-of-health estimation. *J. Power Sources* **2015**, *293*, 416–428. [CrossRef]
13. Li, X.; Jiang, J.; Yi, L.; Chen, D.; Zhang, Y.; Zhang, C. A capacity model based on charging process for state of health estimation of lithium ion batteries. *Appl. Energy* **2016**, *177*, 537–543. [CrossRef]
14. Zhang, C.; Alla, W.; Dinh, Q.; Ascencio, P.; Marco, J. Online estimation of battery equivalent circuit model parameters and state of charge using decoupled least squares technique. *Energy* **2018**, *142*, 678–688. [CrossRef]

15. Du, J.; Liu, Z.; Wang, Y.; Wen, C. An adaptive sliding mode observer for lithium-ion battery state of charge and state of health estimation in electric vehicles. *Control Eng. Pract.* **2016**, *54*, 81–90. [CrossRef]

16. Yang, D.; Zhang, X.; Pan, R.; Wang, Y.; Chen, Z. A novel Gaussian process regression model for state-of-health estimation of lithium-ion battery using charging curve. *J. Power Sources* **2018**, *384*, 387–395. [CrossRef]

17. Sepasi, S.; Ghorbani, R.; Yann, B. Inline state of health estimation of lithium-ion batteries using state of charge calculation. *J. Power Sources* **2015**, *299*, 246–254. [CrossRef]

18. Chen, Z.; Mi, C.C.; Fu, Y.; Xu, J.; Gong, X. Online battery state of health estimation based on Genetic Algorithm for electric and hybrid vehicle applications. *J. Power Sources* **2013**, *240*, 184–192. [CrossRef]

19. Hu, C.; Youn, B.D.; Chung, J. A multiscale framework with extended Kalman filter for lithium-ion battery SOC and capacity estimation. *Appl. Energy* **2012**, *92*, 694–704. [CrossRef]

20. Andre, D.; Appel, C.; Soczka-guth, T.; Uwe, D. Advanced mathematical methods of SOC and SOH estimation for lithium-ion batteries. *J. Power Sources* **2013**, *224*, 20–27. [CrossRef]

21. Li, Y.; Chattopadhyay, P.; Xiong, S.; Ray, A.; Rahn, C.D. Dynamic data-driven and model-based recursive analysis for estimation of battery state-of-charge q. *Appl. Energy* **2016**, *184*, 266–275. [CrossRef]

22. Rahbari, O.; Vafaeipour, M.; Fazelpour, F.; Feidt, M.; Rosen, M.A. Towards realistic designs of wind farm layouts: Application of a novel placement selector approach. *Energy Convers. Manag.* **2014**, *81*, 242–254. [CrossRef]

23. Eddahech, A.; Briat, O.; Bertrand, N.; Delétage, J.; Vinassa, J. Electrical Power and Energy Systems Behavior and state-of-health monitoring of Li-ion batteries using impedance spectroscopy and recurrent neural networks. *Int. J. Electr. Power Energy Syst.* **2012**, *42*, 487–494. [CrossRef]

24. Wu, J.; Wang, Y.; Zhang, X.; Chen, Z. A novel state of health estimation method of Li-ion battery using group method of data handling. *J. Power Sources* **2016**, *327*, 457–464. [CrossRef]

25. El Mejdoubi, A.; Oukaour, A.; Chaoui, H.; Gualous, H.; Sabor, J.; Slamani, Y. State-of-Charge and State-of-Health Lithium-Ion Batteries' Diagnosis According to Surface Temperature Variation. *IEEE Trans. Ind. Electr.* **2016**, *63*, 2391–2402. [CrossRef]

26. Ng, S.S.Y.; Xing, Y.; Tsui, K.L. A naive Bayes model for robust remaining useful life prediction of lithium-ion battery. *Appl. Energy* **2014**, *118*, 114–123. [CrossRef]

27. Baghdadi, I.; Briat, O.; Gyan, P.; Vinassa, J.M. State of health assessment for lithium batteries based on voltage-time relaxation measure. *Electrochim. Acta* **2016**, *194*, 461–472. [CrossRef]

28. Eddahech, A.; Briat, O.; Vinassa, J. Determination of lithium-ion battery state-of-health based on constant-voltage charge phase. *J. Power Sources* **2014**, *258*, 218–227. [CrossRef]

29. Weng, C.; Sun, J.; Peng, H. A unified open-circuit-voltage model of lithium-ion batteries for state-of-charge estimation and state-of-health monitoring. *J. Power Sources* **2014**, *258*, 228–237. [CrossRef]

30. Einhorn, M.; Conte, F.V.; Kral, C.; Fleig, J. A Method for Online Capacity Estimation of Lithium Ion Battery Cells Using the State of Charge and the Transferred Charge. *IEEE Trans. Ind. Appl.* **2012**, *48*, 736–741. [CrossRef]

31. Yang, J.; Xia, B.; Huang, W.; Fu, Y.; Mi, C. Online state-of-health estimation for lithium-ion batteries using constant-voltage charging current analysis. *Appl. Energy* **2018**, *212*, 1589–1600. [CrossRef]

32. De Giorgi, M.G.; Malvoni, M.; Congedo, P.M. Comparison of strategies for multi-step ahead photovoltaic power forecasting models based on hybrid group method of data handling networks and least square support vector machine. *Energy* **2016**, *107*, 360–373. [CrossRef]

33. Fleischer, C.; Waag, W.; Bai, Z.; Uwe, D. On-line self-learning time forward voltage prognosis for lithium-ion batteries using adaptive neuro-fuzzy inference system. *J. Power Sources* **2013**, *243*, 728–749. [CrossRef]

34. Saha, B.G.K. Battery Data Set. NASA Ames Prognostics Data Repository 2007. Available online: https://ti.arc.nasa.gov/tech/dash/groups/pcoe/prognostic-data-repository/ (accessed on 1 May 2007).

35. Shen, W.X.; Member, S.; Chan, C.C.; Lo, E.W.C.; Chau, K.T. Adaptive Neuro-Fuzzy Modeling of Battery Residual Capacity for Electric Vehicles. *IEEE Trans. Ind. Electron.* **2002**, *49*, 677–684. [CrossRef]

36. Liu, D.; Pang, J.; Zhou, J.; Peng, Y.; Pecht, M. Prognostics for state of health estimation of lithium-ion batteries based on combination Gaussian process functional regression. *Microelectron. Reliab.* **2013**, *53*, 832–839. [CrossRef]

 *applied sciences*

Article

Hybrid Battery/Lithium-Ion Capacitor Energy Storage System for a Pure Electric Bus for an Urban Transportation Application

Mahdi Soltani [1,2,*], Jan Ronsmans [3], Shouji Kakihara [4], Joris Jaguemont [1,2], Peter Van den Bossche [1,2], Joeri van Mierlo [1,2] and Noshin Omar [1,2]

[1] Department ETEC Mobility, Logistics and Automotive Technology Research Group (MOBI), Vrije Universiteit Brussel, Pleinlaan 2, 1050 Brussels, Belgium; joris.jaguemont@vub.be (J.J.); peter.van.den.bossche@vub.be (P.V.d.B.); joeri.van.mierlo@vub.be (J.v.M.); noshin.omar@vub.be (N.O.)
[2] Flanders Make, 3001 Heverlee, Belgium
[3] JSR Micro N.V., Technologielaan 8, 3001 Leuven, Belgium; jan.ronsmans@jsrmicro.be
[4] JM Energy Corporation, 8565 Nishi-ide Ooizumi-cho, Hokuto City, Yamanashi 409-1501, Japan; syoji.kakihara@jsrmicro.be
* Correspondence: mahdi.soltani@vub.be; Tel.: +32-(0)26293396

Received: 30 May 2018; Accepted: 17 July 2018; Published: 19 July 2018

Featured Application: A potential application for this research work is the pure electric bus with energy recovery capability. With the hybrid energy storage system based on Lithium-ion battery and Lithium-ion Capacitor, the bus will have a longer range, a higher efficiency and a lower cost in comparison to a bus with non-hybrid energy storage system or a bus with hybrid energy storage based on battery and super-capacitors.

Abstract: Public transportation based on electric vehicles has attracted significant attention in recent years due to the lower overall emissions it generates. However, there are some barriers to further development and commercialization. Fewer charging facilities in comparison to gas stations, limited battery lifetime, and extra costs associated with its replacement present some barriers to achieve better acceptance. A practical solution to improve the battery lifetime and driving range is to eliminate the large-magnitude pulse current flow from and to the battery during acceleration and deceleration. Hybrid energy storage systems which combine high-power (HP) and high-energy (HE) storage units can be used for this purpose. Lithium-ion capacitors (LiC) can be used as a HP storage unit, which is similar to a supercapacitor cell but with a higher rate capability, a higher energy density, and better cyclability. In this design, the LiC can provide the excess power required while the battery fails to do so. Moreover, hybridization enables a downsizing of the overall energy storage system and decreases the total cost as a consequence of lifetime, performance, and efficiency improvement. The aim of this paper is to investigate the effectiveness of the hybrid energy storage system in protecting the battery from damage due to the high-power rates during charging and discharging. The procedure followed and presented in this paper demonstrates the good performance of the evaluated hybrid storage system to reduce the negative consequences of the power peaks associated with urban driving cycles and its ability to improve the lifespan by 16%.

Keywords: hybrid energy storage system; lithium-ion battery; lithium-ion capacitor; lifetime model; power distribution

---

1. Introduction

In recent years, the use of electric vehicles (EVs) has spread widely due to the fewer pollutants they send into the environment. However, there are still some obstacles on the way of their further adoption, including the higher cost in comparison to internal combustion engine vehicles (ICEVs), lower energy density than ICEVs, and their problem in providing the high-power demand during sudden acceleration. To overcome the pricing issue, many Governments encourage car manufacturers and buyers by providing subsidies for further research and development and to reduce the manufacturing cost. Apart from the pricing, a good energy storage system (ESS) capable of providing enough energy for better mileage and enough power for acceleration still needs to be improved in the future. Many analyses in the literature show that the average power demand in vehicular applications is much lower than the peak power demand so that the peak to average power ratio is between 4 and 7 [1]. This requirement raises the need for a special type of EES for EVs. Among the different energy storage systems presented in the market, lithium-ion batteries (LiBs) attract a great deal of attention for their high energy density, however, their low power specification (peak to average ratio between 0.5 and 2 [1]) makes them unfavorable for acceleration purposes [2]. Moreover, high charging/discharging rates when they are used in urban driving cycles has a negative effect on their performance by affecting the efficiency, lifetime, and internal resistance value [3].

Among the different type of LiBs available in the market, lithium iron phosphate (LFP) batteries have shown a significant potential for sudden power consumption and a better cyclability in comparison to the other types of LiBs, but, their lifespan and reliability need to be studied more. Super-capacitors (SCs) are another candidate which can be used as ESS in EVs. They have a higher power density in comparison to the LiBs, as well as a higher cyclability and reliability, nevertheless, they have a lower energy density in the range of 4–7 Wh/kg. As mentioned before, the ESS should be able to fulfill both the power and energy demand for the EVs. One possible solution is to overdesign either of them, which creates an expensive, voluminous, and heavy ESS. Another possibility is to externally combine LiBs as a high-energy system with SCs as a high-power storage system. With this combination, the energy and power density of the entire system can be improved. However, these energy storage systems have different charge and discharge behaviors as LiBs are non-linear devices and SC are rather linear. Moreover, having a lower voltage level than batteries, SCs need to be connected to the DC link through a DC/DC converter, which increases the cost and complexity of the system.

SCs can also be replaced by a hybrid super-capacitor (HSC). HSC is an emerging technology which has attracted a great deal of attention in recent years. Many studies have been presented in the literature aiming to make HSC environmentally friendly and cost effective in comparison to competing technologies while keeping the surface area as high as possible [4,5]. HSC, also known as a lithium-ion capacitor (LiC), is an internal hybrid energy storage device where its structure consolidates SC and LiB technologies. Its basic structure includes a positive electrode with activated carbon, as in super-capacitors, and a negative electrode based on Li-Ion-doped carbon similar to the LiBs [6]. The application of LiCs is increasing quickly due to advantages that they have compared to the SCs, including a high power capability around 10 kW/kg, a higher maximum voltage (3.8 V) and a higher energy density (up to 14 Wh/kg) [7]. However, having a higher voltage level does not necessarily eliminate their need for DC/DC convertor for hybrid application, but it decreases the cost and loss of the convertor significantly since a part of hybridization is done internally in the LiC cells.

In the literature, different methods of hybridization have been presented and their pros and cons from different points of view have been studied. The main topologies examined in the scientific literature can be divided into passive and active topologies. The passive topology which is the simplest hybridization method is achieved by the direct parallel connection of two or more energy storage technology. This topology is cheap, light, easy to implement, highly dynamic, yet has some negative points. Due to the direct connection, power sharing cannot be controlled, and the usable capacity is limited by the operating voltage of the battery. To operate each energy storage system in an optimal way, an active topology is introduced. There are many types of active topologies based on the number of decoupled energy storage systems with convertors, such as DC/DC convertors [8].

Hybrid energy storage systems (HESS) also have been studied from the control strategy point of view in many studies. In [9], a novel controller for a HESS has been proposed which aims to decrease the frequency effect induced on the SC in the process of power sharing. In [10] a new combination of SC and multi-speed transmission system and the usage of regenerative breaking energy was used to increase the energy density of the EVs and make them comparable with ICEVs. In [11], a near-optimal power management strategy was proposed. The presented method was verified for different state of charge and state of health of the battery which can reduce the C-rate of the battery by 10%. In [12], authors investigated the effect of driving cycle characteristics on the optimization result of the HESS and concluded that those results can be generalized to practical bus lines.

As stated before, none of the available researches considered the application of LiC in the HESS and the effect of this combination on the lifetime performance was not studied before.

This paper investigates the effect of hybridization of LiB and LiC on the lifetime performance of a HESS for an urban electric bus. In this regard, a developed lifetime model of a 20 Ah nickel manganese cobalt oxide (NMC) lithium-ion battery in combination with a LiC 2300 F lifetime model developed in our laboratory (mobility, logistics, and automotive technology (MOBI) research center, Vrije Universiteit Brussel (VUB)) was used for simulation purposes. The result approves the effectiveness of hybridization by increasing the lifetime by 16%.

This paper is organized as follows: in Section 2 the electro-thermal and lifetime model of a NMC 20 Ah battery is briefly explained. Section 3 presents a detailed description of the LiC electro-thermal and lifetime model. The hybridization and load sharing methodology is illustrated in Section 4. Further analysis of the driving cycle and power and energy requirements for a pure electric bus is presented in Section 5. Simulation results and discussions are given in Section 6 and, finally, the conclusion is presented in Section 7.

2. Electrothermal and Lifetime Model Explanation for 20 Ah NMC Cell

The model used in this study is a combination of electro-thermal and lifetime model and has been comprehensively explained in [13]. The proposed model is based on the electrical equivalent circuit (EEC) approach which has been presented in [14] and, thus, will not be detailed here.

2.1. Electrical Model

The voltage behavior of the NMC cell was determined by an equivalent circuit model (ECM). The ECM is composed of two parallel RC branches and one ohmic resistor, which is called as a second-order electric model. The ohmic resistance R_o, concentration polarization resistance R_1, and the activation polarization resistance R_2 are estimated based on the experimental results and empirical equations obtained from curve fitting techniques and, thus, the model can accurately simulate the cell's voltage behavior during transient states of charging and discharging. The general equation is defined as [15]:

$$V_{batt} = OCV - R_O I_{batt} - R_1 I_1 - R_2 I_2 \tag{1}$$

where the open circuit voltage is abbreviated as OCV, I_{batt} is the battery current, I_1 is the current in the resistor in the first R-C branch and I_2 is the current in the resistor in the second R-C branch and V_{batt} is the terminal voltage of the battery. The equivalent circuit of the second-order Thevenin model is shown in Figure 1.

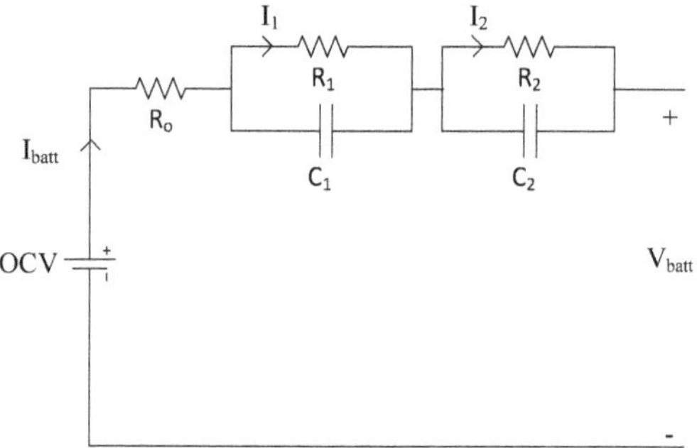

Figure 1. Schematic of the second-order Thevenin model [13].

2.2. Thermal Model

A transient heat equation which is derived from the first law of thermodynamics is used to describe the thermal distribution in the LiB cell, where the amount of generated heat must be stored inside the cell or transferred from the cell to its surroundings [16]:

$$\rho C_p \frac{dT}{dt} = \left[\lambda_x \frac{\partial^2 T}{\partial^2 x} + \lambda_y \frac{\partial^2 T}{\partial^2 y} + \lambda_z \frac{\partial^2 T}{\partial^2 z} \right] + \dot{q} \tag{2}$$

where $\dot{q}$ (W/m^3) and T (K) denote the heat source and the temperature of the cell, respectively. The heat flux transferred from the cell to its surroundings is expressed as follows:

$$-\left[\lambda_x \frac{\partial T}{\partial x} + \lambda_y \frac{\partial T}{\partial y} + \lambda_z \frac{\partial T}{\partial z} \right] \Bigg|_{boundaries} = h(T - Ta)|_{boundaries} \tag{3}$$

where h (W/m·K) and T_a (K) denote the convective heat coefficient, and ambient temperature, respectively. The heat source in the electrode domain is computed from the Bernardi equation [17], where its simplified form can be expressed as follows:

$$\dot{q} = I(U - V) - TI\frac{\partial U}{\partial T} \tag{4}$$

where I (A) is the current flowing through the cell, U (V) the open circuit voltage, V (V) the terminal voltage, and $\frac{\partial U}{\partial T}$ (V/K) is the entropy coefficient.

The heat source of the tabs is computed through this relation:

$$\dot{q} = \frac{R\prime I^2}{V_{tab}}; R\prime = \rho' \frac{l}{S} \tag{5}$$

where R' (Ω), I (A), V_{tab} (m^3), ρ' (Ω m) l (m), and S (m^2) are the electrical resistance, current rate, volume, resistivity, length, and cross-section of the associated tab, respectively.

For more details about this thermal model, readers are referred to [18]. The proposed 0-D thermal model is used to calculate the temperature evolution of the cell based on the current profile and the initial temperature. This temperature change is later used for a more accurate parameter estimation as all parameters shown in Figure 1 are temperature-dependent.

2.3. Lifetime Model

The lifetime model was developed using an empirical approach, as explained in [19]. The capacity degradation trend and internal resistance increase for both the calendaring and cycling aging phenomena were investigated fully and based on the curve fitting techniques. The aging phenomena affects the parameters shown in the ECM in Figure 1. These parameters are adjusted based on the cycling and storage conditions and are stored in the lookup tables for lifime model implementation. Figures 2 and 3 show the internal resistance incremental trend and capacity degradation trend based on the number of cycles, respectively. As it is seen in Figure 2, the internal resistance (R_O, R_1, R_2) increases with the increase of number of cycles. In this figure, the effect of depth of discharge (DoD) on the internal resistance growth is also seen. As shown in Figure 3 the capacity decreases with the increase of number of cycle. This can be translated as a decrease in the capacitance size in the ECM shown in Figure 1. The first step for lifetime estimation is to calculate the equivalent number of cycles. In this regard, the total energy throughput to the cell is divided by the nominal energy of the cell to calculate the equivalent number of cycles. Then, by referring to the look-up tables, according to the DoD, the equivalent number of cycles, the cycling temperature, and the appropriate values for parameters of the ECM shown in Figure 1 are selected.

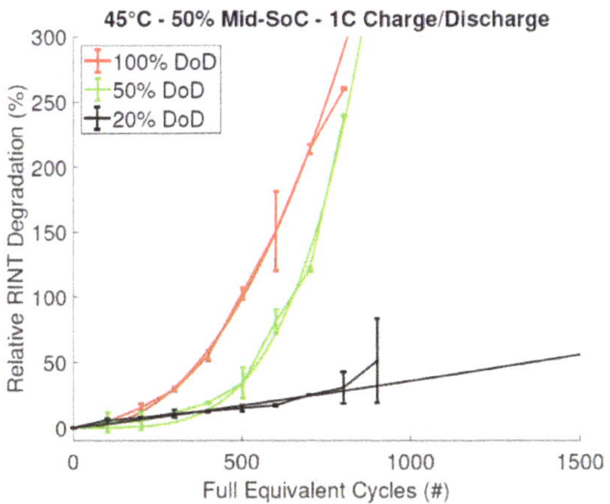

Figure 2. The internal resistance incremental trend based on the equivalent number of cycles [19].

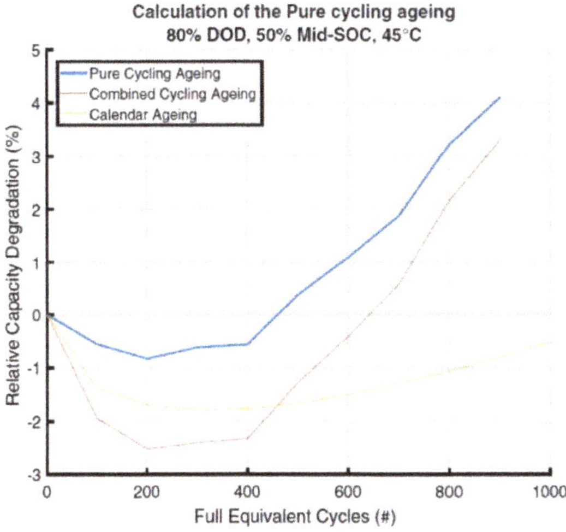

Figure 3. The capacity degradation trend based on the equivalent number of cycles [19].

3. Electrothermal and Lifetime Model Explanation for LiCs

The electrothermal and lifetime model for LiCs have been created based on the empirical approaches as explained in [20]. It consists of three parts similar to the LiB model including: the electrical model, thermal model, and lifetime model.

3.1. Electrical and Thermal Model

The electrical and thermal models as shown in Figure 4, are fully dependent on each other.

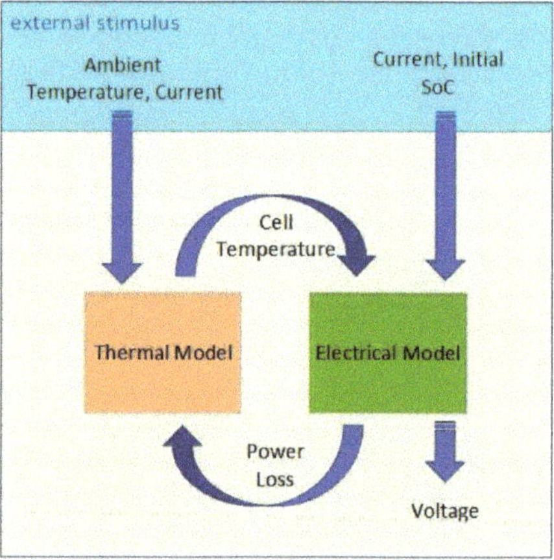

Figure 4. The block diagram of an electrothermal model [20].

The electrical part of the above-mentioned model is a first order electric circuit as shown in Figure 5. The parameters values (OCV, R_O, C_P, and R_P) are extracted during experiments at different conditions, including different temperatures, states of charge (SoC), and current rates. The terminal voltage in the first order electric circuit is calculated based on the Equation (6):

$$V_t = OCV(SoC) - I_L R_0 - V_{cp} \tag{6}$$

where R_o is the ohmic resistance, R_P is concentration polarization resistance, OCV is the open circuit voltage, I_L is the LiC current, I_{CP} is the current in the C_P branch, I_{RP} is the current in the R_P branch, and V_t is the terminal voltage of the LiC.

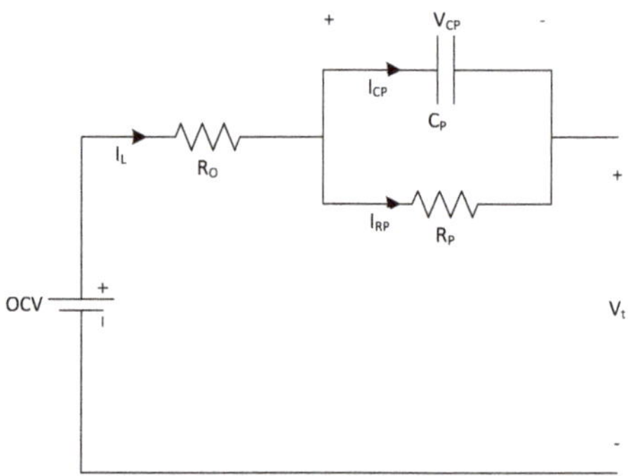

Figure 5. First-order electrical model.

All parameters shown in Figure 5 are temperature-dependent. In order to have a more accurate estimation, the surface temperature of the cell is modeled with a first-order electric circuit. As shown in Figure 6, the heat source, initial temperature, and ambient temperature are essential. The heat source represents power loss in the cell and arises from two references:

1. internal resistance ($R_{int} = R_o + R_p$), which is called "irreversible heat source"; and
2. the entropy changes, which is called "reversible heat source".

The first one is always positive and is due to the ion and electron movement, while the second one can be positive or negative and is generated in the chemical reaction. The power loss is calculated by Equation (7) [20]:

$$P_l = I_{cell}^2 \left(R_{int}\left(I_{cell}, SoC\right)\right) - I_{cell} T_{cell} \frac{\partial V_{OCV}\left(T_{cell,}, SoC\right)}{\partial T_{cell}} \tag{7}$$

where the I_{cell} is the cell's current, R_{int} represents the entire internal resistance, and T_{cell} is the cell's surface temperature. The circuit components shown in Figure 6 are described as follow: the P_g is the power loss generated in the cell, C_{th} is the thermal capacity of the cell, R_{th} represents the thermal conductivity resistance from the cell's center to the surface, R_{con} is the conductivity resistance from the surface to the ambient, and T_a is the ambient temperature.

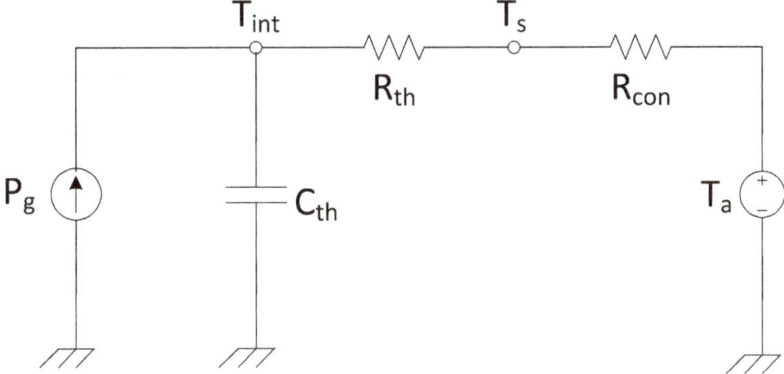

Figure 6. First-order electric circuit for the thermal part of the electrothermal model [20].

Finally, in order to calculate the power loss for charge and discharge conditions, the power loss is calculated by considering the proportional coefficients (α and β) as shown in Equation (8):

$$P_{ltot} = \alpha P_{lch} + \beta P_{ldiss} \tag{8}$$

which P_{lch} represents the power loss during the charge process and P_{ldiss} gives the power loss during the discharge phase. It has been observed in the experiments that LiCs absorb heat while charging and release heat while discharging. However, the absorbed heat and generated heat are not equal when a same value of current is applied. Moreover, at higher value of charge and discharge current, the exothermic part dominates the endothermic one. As shown in Equation (8), α and β are used to consider this phenomenon. In this equation, α represents the portion of the total heat source that is generated during the charge process and is mostly negative, and β donates the portion of the total heat source during the discharge times and is always positive. These parameters are current-dependent and are estimated by applying dynamic and steady state load profiles to the cell at different ambient temperatures. A non-linear least square technique is used for estimation. These values are given in Table 1 and are detailed in [20]. P_{ltot} is the total heat source (P_g) in the presented model in Figure 6.

Table 1. Estimated parameters for Equation (8) at 10 °C.

Parameter	Current (A)								
	10	50	100	150	200	240	300	450	600
α	−4	−1.008	−0.356	−0.134	−0.0426	0.001	0.02	0.054	0.059
β	4	2.27	1.27	0.86	09	0.63	0.54	0.47	0.45

3.2. Lifetime Model

The lifetime of all types of the energy storage technologies is affected by two phenomena:

1. calendar effect; and
2. cycling effect.

The calendar effect occurs while the cell is placed in the storage room for a long time or when it is in the rest period, for example, when the car is not in use for a certain time, the battery system is considered in the calendar mode. Many parameters take part in this phenomenon, like the storage temperature, and the storage SoC. Cycling, on the other hand, is the application in which the cells are used to deliver energy to the load. In these usages, they are charged a discharged fully or partially with different current rates and at different ambient temperatures. Therefore, the lifetime of the cell is affected by different parameters, like the cell temperature, the current rate, the depth of discharge, and the frequency level. In order to study this effect, many experiments were performed in our facilities in the battery innovation center (BIC), MOBI group, VUB University. The lifetime models for LiB and LiC have been explained in [6,13], respectively. Figures 7 and 8 show the capacity and internal resistance trends for the 2300 F LiC cells. As it is seen in Figure 7, the capacity decreases as the number of cycle increases. Moreover, the ambient temperature plays a role in the degradation. The lower the temperature, the lower the capacity degradation. Figure 8 shows the cycling effect on the internal resistance variation. As it is seen, for the cell at 45 °C, after 150,000 cycles, the internal resistance increase tends to a large temperature raise in the cell and, as a result, the test was stopped due to the safety measures. It is also clear in the result that the resistance rise for 45 °C and 0 °C is higher than at 25 °C, which is due to the rapid solid electrolyte interface (SEI) formation on the anode at those temperatures [21].

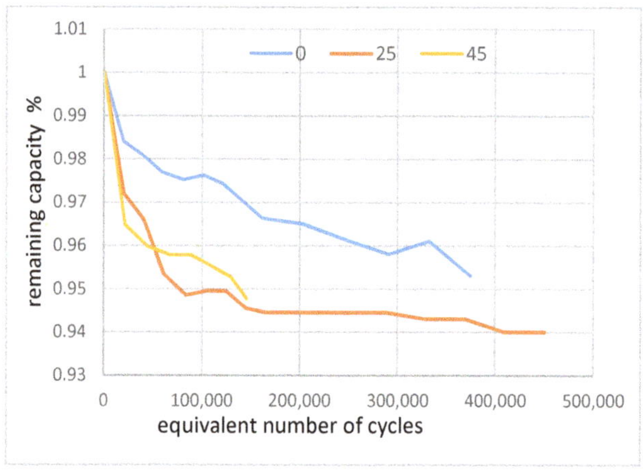

Figure 7. Capacity evolution of LiC 2300F with a heavy-duty load profile.

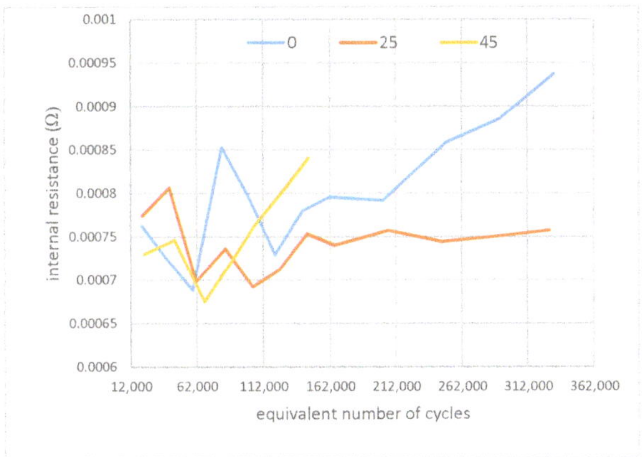

Figure 8. Internal resistance evolution of LiC 2300 F with a heavy-duty load profile.

4. Hybridization and Load Sharing Methodology

As briefly mentioned in the introduction, there are two main topologies to realize a hybrid unit with a high power (HP) storage system which is realized by means of super-capacitor or lithium-ion capacitors and a high energy (HE) storage system which is realized by means of a battery:

1. active topology [8], and
2. passive topology [8].

There is also a combination of those two methods which is called the semi-active topology [22]. A review article [23] validates the performance and cost efficiency of the semi-active topology in comparison to the passive and fully-active topology. Since the purpose of this study is to investigate the hybridization effect on the lifetime improvement of the energy storage system and to extend the range of the urban electric bus with one fully-charged ESS, a semi-active topology of hybridization is used herein. Some of the semi-active topologies are shown in Figure 9.

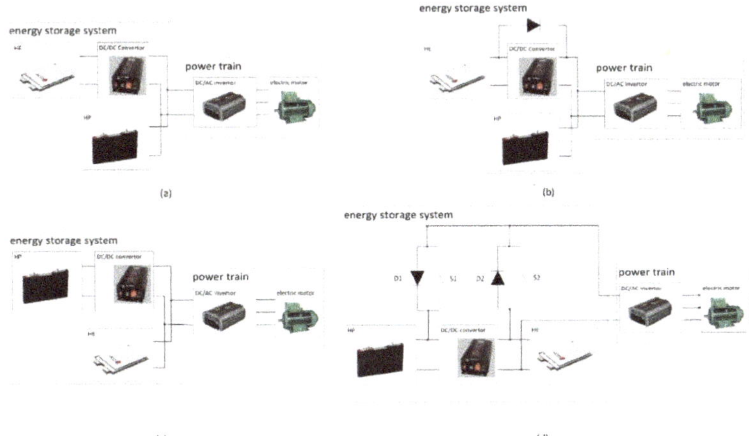

Figure 9. Semi-active hybrid topologies: (**a**) battery semi-active hybrid energy storage topology, (**b**) extended battery semi-active hybrid energy storage topology, (**c**) LiC semi-active hybrid energy storage topology, and (**d**) advanced LiC semi-active hybrid energy storage topology [8].

As the voltage level of the LiC module is different than the LiB module, the selected semi-active topology in this study, as shown in Figure 10, only utilizes a bi-directional DC/DC convertor to adjust the voltage level of the HP source with the voltage level of the HE source. In order to maintain this, a low-complexity control system with less computational demanding is proposed. The simplified diagram of the control unit is shown in Figure 10.

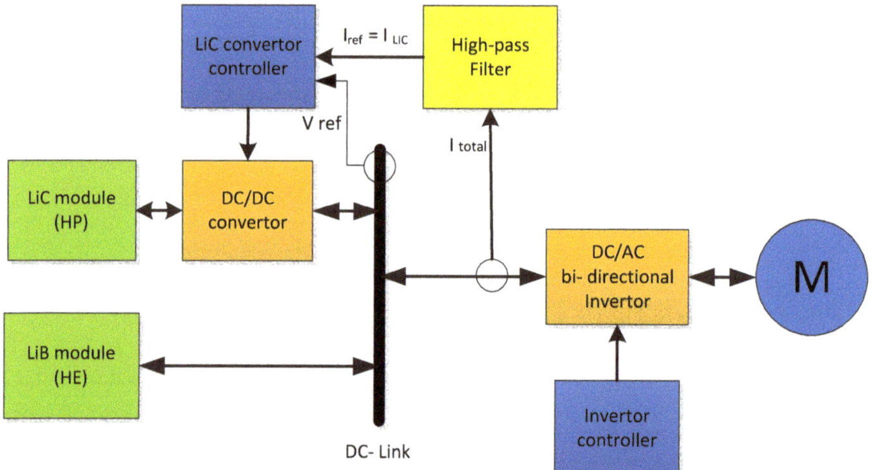

Figure 10. Simplified diagram of the hybrid battery-LiC power source and the proposed control system [22].

The controller also aims to limit the LiC current below the maximum value, which is 1200 A, and is given in the datasheet [24]. Moreover, it aims to keep the SoC of the LiC within the acceptable range. The maximum value of the current is also limited by the safety measures to keep the cells' temperature below 70 °C. The acceptable range of SoC is defined by the control strategy. In this study, the initial SoC of the LiCs is kept at 50% in order to have enough stored energy to deliver to the bus during acceleration, as well as to have free capacity to store energy which goes back to the cell during the regenerative braking. Clearly, the SoC cannot go over 100% or below 0%, however, by limiting the SoC of LiCs to a narrower window (for example, 20% to 80%), LiB-ESS is asked to work with a higher current which, as a result, is imposed to a high-frequency current component. The range of the SoC needs to be defined so as to optimize the cost, size, volume, and lifetime of the entire system.

The control system is composed of a high-pass filter which is used to separate the high-frequency current components. These high-frequency components of the load are asked to be delivered by the LiC module. As a consequence, the high-current stress factor is removed from the battery and its lifetime is improved. As it is seen in Figure 2 and is explained in [13], a higher DoD tends to a higher increase in the internal resistance and a faster degradation in the capacity of the cell. By eliminating the high-frequency current components from the LiBs, these cells are imposed to a lower DoD which consequently improves the lifetime. In this study, the high-frequency current component (I_{ref}) is used as a reference for the LiC control unit. The DC-link voltage is also used as a reference voltage (V_{ref}) for the LiC controller. Figure 11 shows the LiC and LiB current component for the hybrid bus extracted from the London driving cycles.

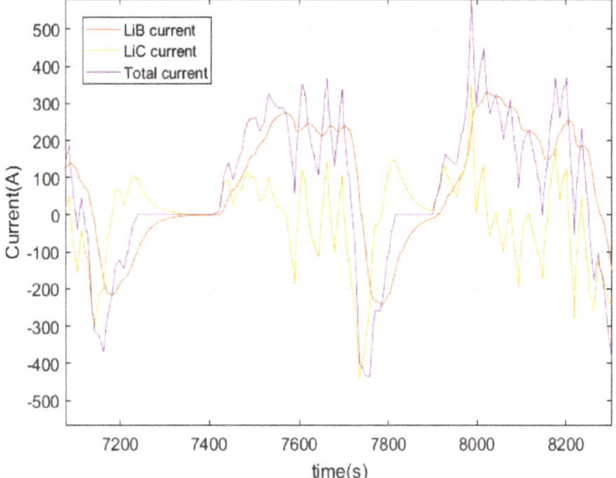

Figure 11. The LiB and LiC current component for the proposed electric bus based on the London driving cycle.

To simplify the lifetime improvement calculation, a proportional equation for the DC/DC converter with the efficiency of 85% [25], as shown in Equation (9), is used:

$$P_{conv-out} = \eta * P_{conv-in} \tag{9}$$

where the $P_{conv-out}$ is a portion of power demand requested by the driver and the driving path which needs to be delivered by LiC module. However, this power demand must obey the reference values given by control unit, (reference current (I_{ref}) which is separated by filter from the total current and the reference voltage (V_{ref}) which comes from the DC-link). The $P_{conv-in}$ on the other hand, is the real value of the LiC module and is calculated as shown in Equation (10):

$$\left(V_{ref} * I_{ref} = V_{batt} * I_{LiC-req} \right) = \eta * V_{LiC-real} * I_{LiC-real} \tag{10}$$

5. Electric Bus Specification and Energy Storage Unit Sizing

In this research, the authors intend to design a HESS for a pure electric bus for an urban public transportation application. The very first step for this design is to specify the bus requirements, including the electric motor power, energy storage capacity, passenger capacity, and requested range. The bus specifications are presented in Table 2, which is based on a commercial electric bus (Solaris Urbino 12 Electric) produced in Poland.

Table 2. Electric bus specification.

Parameter	Value
Length	12 m
Net Weight plus passengers	13.6 t + 6.4 t
Electric motor power	120 kW (two motors of 60 kW)
Range	240 km
Battery capacity	240 kWh

In order to calculate the peak current demand and the requested energy to travel for 240 km with one charge, the Millbrook London Transport Bus (MLTB) driving cycle is selected. As shown in Figure 12, the driving cycle is composed of Inner and Outer London to have a more realistic load profile.

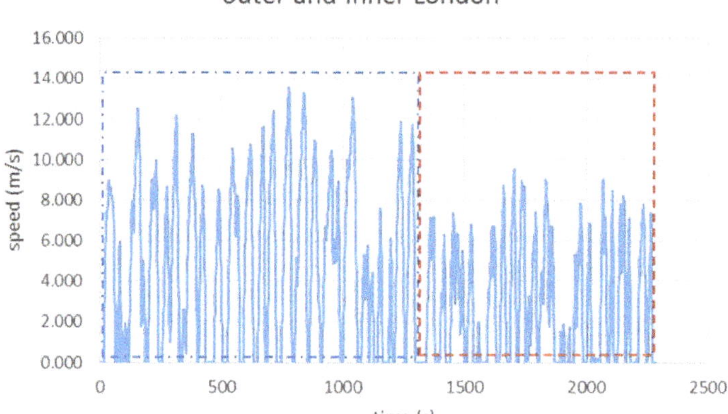

Figure 12. Speed profile of the MLTB driving cycle.

Table 3 gives a summary of this driving cycle.

Table 3. Summary of the MLTB driving cycle.

Parameter	Unit	Ph1 Outer London	Ph2 Inner London	Overall
Total time	s	1381	901	2282
Time at idle	s	1381	901	2282
Distance	km	6.46	2.50	8.96
Average speed	km/h	22	15.9	20.5
Max speed	km/h	47	34.4	48.7
Max acceleration	m/s^2	5	1.5	1.5
Max deceleration	m/s^2	−2.1	−2.2	−2.2
No of stops ≥15 s		11	8	19

The speed load profile should be converted to the power and current profile for further calculation. Based on the simple physics equations we have:

$$a = \frac{V_2 - V_1}{t_2 - t_1} \tag{11}$$

where $V_2\left(\frac{m}{s}\right)$ and $V_1\left(\frac{m}{s}\right)$ are the speed at t_2 (s) and t_1 (s) time sequences, respectively, and a $\left(\frac{m}{s^2}\right)$ is the acceleration. The required force F (N) to reach this acceleration is calculated based on the total mass m (kg) of the vehicle as given in the Equation (12):

$$F = m \times a \tag{12}$$

Having the average speed in each time interval, the average power can be calculated based on the Equation (13):

$$P_{ave} = V_{ave} \times F \tag{13}$$

where V_{ave} is the average speed in one sampling period. Then the power transformation efficiency from mechanical form to the electrical form and from AC mode to the DC mode, and vice versa, is considered to calculate the required power profile. In this study, the transformation efficiency is calculated as 80%. Figure 13 shows the power profile of the MLTB driving cycle.

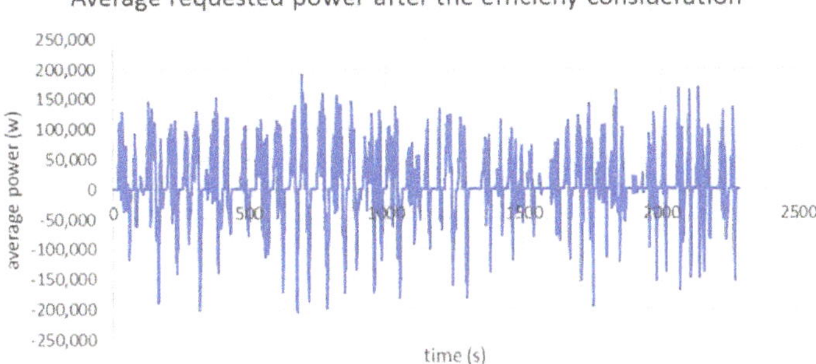

Figure 13. Power profile of the MLTB driving cycle.

As it can be seen in Figure 13, there are positive and negative power peaks. Negative power peaks mean that the power returns to the HESS. To have a more efficient electric bus and to improve the energy density of the vehicle, the negative power should be restored in the ESS.

As mentioned before, in this study the NMC 20 Ah cell is used as a HE storage unit. The specifications are given in [13]. The discharge and charge current for this cell are limited to the 5 C and 0.5 C, respectively. Specific measures should be taken in these cases to avoid extra current extraction and insertion which can negatively affect the lifetime of the cell.

5.1. Battery Bank Configuration for the MLTB Load Profile

The battery bank configuration is given in Table 4.

Table 4. The configuration of the LiB battery bank.

Battery Type	Voltage Range	Energy
NMC 20 Ah	250–400 V	240 kWh

To meet the requirement, the maximum voltage is used to calculate the number of cells in series connection. Other calculation is given in Table 5.

Table 5. The specifications of the LiB battery bank.

Parameter	Value
capacity of each cell @ 100% DoD	20 Ah
Voltage @ 0% SoC	3 V
Nominal voltage	3.65 V
Voltage @ 100 SoC	4.15 V
number of cells in series	$400/4.15 = 96$
Nominal voltage of battery bank	$96 \times 3.65 = 350.4$ V
Required capacity in Ah	240 kWh/350.4 V = 685 Ah
Number of stacks in parallel	$N = 685/20 = 35$ [1]
Stored energy in one stack	$350.4 \times 20 = 7$ kWh
Maximum continuous current (nominal power of electric motor divided by nominal voltage)	$120,000$ W$/350.4$ V $= 342.46$ A $\xrightarrow{yields}$ current of each stack $= 342.46/35 = 9.78$ A

[1] In the simulation environment, 35 stacks in parallel do not provide 240 kWh energy (due to the internal losses in each cell) which results in a shorter mileage, as a result, we increased the number of stacks to 37.

The battery module is based on the electrothermal model of the NMC 20 Ah cell introduced in Section 3. By applying the power profile to the model as shown in Figure 14, at low SoC, the charge current will increase up to 15 A, which is higher than the maximum charge current limit (10 A) for this battery. To avoid this extra current, there are two possibilities: either to increase the number of parallel branches by 50% or to use the HESS.

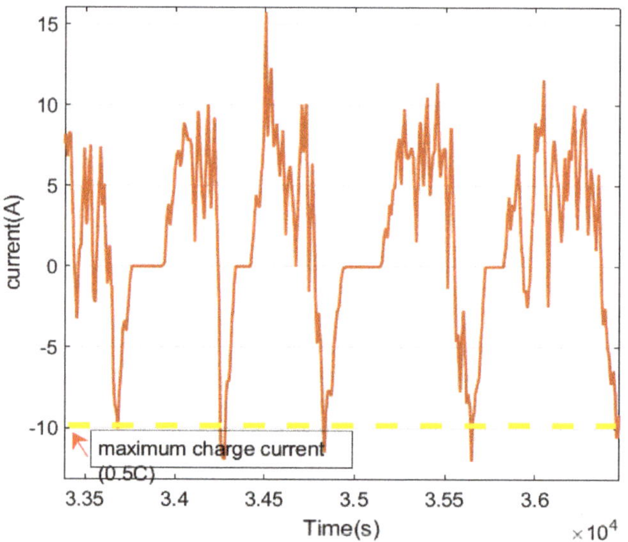

Figure 14. The simulation result for the HE storage system.

5.2. Lithium Ion Capacitor (LiC) Pack Configuration

Table 6 shows the LiC pack requirement. The number of cells and branches are given in Table 7.

Table 6. The LiC bank configuration.

Battery Type	Voltage Range	Energy
LiC 2300 F (1 Ah)	400–700 V	572.9 Wh

Table 7. The specification of LiC bank.

Parameter	Value
Capacity of each cell @ 100% DoD	1 Ah
Voltage @ 0% SoC	2.2 V
Voltage @ 100 SoC	3.8 V
number of cells in series	$700/3.8 = 184$
Stored energy in one string	0.5729 kWh

6. Simulation Results and Discussion

As explained in the previous section, to limit the charge current of the LiB cell, there are two possibilities:

1. To use the HESS.
2. To increase the number of stacks by 50%.

Here, in the simulation result, the two possibilities are compared from the cost, volume, weight, lifetime, and range points of view. Figure 15 shows the simulation environment for both the LiB-based ESS and HESS.

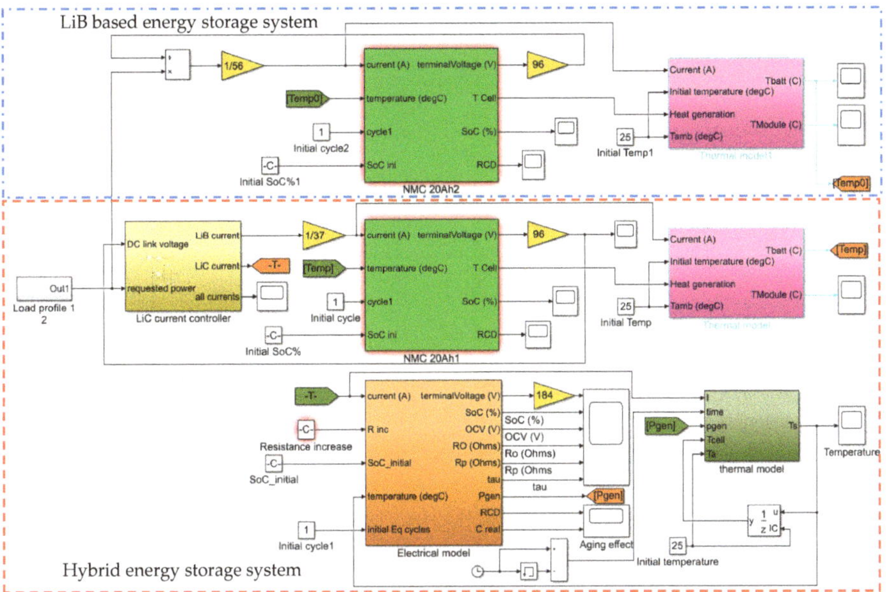

Figure 15. The simulation environment for both ESS and HESS.

6.1. Hybrid Configuration

As mentioned before, the novelty of this research is using LiC instead of SC in the HESS. Using the LiC bank instead of SC with the same capacity and voltage level is lighter and less voluminous. These advantages increase the efficiency of the bus and free up more space for other applications.

As explained in Section 5, the hybrid configuration is composed of LiB pack (37 parallel stacks of LiB cells where each stack has 96 cells in series) in parallel with a LiC pack (one stack of LiC which has 184 cells in series). In this configuration, the total distance which the bus can travel with and without LiC is calculated. As shown in the Table 8, with the HESS, the bus can traverse about 16% more than LiB-EES. As it is seen in Table 8, the maximum charge current for the LiB-ESS is 50% more than the nominal value of the current although these charging pulses are applied in a very short duration, they are repeated thousands of times over the life period of the LiB EES. As these charge pulses are beyond the nominal charge C-rate value based on the datasheet, they may have a negative impact on the lifetime and efficiency of the NMC cell and should be avoided.

In Figure 16 the capacity degradation trend for LiB-ESS is compared with the HESS. In the LiB system, the charge over-current and high-frequency current pulses are applied to the cells while, with the application of HESS, the high-frequency stress, as well as the charge over-current are lifted from the cell resulting a longer life time and lower temperature raise in the cells. As explained in Section 2.3, the capacity degradation for LiBs is highly dependent to the temperature, DoD, and number of cycles. With the application of the HESS, as shown in Table 8, the cell temperature decreases slightly, but the discharge energy from the cell, which is represented by the number of cycles, decreases by 14%. Moreover, by lifting the peak current pulses over the LiB cells, the effect of high DoD is eliminated. Due to the above-mentioned reasons, the mileage is extended by 70,000 km.

Table 8. Obtained results for both the LiB-ESS and HESS.

Parameter	LiB-ESS	HESS	
Time for one discharge (s)	66,259	66,794	
Range for one discharge (km)	260.37	262.48	
Cell temperature (°C)	25.65	LiB: 25.4	Lic: 35
Number of Eq cycles for one discharge	1.9	1.64	625.4
Maximum discharge current	20.15	12.26	451
Maximum charge current	15.16 > 10	9.7	572
Range until end of life (km)	410,476.15	480,148.35	
Driving time until end of life (h)	29,015.15	33,940.04	

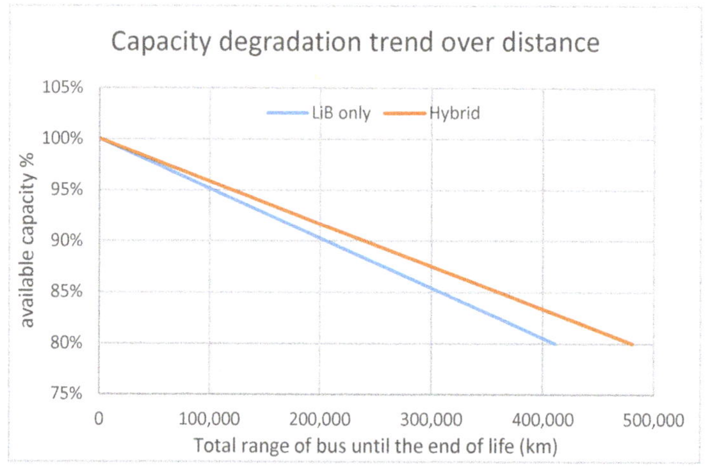

Figure 16. Capacity degradation trend versus the passable distance for LiB-ESS and HESS.

6.2. LiB Configuration with 50% Over-Design

As explained in Section 6, another possibility to limit the charge over-current of the LiB module, is to increase the initial number of parallel stacks by 50%. With this change, as shown in Table 9, the charge current stays within the limit, but the stored energy, volume, weight, and cost increase significantly.

Table 9. Obtained results for the LiB-ESS with 50% over-design.

	Parameter						
	Time for One Discharge (s)	Range for One Discharge (km)	Cell Temperature (°C)	Number of Eq Cycle for One Discharge	Maximum Discharge Current (A)	Maximum Charge Current (A)	Range until End of Life (km)	Driving Time until End of Life (h)
LiB-ESS	100,549	395.12	25.25	2	13	10	592,692.73	41,895.41

As it is shown in Figure 17, with this configuration, the range is improved noticeably, which is due to the greater amount of energy stored in the batteries. Having a lower current passing through the LiB-ESS in comparison to the HEES, the cells are degraded slower and the cell temperature is also much lower. As a result, the degradation process will be much lower than the HESS.

To compare both configurations from the volume, weight, range, and lifetime points of view, in the Tables 10 and 11 the result for the HESS and LiB-ESS are given, respectively.

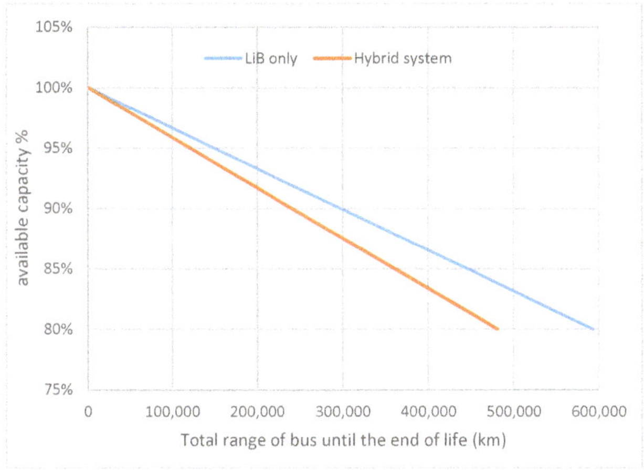

Figure 17. Capacity degradation trend versus the passable distance for the over-designed LiB-ESS and HESS.

Table 10. Size and cost of the HESS.

Parameter	LiC		LiB		Total
	Single Cell	Module	Single Cell	Module	
Weight (kg)	0.365	67.16	0.428	1520.25	1587.41
Volume (l)	0.216	39.74	0.196	696.19	735.93
Cost (%)		5		95	100

Table 11. Size and cost of the energy storage system based on LiB with 50% oversizing.

Parameter	Single Cell	Oversized LiB Module
Weight (kg)	0.428	$0.428 \times 96 \times 37 \times 1.5 = 2280.38$
Volume (l)	0.196	$0.196 \times 96 \times 37 \times 1.5 = 1044.28$
Cost (%)		144%

To summarize the obtained results for both the LiB-ESS and HESS, the results are summarized in Table 12.

It is clear from the obtained results that using a HESS downsizes the energy storage system, yet is capable of providing enough energy and power to the vehicle to meet its demand. It is also seen that with the hybrid configuration, the lifetime is improved by more than 16%. The volume and weight decrease by 30% in comparison to the over-designed LiB-ESS.

Table 12. A summary of Tables 10 and 11.

Parameter	96S56P	96S37P + 184S1P
weight (kg)	2300.92	1587.41
volume (l)	1053.69	735.93
Cost (%)	144	100
energy (kWh)	392 [1]	259.57 [2]
range (km)	592,692.73	480,148.35
continuous driving time (h)	41,895.41	33,940.04

[1] 7 kWh $\times$ 56 = 392 kWh, [2] 7 kWh $\times$ 37 + 0.5729 kWh = 259.57 kWh.

7. Conclusions

In this paper, a HESS, which is a combination of a lithium-ion battery and lithium-ion capacitor, was presented. The aim of this study is to show the effectiveness of hybrid systems to downsize the ESS and to decrease the volume and weight, as well as life-time improvement. At the first step, the electrothermal model and lifetime model for both the LiB and LiC units are explained. Then the hybridization method, which is a semi-active hybrid topology is briefly explained. The advantage of this topology, which has been presented in some of the previous studies, led us to this selection. The low complexity control strategy for load sharing between the LiB and LiC was explained briefly and the advantages were shortly explained. In addition to a lower computational cost, which it requires, the simplicity in load sharing based on a low-pass filter was considered for this selection. In the end, the unit sizing for both the energy storage units was thoroughly explained and was validated with the simulation results. Lastly, based on the calculation and simulation results, a comparison between the LiB-ESS and HESS was presented. This study shows that with the HESS, the lifetime is improved by 16% and the size is lowered by 30%. As a future work, the authors intend to optimize the LiC and LiB unit for a lower cost and size and with a higher energy and power density.

Author Contributions: M.S. has written the original draft and presented the hybridization methodology and the simulation environment; J.R. and S.K. provided technical data and practical information to make this study applicable for the industry and they also investigated the result from the technical point of view; and J.J., N.O., P.V.d.B. and J.v.M. reviewed and edited the manuscript and they also provided supervision guidance to this research.

Funding: This research received no external funding.

Acknowledgments: We acknowledge Flanders Make for the support to our research group. We also acknowledge JSR Micro NV for their support for this research.

Conflicts of Interest: The authors declare no conflict of interest.

References

1. Schupbach, R.M.; Balda, J.C.; Zolot, M.; Kramer, B. Design methodology of a combined battery-ultracapacitor energy storage unit for vehicle power management. In Proceedings of the IEEE 34th Annual Conference on Power Electronics Specialist, Acapulco, Mexico, 15–19 June 2003; pp. 88–93. [CrossRef]
2. Masih-Tehrani, M.; Ha'iri-Yazdi, M.R.; Esfahanian, V.; Safaei, A. Optimum sizing and optimum energy management of a hybrid energy storage system for lithium battery life improvement. *J. Power Sources* **2013**, *244*, 2–10. [CrossRef]
3. Veneri, O.; Capasso, C.; Patalano, S. Experimental investigation into the effectiveness of a super-capacitor based hybrid energy storage system for urban commercial vehicles. *Appl. Energy* **2017**. [CrossRef]
4. Repp, S.; Harputlu, E.; Gurgen, S.; Castellano, M.; Kremer, N.; Pompe, N.; Wörner, J.; Hoffmann, A.; Thomann, R.; Emen, F.M.; et al. Synergetic effects of Fe^{3+} doped spinel $Li_4 Ti_5 O_{12}$ nanoparticles on reduced graphene oxide for high surface electrode hybrid supercapacitors. *Nanoscale* **2018**. [CrossRef] [PubMed]
5. Genc, R.; Alas, M.O.; Harputlu, E.; Repp, S.; Kremer, N.; Castellano, M.; Colak, S.G.; Ocakoglu, K.; Erdem, E. High-Capacitance hybrid supercapacitor based on multi-colored fluorescent carbon-dots. *Sci. Rep.* **2017**, *7*, 1–13. [CrossRef] [PubMed]
6. Soltani, M.; Jaguemont, J.; Omar, N.; Ronsmans, J.; van den Bossche, P.; Van Mierlo, J. Cycle Life Evaluation for Lithium-Ion Capacitors. In Proceedings of the Evs30 the 30th International Electric Vehicle Symposium and Exhibition, Stuttgart, Germany, 9–11 October 2017; pp. 1–7.
7. Ultimo Lithium Ion Capacitor Prismatic Cells. Available online: www.jsrmicro.be (accessed on 18 July 2018).
8. Zimmermann, T.; Keil, P.; Hofmann, M.; Horsche, M.F.; Pichlmaier, S.; Jossen, A. Review of system topologies for hybrid electrical energy storage systems. *J. Energy Storage* **2016**, *8*, 78–90. [CrossRef]
9. Wang, X.; Yu, D.; Le Blond, S.; Zhao, Z.; Wilson, P. A novel controller of a battery-supercapacitor hybrid energy storage system for domestic applications. *Energy Build.* **2017**, *141*, 167–174. [CrossRef]
10. Ruan, J.; Walker, P.D.; Zhang, N.; Wu, J. An investigation of hybrid energy storage system in multi-speed electric vehicle. *Energy* **2017**, *140*, 291–306. [CrossRef]

11. Zhang, S.; Xiong, R.; Cao, J. Battery durability and longevity based power management for plug-in hybrid electric vehicle with hybrid energy storage system. *Appl. Energy* **2016**, *179*, 316–328. [CrossRef]

12. Song, Z.; Hou, J.; Xu, S.; Ouyang, M.; Li, J. The influence of driving cycle characteristics on the integrated optimization of hybrid energy storage system for electric city buses. *Energy* **2017**, *135*, 91–100. [CrossRef]

13. De Hoog, J.; Timmermans, J.M.; Ioan-Stroe, D.; Swierczynski, M.; Jaguemont, J.; Goutam, S.; Omar, N.; Van Mierlo, J.; Van Den Bossche, P. Combined cycling and calendar capacity fade modeling of a Nickel-Manganese-Cobalt Oxide Cell with real-life profile validation. *Appl. Energy* **2017**, *200*, 47–61. [CrossRef]

14. Nikolian, A.; Firouz, Y.; Gopalakrishnan, R.; Timmermans, J.M.; Omar, N.; van den Bossche, P.; van Mierlo, J. Lithium ion batteries—Development of advanced electrical equivalent circuit models for nickel manganese cobalt lithium-ion. *Energies* **2016**, *9*, 360. [CrossRef]

15. Nikolian, A.; De Hoog, J.; Fleurbay, K.; Timmermans, J.; Van De Bossche, P.; Van Mierlo, J. Classification of Electric modelling and Characterization methods of Lithium-ion Batteries for Vehicle Applications. In Proceedings of the European Electric Vehicle Congress, Brussels, Belgium, 13–16 May 2014; pp. 1–15.

16. Berckmans, G.; Ronsmans, J.; Jaguemont, J.; Samba, A.; Omar, N.; Hegazy, O.; Soltani, M.; Firouz, Y.; van den Bossche, P.; Van Mierlo, J. Lithium-Ion Capacitor: Analysis of Thermal Behavior and Development of Three-Dimensional Thermal Model. *J. Electrochem. Energy Convers. Storage* **2017**, *14*, 041005. [CrossRef]

17. Liu, Z.; Li, H.X. Thermal modeling for vehicle battery system: A brief review. In Proceedings of the 2012 International Conference on System Science and Engineering, Dalian, Liaoning, China, 30 June–2 July 2012; pp. 74–78.

18. Jaguemont, J.; Nikolian, A.; Omar, N.; Goutam, S.; Van Mierlo, J.; Van den Bossche, P. Development of a 2D-thermal model of three battery chemistries considering entropy. *IEEE Trans. Energy Convers.* **2017**, *32*, 1. [CrossRef]

19. De Hoog, J.; Jaguemont, J.; Abdel-Monem, M.; Van Den Bossche, P.; Van Mierlo, J.; Omar, N. Combining an electrothermal and impedance aging model to investigate thermal degradation caused by fast charging. *Energies* **2018**, *11*. [CrossRef]

20. Firouz, Y.; Omar, N.; Van Den Bossche, P.; Van Mierlo, J. Electro-thermal modeling of new prismatic lithium-ion capacitors. In Proceedings of the 2014 IEEE Vehicle Power and Propulsion Conference, Coimbra, Portugal, 27–30 October 2014; Volume 2, pp. 1–6.

21. Sivakkumar, S.R.; Pandolfo, A.G. Evaluation of lithium-ion capacitors assembled with pre-lithiated graphite anode and activated carbon cathode. *Electrochim. Acta* **2012**, *65*, 280–287. [CrossRef]

22. Cao, J.; Emadi, A. A new battery/ultracapacitor hybrid energy storage system for electric, hybrid, and plug-in hybrid electric vehicles. *IEEE Trans. Power Electron.* **2012**, *27*, 122–132. [CrossRef]

23. Song, Z.; Hofmann, H.; Li, J.; Hou, J.; Han, X.; Ouyang, M. Energy management strategies comparison for electric vehicles with hybrid energy storage system. *Appl. Energy* **2014**, *134*, 321–331. [CrossRef]

24. LiC Cell 2300F, Prismatic (JSR CPQ2300S). Available online: https://www.electrostandards.com/ProductDetail/?productid=2446 (accessed on 18 July 2018).

25. Shrivastava, A.; Calhoun, B. A DC-DC converter efficiency model for system level analysis in ultra low power applications. *J. Low Power Electron. Appl.* **2013**, *3*, 215–232. [CrossRef]

Article

In-Life Range Modularity for Electric Vehicles: The Environmental Impact of a Range-Extender Trailer System

Nils Hooftman [1,2,*], Maarten Messagie [1,2], Frédéric Joint [3], Jean-Baptiste Segard [3] and Thierry Coosemans [1,2]

[1] Electrotechnical Engineering and Energy Technology, MOBI Research Group (VUB-MOBI Group is member of Flanders Make.), Vrije Universiteit Brussel, Pleinlaan 2, 1050 Brussels, Belgium; maarten.messagie@vub.be (M.M.); thierry.coosemans@vub.be (T.C.)
[2] Flanders Make, 3001 Heverlee, Belgium
[3] EP Tender, 22 rue Gustave Eiffel, 78300 Poissy, France; frederic.joint@eptender.com (F.J.); jean-baptiste.segard@eptender.com (J.-B.S.)
*   Correspondence: nils.hooftman@vub.be; Tel.: +32-(2)-629-37-67

Received: 28 April 2018; Accepted: 17 June 2018; Published: 21 June 2018

Abstract: *Purpose*: In the light of decarbonizing the passenger car sector, several technologies are available today. In this paper, we distinguish plug-in hybrid electric vehicles (PHEV), electric vehicles (EV) with a modest battery capacity of 40 kWh, and long-range EVs with 90 kWh installed. Given that the average motorist only rarely performs long-distance trips, both the PHEV and the 90 kWh EV are considered to be over-dimensioned for their purpose, although consumers tend to perceive the 40 kWh EV's range as too limiting. Therefore, in-life range modularity by means of occasionally using a range-extender trailer for a 40 kWh EV is proposed, based on either a petrol generator as a short-term solution or a 50 kWh battery pack. *Method*: A life cycle assessment (LCA) is presented for comparing the different powertrains for their environmental impact, with the emphasis on local air quality and climate change. Therefore, the combination of a 40 kWh EV and the trailer options is benchmarked with a range of conventional cars and EVs, differentiated per battery capacity. Next, the local impact per technology is discussed on a well-to-wheel base for the specific situation in Belgium, with specific attention given to the contribution of non-exhaust emissions of PM due to brake, tyre, and road wear. *Results*: From a life cycle point of view, the trailer concepts outperform the 90 kWh EV for the discussed midpoint indicators as the latter is characterized by a high manufacturing impact and by a mass penalty resulting in higher contributions to non-exhaust PM formation. Compared to a petrol PHEV, both trailers are found to have higher contributions to diminished local air quality, given the relatively low use phase impact of petrol combustion. Concerning human toxicity, the impact is proportional to battery size, although the battery trailer performs better than the 90 kWh EV due to its occasional application rather than carrying along such high capacity all the time. For climate change, we see a clear advantage of both the petrol and the battery trailer, with reductions ranging from one-third to nearly sixty percent, respectively. *Conclusion*: Whereas electrified powertrains have the potential to add to better urban air quality, their life cycle impact cannot be neglected as battery manufacturing remains a substantial contributor to the EV's overall impact. Therefore, in-life range modularity helps to reduce this burden by offering an extended range only when it is needed. This is relevant to bridge the years up until cleaner battery chemistries break through, while the energy production sector increases the implementation of renewables. Petrol generator trailers are no long-term solution but should be seen as an intermediate means until battery technology costs have further dropped to make it economically feasible to commercialize battery trailer range-extenders. Next, active regulation is required for non-exhaust PM emissions as they could dominate locally in the future if more renewables would be applied in the electricity production process.

Keywords: range-extender; CO_2; air quality; mobility needs; LCA; Paris Agreement

1. Introduction

Air quality levels across Europe remain problematic, especially in urban regions [1]. These are found to be hotspots for nitrogen oxides (NO_x) and particulate matter (PM), for which road transport has a substantial contribution [2]. In the light of mitigating local air quality levels, pollutant emissions from both passenger cars, light-commercial, and heavy-duty vehicles have been regulated by the so-called Euro emission standards [3]. For cars, these progressive reduction targets have proven their efficacy for bringing down the emissions of exhausted PM, carbon monoxide (CO), unburned hydrocarbons (HC), and petrol NO_x. For diesel cars, however, substantial exceedances with the imposed NO_x limits are found. Evidence results from both rigorous chassis dynamometer testing [4,5], real-world driving tests using portable emissions measurement systems (PEMS) [6,7], and roadside remote sensing campaigns [8,9]. For heavy-duty vehicles, which are no further discussed in the presented paper, substantial NO_x reductions for heavy-duty vehicles have been realized since Euro VI. As such, they produce only half the NO_x emissions compared to the average Euro 6 diesel car, when compared on a kilometer basis [10].

Concerning transport's impact on climate change, the European light-duty vehicle sector is imposed to a target for carbon dioxide (CO_2) emissions. Therefore, each car manufacturer is required to obtain a corporate average fleet fuel economy of 95 g of CO_2 per kilometer by 2021, as described in Regulation (EC) 2009/443 [11]. Electric vehicles (EVs) are key assets in reaching this target, as they are given extra weight in the balance using so-called *super-credit factors* [12]. These allow EVs to count for more than one car in the fleet average calculation. An indirect effect of Regulation 443 is that EV-producing car manufacturers are less incentivized to lower the CO_2 emissions of their remaining conventional models, as electric vehicles bring down their fleet's average [13]. The importance of the 2021 and future CO_2 targets in Europe is strengthened by the 2015 Paris Agreement, in which the majority of the world's nations agreed to strictly reduce greenhouse gas emissions (GHG, represented by CO_2-equivalent gasses), in order to keep the global temperature increase well below 2 °C, relative to pre-industrial levels [14]. Despite the current lack of a well-defined roadmap towards this goal, a net-zero GHG economy is required by 2050 or shortly after that [15]. For this reason, the European Union strives to a minimum GHG reduction by 60 percent for its transport sector by 2050, while its entire economy is bound to reduce its GHG contribution by 80 to 95 percent [16]. In the light of decarbonizing the light-duty fleet, electrification is believed to play a major role if the electricity production sector follows the same decarbonization trend. Notwithstanding, we see that EVs applied in Europe already produce less CO_2 than a diesel car of the same segment on a well-to-wheel basis, even when applying electricity from a coal-intensive production like in Poland [17]. Besides the CO_2 reduction potential, however, current battery technologies have a significant environmental impact when their manufacturing process is considered, and this is primarily due to mining practices [18]. This indicates the need for cleaner technologies and a ramping-up of recycling used batteries. Given an average lifetime of a battery applied in an automobile of ten years, a substantial stream of used batteries is yet to come. Moreover, second-life application in stationary power storage can extend the useful life of batteries significantly, indicating a further shift in time before recycling the current generation of batteries becomes economically viable. Disregard this fact, precious metals like cobalt are already being recovered from battery waste, whereas the increase of prices raw lithium has led to the start of industrializing its recovery as well [19]. Large-scale recycling is nonetheless to be expected beyond 2025 [20].

Widespread adoption of EVs requires a paradigm shift in the mind of consumers, as the technology is characterized by limited onboard energy storage. Despite substantially lower operational expenses (OPEX), both fiscal and financial incentives remain a necessity to bring down the capital expenditure (CAPEX). Next, there is the need for a widely available network of public charging infrastructure.

Incentivization will, therefore, be required (at least) up until the point where cost parity is reached with a conventional car of the same segment. This moment is forecast to arrive between 2022 and 2026 [21]. Whereas the electric range is repeatedly indicated as one of the significant hurdles for EV breakthrough, it is a fact that the technology cannot cover the mobility needs of every consumer. Travel surveys are an essential source for estimating real-world needs for the daily range for passenger car users. Examples of such studies can be found in Pearre et al. [22] and Needell et al. for North-American statistics [23]; and Pasaoglu et al. [24] and Corchero et al. [25] for European variants. These different sources confirm that most of the daily range needs are in the 0 to 80 km range, while ranges exceeding 150 km/day are found to occur only on a limited number of days per year, i.e., for only 5 percent of the daily trips [26,27]. An exemplary distribution of the daily driven distances is given in Figure 1. These examples show the potential for substituting conventional cars with EVs for a substantial part of the year while relying on alternatives for the days when more considerable distances are traveled. In the absence of a sufficiently deployed charging infrastructure, this statement firstly targets consumers that have private parking places and/or a garage that can be equipped with a charger, i.e., the early adopters.

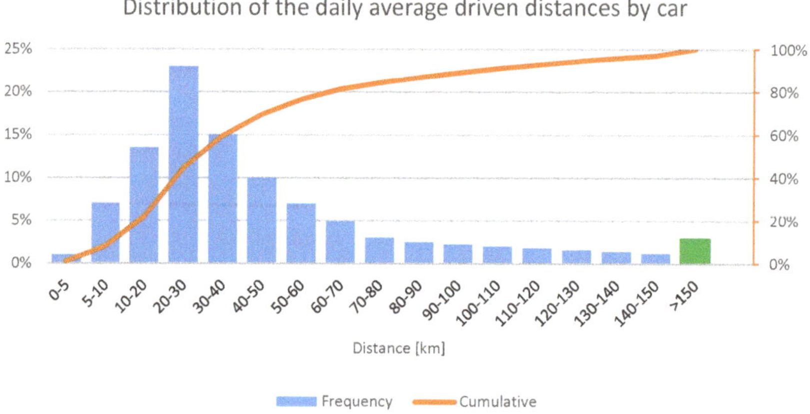

Figure 1. Example of an average daily driven distance distribution (based on Redelbach et al. [28]).

Options for covering these few percent of long daily distances could either be a conventional car, a plug-in hybrid electric vehicle (PHEV), or an EV with a high-capacity battery pack. PHEVs have the potential to mitigate both greenhouse gas and pollutant emissions as they typically allow an all-electric distance of 30–50 km. This nonetheless implies PHEV users must recharge on a daily basis, which is not always the case, as Ligterink et al. indicated by concluding only 30 percent of the Dutch PHEV company cars' kilometers were covered electrically [29]. The Dutch example shows how a large-scale market adoption of PHEVs was rather a consequence of a favorable tax regime than of an environmental motivation. Due to the incentives for car manufacturers to market PHEVs, European variants typically combine powerful engines with modest battery packs, which allow unrealistically low type-approval CO_2 emissions [30]. Nonetheless, a significant amount of attention has been given to PHEVs in the scientific literature. Popular topics are the optimization of battery capacity [28,31–33], the environmental impact of the dual technology [34,35], and total cost of ownership [36–39]. Whereas the automotive industry regards PHEVs as an important asset for reaching future GHG and pollutant emission targets [40], the question remains which future market share this expensive dual technology is destined for, as EVs are gaining momentum and battery innovation aims for driving ranges equaling those of conventional petrol cars, while the cost difference decreases as well. Despite current EV autonomies ranging from 250 to 400 *real* kilometers, manufacturers claim ranges up to 600 km and more for models that will be introduced in short-term. Given the limited adoption of EVs to date, we can assume those consumers that are buying such cars are early adopters with the possibility to recharge on a daily basis. In this case, however, the full potential of such batteries will potentially only be addressed in rare cases. Thus, the respective EV will have to move a 'dead mass' for most of the time, diminishing its inherent environmental impact due to high energy consumption, a higher manufacturing impact and higher tyre and road wear [41].

Another solution might be a roll-out of fast-chargers and, parallel to this, the development of battery chemistries that can be charged at higher currents. Although this option creates a lesser burden for the EV user in terms of charging time, the challenge remains to limit the impact of charging rates >50 kW have on the cycle-life of the respective battery packs. Moreover, fast-chargers put a substantial strain on the available power grid [42,43] , if they are not coupled to a local storage system, consisting of second-life batteries or supercapacitors that are charged when power demand is low. For these reasons, a concept of in-life modularity is proposed in this paper, consisting of a range-extender trailer that can be connected to an EV. Both a petrol generator and a battery pack trailer are shown in Figure 2.

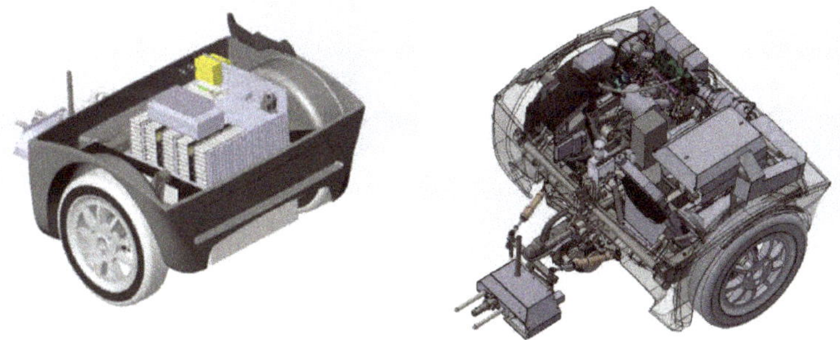

Figure 2. Graphical representation of the generator trailer (**left**) and the battery trailer (**right**).

Ideally, this concept consists of a trailer fitted with an extra battery pack, allowing the EV's battery to be used in a charge-sustaining mode for an additional 300 km. Whereas extra range is typically required to cover long distances over highways, this range-extender concept (from now on abbreviated to 'ReX') could complement fast-charging stations located near highways and would be offered on a rental basis, to avoid a high upfront cost for its sporadic use. As an intermediate towards the ideal situation, a petrol generator could serve as a power source for a generator to supply energy to the EV. As battery costs continue to decrease with increased production and a further fine-tuning of the production process, the petrol generator could be phased-out on the short to mid-term to maximize the environmental potential of in-life range modularity. Therefore, the objective of the presented paper is first to assess the environmental impact of this setup considering both climate change and air quality, by comparing a 40 kWh EV + ReX trailer combination to a range of mid-sized family cars. These are based on either petrol, diesel, petrol hybrid, or petrol PHEV powertrains. Also, the comparison is made with four existing EV models, characterized by battery capacities of 30, 60, and 90 kWh, respectively. The main focus is on the results for a petrol generator and a 50 kWh battery trailer, which are discussed in relation to the PHEV and the 90 kWh EV, as these are the technologies we expect to compete in the coming decade.

2. Methodology

2.1. Life Cycle Assessment

An environmental Life Cycle Assessment (LCA) is applied to compare the impacts, damages, and benefits of the combination of an EV + ReX trailer while considering all the associated emissions, both direct and indirect. An LCA consists a four-step approach, including a definition of a goal and scope, a life cycle inventory, an impact assessment, and the interpretation of the results, following the methodology according to ISO 14040 [44] and ISO 14044 [45].

2.1.1. Goal and Scope

For the presented paper, different powertrain technologies are compared to the combination of a 40 kWh EV and a range-extender trailer for their impact on both climate change and human health. Despite the technological differences between the discussed powertrains regarding, for instance, their nominal driving range, they all provide the same function of mobility. Therefore, the functional unit for comparing the different powertrains is their respective impact per kilometer driven. The entire life cycle impact is calculated by considering a lifespan of 210,000 km, which represents the European average end-of-life age for passenger cars of 15 years and an assumed annually driven distance of 14,000 km. For the discussed EVs, we assume a battery pack replacement after 150,000 km.

Whereas the scope of this assessment concerns the global impact of the different technologies, seen over their entire life cycle, we also zoom in on the specific case for Belgium to assess the impact on human health locally. Therefore, a full-scale LCA for both the impacts on climate change and human health is complemented with a well-to-wheel emissions analysis for which all emissions produced outside of Belgium's national borders are excluded from the scope. Therefore, refinery-to-tank (RTT) emissions are applied instead of the conventional well-to-tank (WTT) emissions, as discussed earlier in Hooftman et al. [41]. For a graphical overview of the methodology, please refer to Figure 3. Addressing local impacts is relevant in the light of improving air quality levels in European cities, as virtually every city dweller is deemed to breathe air that is found harmful to human health.

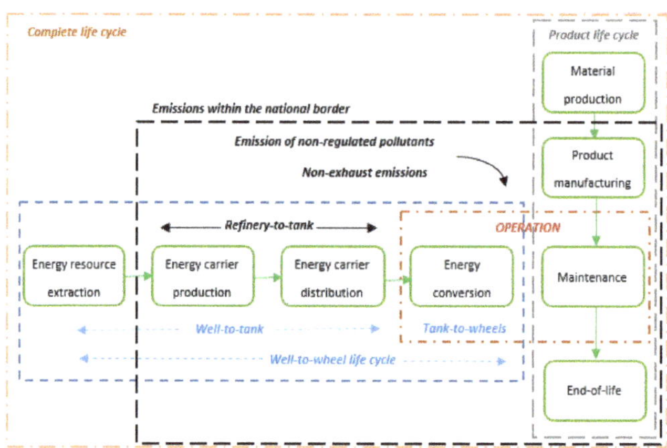

Figure 3. Flowchart of the LCA approach (based on [36,41]).

2.1.2. Lifecycle Inventory

For the different powertrains, life cycle inventories (LCI) were produced based on Messagie et al. in [46], and cover both emissions related to the well-to-tank and tank-to-wheel phase, as well as those related to the manufacturing of the bodywork and powertrain. Concerning end-of-life recycling, we accredit the benefit of recovering materials from the recycling processes to the manufacturing phase, avoiding the material production from virgin ores. The advantage of separating the different product life stages enables the identification of the causes of specific impacts and emissions per stage in the product's value-chain. The original inventories have been updated to reflect the real-world fuel consumption and regulated emission data based on the discrepancies found in literature, and have been supplemented with emission factors for the most important non-regulated pollutants based on Hooftman et al. in [41]. A deliberate divergence from the official emission factors was chosen to allow a fair comparison between the different technologies. For the EVs, the energy consumption data published by the U.S. Department of Energy (DOE) was used [47]. The specifications of the assessed EVs are given in Table 1, while the assumed energy consumption factors for the discussed ICE cars are given in Table 2. The energy consumption of the 40 kWh EV combined with a trailer is assumed to increase by five percent, Finally, whereas the original powertrains database only described a 24 kWh EV, the LCI for the 30, 40, 60, and 90 kWh EVs discussed in this paper result from a parametrization exercise for which vehicle mass, battery mass, and electric consumption served as determinants.

Table 1. Specifications of the discussed electric vehicles (based upon [47]).

Parameter [Unit]	30 kWh EV (Nissan Leaf)	40 kWh EV (Renault Zoe)	60 kWh EV (Chevrolet Bolt)	90 kWh EV (Tesla Model S90)
Capacity [kWh]	30	40	60	90
Mass in Running Order [kg]	1591	1450	1624	2200
Weight battery [kg]	272	305	435	540
Average consumption [kWh/km]	0.15	0.15	0.15	0.25
Highway consumption [kWh/km]	0.20	0.20	0.20	0.26
Highway consumption with ReX trailer [kWh/km]		0.21		
EU electricity mix in g/kWh			276	

Table 2. Overview of the real-world tank-to-wheel fuel consumption indicators per technology (based upon [29,30]).

Unit	Petrol	Petrol Hybrid	Diesel	Petrol Plug-in Electric Vehicle
[l/100 km]	6.8	5.6	5.3	3.4
gCO$_2$/km	162.7	134.0	140.0	81.3

3. Life Cycle Inventory

3.1. Impact Assessment

The selected impact assessment methodology that was applied in the SimaPro 8.3 software is ReCiPe midpoint (H) [48]. Out of a set of eighteen midpoint impact categories, four are discussed in this paper as they represent both GHG emissions and air quality in urban environments. These are Climate Change (CC), Photochemical Oxidant Formation (POF), Human Toxicity (HT), and Particulate Matter Formation (PMF). These midpoint indicators serve as an intermediate between the emission source and the 'endpoint', representing the recipients of the environmental effects caused by anthropogenic activities; these are Human Health, Ecosystem Quality, and Natural Resources [39]. No endpoint indicators are discussed in this paper.

Concerning the midpoint indicator for climate change, CO$_2$-equivalents represent the group of greenhouse gasses. Primary drivers of POF are elements from the family of benzenes, nitrogen oxide(s), and other non-methane organic compounds, which are precursor gasses for ground-level ozone (O$_3$) formation. POFs are in this paper represented by the group of non-methane volatile organic compounds (NMVOC). As for HT, copper, dioxins, cadmium, silver, and zinc among others contribute to the impacts in this category, grouped as 1,4-dichlorobenzene equivalents (1,4-DB). Particulate Matter Formation highlights the impacts of primarily formed particulates as well as particulates formed by the condensation of nitrogen oxides, sulphur oxides, ammonia, and non-methane volatile organic compounds (secondary PM). PMF is represented by the emission of PM10-equivalents, i.e., particles with an aerodynamic diameter smaller or equal to 10 micrometers. Both PMF from combustion and from tyre, brake, and road wear are considered to offer a maximal scope on the most relevant pollutants.

3.2. Assumptions

The discussed EVs are assumed to be charged with the average European energy mix, characterized by an average CO$_2$ emission intensity of 276 g per kWh of electricity produced [49]. The authors of this paper deliberately chose not to address marginal energy production for generating the electricity for EVs, as EVs are thought to be part of the total load system, confirming the viewpoints found in Refs. [35,50–52].

For the comparison of the different powertrain technologies, the 40 kWh EV is chosen as the reference to be combined with the range-extender. Also, the marginal application of the trailer is assumed to be 5 percent of the vehicle's lifetime driven distance, reflecting the outcome of the aforementioned travel surveys and the fact that long distances are in general only seldom performed. Next, one trailer is assumed to be shared by a maximum of 15 users on a rental basis, resulting in the fact that its manufacturing impact is subsequently shared over these 15 users. In the sensitivity analysis in Section 5, we assess the impact of a higher trailer uptake and thus a lower number of users per unit. For the remaining 95 percent of the EV's lifetime, it is assumed to drive purely electric. Equation (2) represents how well-to-tank emissions are based on both the EV's average consumption (kWh/km) and the increased consumption while towing the trailer. This increase is assumed to be capped at 5 percent as the trailer is closely coupled behind the EV and therefore has a minimal influence on its aerodynamics. In case of towing the battery trailer, an extra consumption of 10 percent is considered. The specifications for the trailers per power source concept are given in Table 3, while Table 4 presents the use stage emissions for a suitable generator.

$$WTT_{EV40+ReX} = \left(0.95 \times WTT_{EV40,\ avg}\right) + \left(0.05 \times \left(WTT_{EV40,\ highway} + WTT_{ReX}\right)\right) \qquad (1)$$

Table 3. Overview of the trailer characteristics for a petrol generator and an additional battery pack.

Parameter	Petrol Generator	Battery Pack
Rated power [kW]	25	50
Mass [kg]	265	480
Fuel tank [L]	35	/
Fuel type	Petrol	Electric
Range [km]	300	300
Average consumption [L/kWh]	0.44	/
Average consumption [L/100 km]	7.5	/

Table 4. Emission factors for the generator ReX trailer [53].

Unit	Average Generator Emissions				
	HC	CO	NO$_x$	CO$_2$	HC + NO$_x$
[g/kWh]	2.5	40.5	1.1	999.4	3.6

Based on a life cycle inventory (LCI), the elementary flows which are linked to the various vehicle technologies need to be converted to the different impact categories. These allow quantification and a comparison between the potential impacts. This step is referred to as the life cycle impact assessment (LCIA). An exemplary full LCIA of a 30 kWh EV is given in Table A1 in the Appendix A. Concerning the environmental performance of both the generator and the battery ReX trailer concepts, please refer to Tables 5 and 6, respectively. These tables show a deliberate distinction between the production of the 'trailer body' and the production of the power source, while the operation of the latter was analyzed during its use phase. This choice was made to allow better insight into the allocation of their respective contribution to the midpoint indicators. The total lifetime of the trailer was chosen to be identical to that of a passenger car itself, namely 210,000 km. The European electricity mix is included, representing the 95 percent of the time during which the generator trailer is decoupled from the EV. For all four midpoint indicators discussed in Table 5, it is this 'EV part' which is responsible for approximately three-quarters of the respective impact. The fact that the generator ReX trailer has a significant impact for being active only 5 percent of the time emphasizes the potential environmental improvements if the generator would be substituted by a battery pack, as indicated in Table 6. Keep in mind that in the remainder of this paper, the impacts of the trailer's assembly for both the bodyworks and the generator are divided by 15, as the product is developed to be shared by the same number of users.

Table 5. LCIA of the 40 kWh EV + generator ReX for the various impact categories

Impact Category	Unit	Total	Trailer Assembly Excl. Generator	Generator Manufacturing	Generator Operation	EV Electricity (EU Mix)
Climate change	gCO_2-eq./km	109.31	2.12	3.67	12.66	90.85
Human toxicity	g1,4-DB-eq./km	63.44	2.33	9.94	0.45	50.72
Photochemical oxidant formation	gNMVOC/km	0.26	0.01	0.01	0.04	0.19
Particulate matter formation	gPM_{10}-eq./km	0.16	0.01	0.01	0.01	0.13

Table 6. LCIA of the 40 kWh EV + battery trailer.

Impact Category	Unit	Total	Trailer Assembly Excl. Battery	Battery Manufacturing	Battery Operation	EV (EU Mix)
Climate change	gCO_2-eq./km	102.5	3.2	18.0	0	81.3
Human toxicity	g1,4-DB-eq./km	36.3	3.5	31.5	0	1.27
Photochemical oxidant formation	gNMVOC/km	0.62	0.02	0.05	0	0.10
Particulate matter formation	gPM_{10}-eq./km	1.16	0.02	0.07	0	0.02

4. Results and Discussion

4.1. Climate Change

Figure 4 presents the life cycle impact on climate change for the different powertrain technologies. Starting from the left-hand side, the impact of the use phase is emphasized for the ICE-based powertrains. For the plug-in hybrid, we report a similar impact than for diesel, and this is primarily due to the high real-world tank-to-wheel emissions of the former. Moving over to the EVs, an unmistakable difference gap is caused by the absence of tank-to-wheel emissions for the electric models, resulting in a lesser overall CO_2 impact, while well-to-tank emissions result from the average energy consumption and the characteristic CO_2-intensity of the European energy mix. The fact that battery size is proportional to energy consumption is highlighted for the 90 kWh EV. Thus, it represents nearly twice the climate change impact of the 30 kWh variants. Regarding the impact of the EV + trailer combinations, the battery and the generator trailer prove to have a favorable effect when compared to the 90 kWh EV. While offering the same range as the latter, the generator trailer combination allows a 33 percent lower CO_2 emission. In the case of the battery-equipped trailer, this gap widens to over 40 percent. When comparing the trailer combinations to the 60 kWh EV, only a small difference is reported for both concepts. Regarding the potential for carbon savings by substituting PHEVs by a range-extender trailer that is only occasionally coupled to a 40 kWh EV, a reduction exceeding 50 percent is offered by the petrol trailer and over 58 percent in the case of the battery trailer.

The results shown in Figure 4 confirm both the benefits of electric vehicles over internal combustion engine vehicles when climate change is concerned, and the energy mix as a strong determinant of the EV's well-to-tank emission profile. This energy mix could be managed intelligently if smart charging mechanisms would allow EVs to recharge only when solar and wind energy is abundant. Thus, EVs could be the intermediary solution until battery storage of renewables becomes economically viable. In the absence of smart meters, an uncontrolled grid connection of a significant EV fleet could nonetheless result in additional peak loads. Responding to this power demand using coal-fired power plants results in high CO_2 emissions and adversely impacts air quality. Merely allocating these excess emissions to EVs alone should be avoided as they are deemed to be part of the total load system. These factors indicate that the environmental potential of EVs goes hand in hand with how one produces the electricity they charge with.

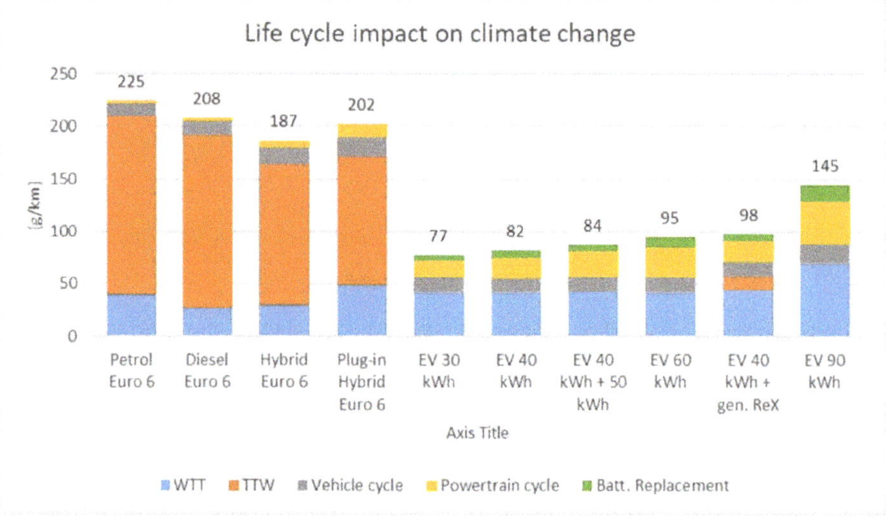

Figure 4. Impact per kilometer on climate change considering the entire LCA.

4.2. Photochemical Oxidant Formation

Concerning photochemical oxidant formation (POF) seen over the vehicle's lifetime, Figure 5 shows the significant impact of diesel technology originating from the tank-to-wheel phase, as its combustion inherently causes high emissions of NO_x. By including real-world driving emissions (RDE), the 'struggle' for European car manufacturers to control NOx emissions is emphasized. Disregarding the dominance of diesel powertrains for POF, we see that the higher the battery capacity becomes, the higher the impact on POF gets. Thus, the petrol-fueled plug-in hybrid performs worse than the conventional petrol car, although the latter's impact would be far higher if it were diesel-fueled. Linked to the battery size are the vehicle weight and the electric consumption, which is reflected in the well-to-tank emissions for the electrified powertrains. Thus, the 90 kWh EV once more presents a POF contribution that is nearly twice what is reported for the 30 kWh model. Principal sources for POF among the feedstocks for electricity production are coal, gas, and oil, indicating the potential reductions if the more renewable energy sources would be applied. Considering the trailer combinations, we report a very similar POF contribution when compared to the 60 kWh EV, while representing about 35 percent lower POF emissions when compared to the 90 kWh EV. Compared to the PHEV, the occasional application of a generator or a battery trailer generates 9 to 14 percent higher POF contributions, respectively.

If we were to assess only the POF emissions caused locally, the scope narrows down to a well-to-wheel analysis for Belgium, for which we consider the earlier mentioned refinery-to-tank (RTT) emissions for the contributions occurring upstream the use phase. Figure 6 represents the results of this exercise and shows the potential for EVs in strategies to bring down local emissions without ignoring the impact of manufacturing their powertrains. Locally, the TTW phase of the generator trailer makes it perform like the PHEV while contributing roughly 60 percent more to photochemical oxidant formation when compared to the 90 kWh EV. Substituting the generator with an additional battery pack would level its impact with that of the 60 kWh EV, or about 40 percent less than what is reported for the 90 kWh EV. If we compare the battery trailer combination to the PHEV, local POF contributions for the former end up to be two-thirds less. No POF originates from tyre, brake or road abrasion.

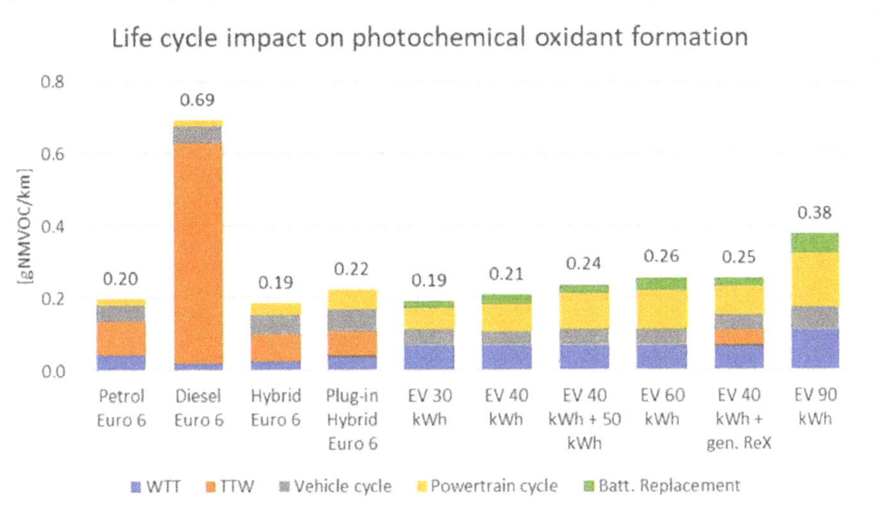

Figure 5. Impact per kilometer on photochemical oxidant formation.

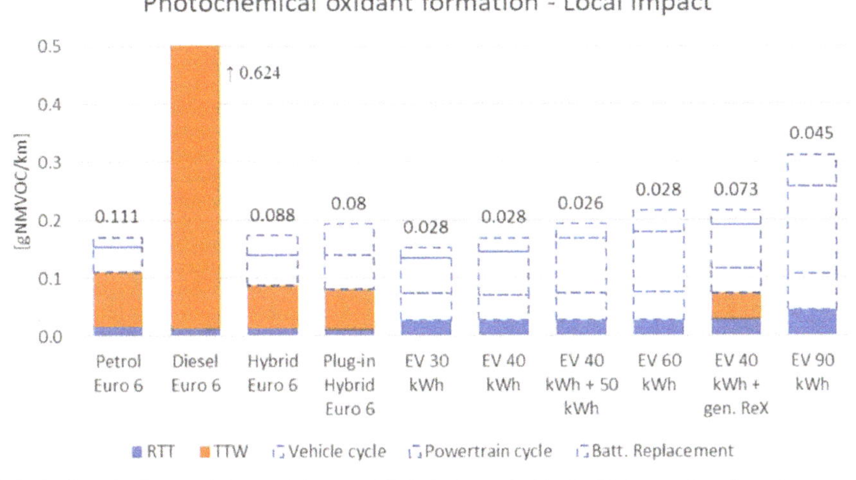

Figure 6. Photochemical oxidant formation at a local level per powertrain technology.

4.3. Particulate Matter Formation

For particulate matter formation (PMF), Figure 7 shows the contribution of the different powertrain technologies per kilometer driven. For this impact category, non-exhaust emissions are distinguished from the total TTW emissions, to highlight their significance relating to battery size. The specific impact regarding PMF from non-exhaust sources ranges from roughly one-fifth of the total PM emissions (petrol) to about one-tenth in case of the 90 kWh EV and the diesel car. Keep in mind that the non-exhaust fraction originates from brake, tyre, and road wear and currently is not regulated within Europe. Again, we see the inherently high contribution of PM emissions during the use phase for diesel cars, while for the other ICEVs and EVs this impact increases with the applied battery capacity. What catches the eye is that for the entire life cycle, the 90 kWh EV's PM emissions actually exceed those of a diesel car, particularly if a battery replacement is considered.

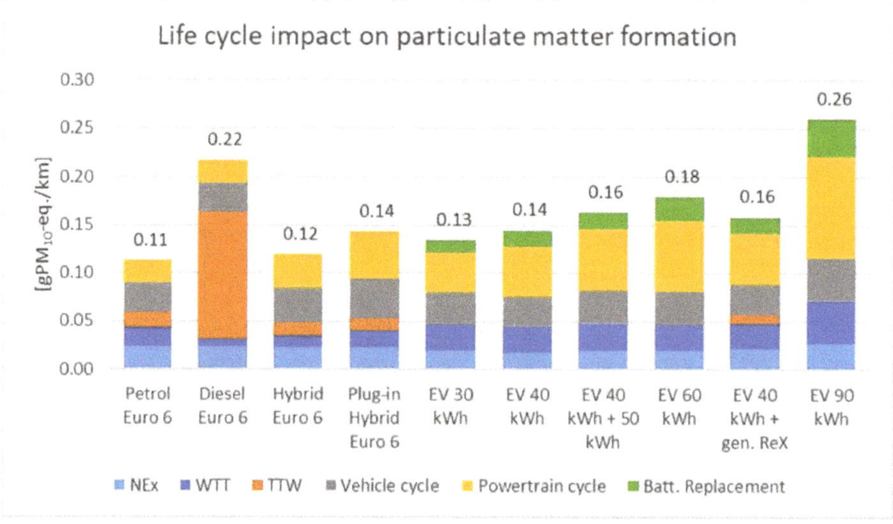

Figure 7. The contribution of particulate matter per technology.

For the range-extender trailer combinations, brake wear is excluded as the trailer is assumed not to be provided with a braking system. The effect of towing a trailer—and thus an extra weight—on the non-exhaust sources is nonetheless encompassed, albeit that due to the occasional nature of using the trailer, these impacts are negligible. Notice how both the trailer combinations and the 60 kWh EV have a similar impact over the full life cycle, which is about 40 percent lower than for the 90 kWh EV. Compared to a PHEV, the trailers have a 14 percent higher contribution to PMF, although one must keep in mind that the discussed PHEV is based on petrol technology. In case of a diesel PHEV, the tank-to-wheel contribution would be significantly higher.

If we look at the local emissions considering a refinery-to-wheel analysis, Figure 8 stresses the importance of regulation of non-exhaust emissions as they now have relative shares ranging up to 75 percent in case of the 40 kWh EV. Locally, the battery trailer performs about 30 percent better than the generator trailer and the 90 kWh EV, which have an equal contribution. Compared to the other EVs, the 90 kWh EV does represent a substantially higher local impact, which is mostly because of the higher vehicle mass' influence on non-exhaust emissions. Thus, we can confirm that the extra weight of the 90 kWh EV's battery has a negative effect seen for both its lifetime and local impact, although for the latter this effect is similar to the contribution of a hybrid powertrain. Whereas the trailer concept is designed in the light of a sharing economy and for covering longer distances, it is likely the trailer will be dropped off and picked up near highways. Therefore, the impact of the trailer combination is less relevant concerning local air quality levels.

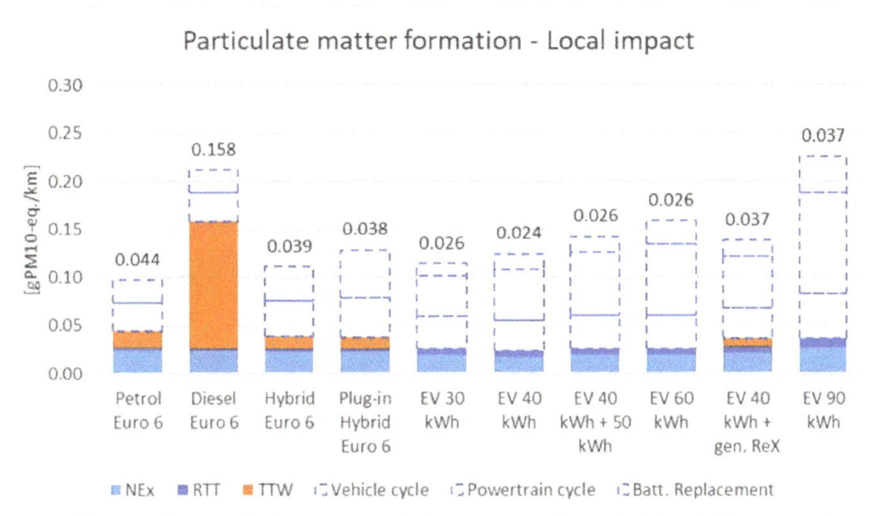

Figure 8. Impact per kilometer on particulate matter formation during use of the vehicle (including the refinery-to-tank (RTT) emissions upstream). Tank-to-wheel (TTW) emissions are further disaggregated to exhaust TTW and non-exhaust PMF (NEx).

4.4. The Impact on Human Toxicity

Figure 9 presents the impact of the different powertrains on human toxicity. As can be seen, the contribution of the well-to-wheel phase (i.e., non-exhaust, WTT, and TTW) is only marginal for the conventional powertrains, while it increases with battery capacity. Also, the contribution by the electricity production is emphasized in this exercise, indicating the reduction margin for changing the feedstock with renewable sources. What must be noted here is that the HT impact of EVs is far more researched than for the ICE-based technologies. For conventional petrol and diesel cars, for instance, an update of the current situation is needed as the substantial increase in sensor applications over the last decade resulted in more copper use, the 90 kWh performs worst of the given options, followed by the battery trailer combination due to the high impact of the battery production process. Locally, the impact of the manufacturing phase is filtered out of scope in Figure 10. Thus, the impact of non-exhaust PM emissions is emphasized, for which EVs have the benefit of relying less on the conventional brakes. Notice the little difference between the presented powertrains based on internal combustion engines, the generator trailer combination, and the 90 kWh EV.

What can be concluded from these figures is that the manufacturing phase of EVs should not be neglected in the debate on which powertrain causes the least environmental impact. If we use the petrol car's powertrain as a reference, the impact of EVs on human toxicity ranges from double to nearly four times, although this difference would reduce if more recent LCI data on conventional cars were available. Next, we highlight the need for both more sustainable mining techniques, battery chemistries that rely on more locally and abundantly available feedstocks and an increased uptake of recycled materials in the battery manufacturing process. Nonetheless, if local policy is concerned, attention should primarily go to the impact different technologies have locally. In this regard, neither the PHEV nor the 90 kWh EV prove to be the optimal solution. Instead, a reduced impact on human toxicity is reached by addressing a battery trailer on those occasions when the nominal range of a 40 kWh EV is not sufficient.

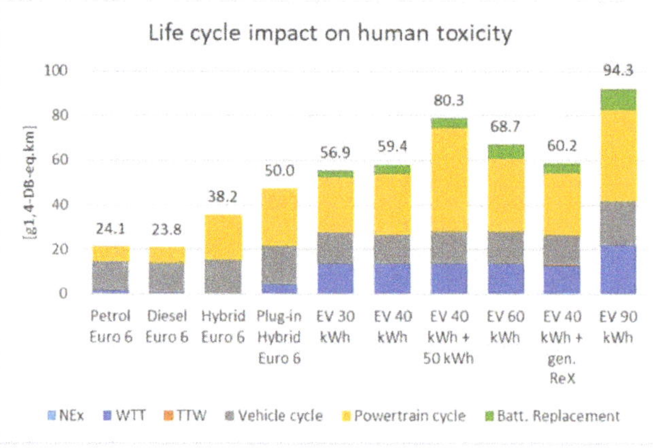

Figure 9. An overview of the well-to-wheel impact on human toxicity per powertrain.

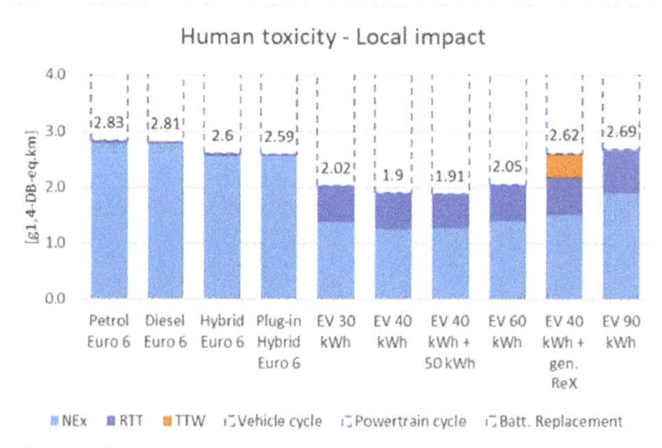

Figure 10. The impact on human toxicity at a local level per powertrain technology, considering refinery-to-tank (RTT) instead of well-to-tank emissions.

4.5. Sensitivity Analysis

Whereas the results presented in this paper are based on certain assumptions, a sensitivity analysis is recommendable to highlight possible weaknesses and/or potentials for the trailer combinations, combined to an EV. An overview is given of examples which are thought to influence the results of this exercise.

4.5.1. The Electricity Production Mix

For the comparisons made in this paper, the European electricity mix was applied, representing 276 g of CO_2 produced per kWh. As EVs are only exploited to their full environmental potential when its electricity originates from renewable sources, this is where the focus of European policy concerning energy production should be. Thus, we emphasize the agreed commitments towards carbon-neutral economy by 2050. As indicated by Messagie et al., the climate change impact of EVs powered by energy from renewable sources could be reduced to approx. 40 g of CO_2-equivalents per kilometer driven. The other way around; if energy production would shift to fossil fuels such as oil or coal, the impact per kilometer would significantly exceed the impact of conventional cars. This is shown in Figure 11, for which the 40 kWh EV's impact is distinguished per energy production source (based on [42]). A drastic shift to renewable sources would thus mean the well-to-tank emissions for EVs could be marginalized, which will also have a substantial impact on the other three discussed midpoint indicators.

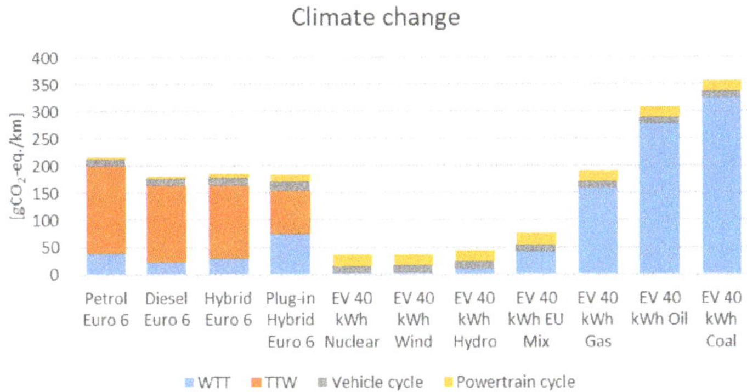

Figure 11. Impact of the energy production mix in grams per kilometer (based on 46).

In the current absence of large-scale power storage facilities for renewable sources such as wind and solar power, EVs might prove to be a solution as well. Slow-charging facilities could, therefore, be maximally provided with green power, although this is far less likely for fast-charging facilities unless ultracapacitors are applied. In the case of fast-chargers, energy is needed immediately at high power rates, which is why the share of renewable energy is likely to be on the low side.

4.5.2. The Use Pattern for the ReX Trailer

An important aspect of this sensitivity analysis is the marginal application factor of the ReX trailer. If one would apply the ReX trailer for one-tenth of the driven distance rather than one-twentieth, this could have a significant influence on the comparison of the different midpoint indicators. For climate change, applying the ReX trailer for 10 percent of the time would result in a relative increase by 17 percent.

Another aspect of the use pattern is the central issue of people tending to cover longer trips during weekends, e.g., to visit relatives. This can result in distorted availability for the ReX concept and thus, at peak demand in a lower number of users for one trailer. This sensitivity has been calculated for five users and ten users concerning the impact on climate change. Results showed no significant impact on either the vehicle cycle or the powertrain cycle, as the increase was found to remain below 2.5 percent in the worst case of 5 users.

5. Conclusions

In the light of decarbonizing society by mid-century, electrification of passenger cars is imminent. Therefore, several alternatives for the conventional ICE-based passenger car are available today, although their sustainability varies. Whereas the average motorist mostly covers short distances on a daily basis, one can either opt for a plug-in hybrid electric vehicle, an EV with a large battery pack, or a small (40 kWh) EV and the option of fast-charging. A fourth alternative is proposed in this paper by means of occasionally coupling the 40 kWh to a range-extender trailer that is shared with other EV users. Here, we discussed a petrol generator trailer for application in the short-term, while substituting the generator on the mid-term with a 50 kWh battery pack as battery prices further decrease. While the impact of fast-charging exceeds the scope of the presented paper, we did compare the environmental impact for the PHEV, 90 kWh EV, and the 40 kWh EV + range-extender for their contribution to climate change, photochemical oxidant formation, particulate matter formation, and human toxicity. The European electricity mix was considered for charging the electrified powertrains, characterized by a carbon-intensity of 276 g CO_2/kWh, while an end-of-life range of 210,000 km was assumed for each technology. Finally, we included an EV battery replacement after 150,000 km.

Results show that seen over the different life stages of a passenger car, the trailer concepts outperform the 90 kWh EV's contributions for the discussed midpoint indicators. Compared to the petrol-fueled PHEV, both trailers are found to exceed the impact on POF and PMF by up to 15 percent, whereas this situation would be different if the PHEV would be based on diesel technology, given the impact during the latter's use phase. Concerning HT, results show that the higher the larger the batteries are dimensioned, the bigger the impact gets. Thus, the 40 kWh EV with a battery trailer performs significantly worse than the PHEV, while remaining a less toxic solution compared to the 90 kWh for offering the same range. For climate change, we see a clear benefit of driving all-electric most of the time with a modest battery capacity of 40 kWh, even when addressing a petrol-based range-extender. These life cycle results indicate the potential for substituting the internal combustion engine with an electrified powertrain, although there are limits to the sustainability of the selected battery pack size. Nonetheless, the results also indicate the need for cleaner battery technologies, as significant contributions to the discussed impact factors can be linked to the battery manufacturing process. In most cases, it is the impact of mining for specific metals that is expressed the most.

Next to the life cycle impact, we also addressed to the potential of electrifying powertrains on local air quality levels, which remain poor throughout Europe's cities. Therefore, a well-to-tank assessment for all emissions occurring within Belgium was presented. Thus, the significance of non-exhaust PM emissions are highlighted, as the mass penalty for large battery packs translates to a substantial impact on both PMF and HT. The petrol trailer was found to have a slightly worse impact locally, compared to the petrol PHEV, while performing significantly worse on POF when compared to the 90 kWh EV. This indicates the potential effect the substitution of combustion-based technologies in cities can have on local air quality.

The comparison of the PHEV, an EV with a large battery pack and an EV relying on in-range range modularity is relevant as our personal transportation system drastically needs to shift away from the monopoly of conventional internal combustion engine vehicles. This shift should be as durable as possible, which is why different midpoint indicators have been compared for the mentioned technologies. The specific impact of fast-chargers on the electricity grid has not been modeled for this exercise, although its potential cannot be neglected in case the recharging power is sourced from a durable source through energy storage locally. In the light of improving the presented model, more insights into the human toxicity impact of both the most recent conventional cars as in upcoming battery chemistries are important. Next, the inclusion of a diesel PHEV into the benchmark could indicate to what extent the 40 kWh EV + range-extender could be beneficial in terms of the impact on air quality, both locally and seen over the entire life cycle. Finally, this work could be complemented with a one-on-one comparison of the EV + trailer combination and the same EV relying on fast-chargers. For the latter, the focus should be expanded to the environmental impact of the uncontrolled demand for high-power during the charging session, while solutions must be sought to maximize the implementation of renewable energy, for instance by locating fast-chargers next to renewable power facilities. The presented paper aims to add to the potential of electric vehicles, by offering a solution bridging current technologies to future generations.

Author Contributions: N.H. created the models with the assistance of M.M., F.J., J.-B.S. and T.C. contributed by scientifically validating the various iterations of the presented work with consecutive reviews.

Funding: This project was funded by the European Horizon 2020 project ID 684085.

Acknowledgments: We acknowledge Flanders Make for the support to our research group.

Conflicts of Interest: The authors declare no conflicts of interest.

Appendix A

Table A1. LCI of a 30 kWh EV for the European electricity mix (276 g CO_2/kWh) (based on 46).

Midpoint Indicator	WTT		TTW				Vehicle Cycle					Powertrain Cycle					
	WTT	Public Charging Station	Tire Abrasion	Road Abrasion	Brake Abrasion	TTW	Body Shell	Lead Battery	Maintenance	Li Battery	Electric Motor	AC/DC Converter	DC/DC Converter	Onboard Charger	Catalytic Converter	Starter and Generator	Engine Control Unit
CC [kgCO$_2$/km]	4.14×10^{-2}	2.95×10^{-4}	0.00	0.00	0.00	0.00	1.30×10^{-2}	6.29×10^{-5}	1.52×10^{-3}	1.24×10^{-2}	1.19×10^{-3}	1.35×10^{-3}	5.08×10^{-4}	4.14×10^{-4}	0.00	0.00	0.00
POF [kgNMVOC/km]	2.68×10^{-5}	7.42×10^{-7}	0.00	0.00	0.00	0.00	4.08×10^{-5}	2.56×10^{-7}	4.94×10^{-6}	4.54×10^{-5}	4.68×10^{-6}	5.66×10^{-6}	1.99×10^{-6}	1.98×10^{-6}	0.00	0.00	0.00
PMF [kgPM$_{10}$/km]	5.83×10^{-6}	4.51×10^{-7}	7.05×10^{-6}	1.00×10^{-5}	2.46×10^{-6}	0.00	3.12×10^{-5}	1.91×10^{-7}	2.46×10^{-6}	3.16×10^{-5}	4.77×10^{-6}	3.21×10^{-6}	1.19×10^{-6}	1.07×10^{-6}	0.00	0.00	0.00
HT [kg1,4-DB/km]	2.03×10^{-4}	4.43×10^{-4}	6.37×10^{-4}	4.01×10^{-6}	7.35×10^{-4}	0.00	1.34×10^{-2}	3.03×10^{-4}	5.86×10^{-4}	8.22×10^{-3}	5.72×10^{-3}	6.12×10^{-3}	2.31×10^{-3}	1.97×10^{-3}	0.00	0.00	0.00

References

1. European Environment Agency (EEA). *Air Quality in Europe—2017 Report*; European Environment Agency: Copenhagen, Denmark, 2017.
2. European Environment Agency (EEA). National Emission Ceilings Directive Emissions Data Viewer—European Environment Agency. 2017. Available online: https://www.eea.europa.eu/data-and-maps/dashboards/necd-directive-data-viewer#tab-related-briefings (accessed on 31 October 2017).
3. Dieselnet. EU: Cars and Light Truck. 2016. Available online: https://www.dieselnet.com/standards/eu/ld.php (accessed on 6 May 2017).
4. Fontaras, G.; Franco, V.; Dilara, P.; Martini, G.; Manfredi, U. Development and review of Euro 5 passenger car emission factors based on experimental results over various driving cycles. *Sci. Total Environ.* **2014**, *468–469*, 1034–1042. [CrossRef] [PubMed]
5. Yang, L.; Franco, V.; Mock, P.; Kolke, R.; Zhang, S.; Wu, Y.; German, J. Experimental Assessment of NOx Emissions from 73 Euro 6 Diesel Passenger Cars. *Environ. Sci. Technol.* **2015**, *49*, 14409–14415. [CrossRef] [PubMed]
6. Weiss, M.; Bonnel, P.; Hummel, R.; Manfredi, U.; Colombo, R.; Lanappe, G.; Le Lijour, P.; Sculati, M. Analyzing on-road emissions of light-duty vehicles with Portable Emission Measurement Systems (PEMS). *Environ. Sci. Technol.* **2011**, *45*, 8575–8581. [CrossRef] [PubMed]
7. Demuynck, J. Real-driving emission results from GDI vehicles with and without a GPF Association for Emissions Control by Catalyst (AECC). In Proceedings of the IQPC 4th International Conference Advanced Emission Control Concepts for Gasoline Engines, Bonn, Germany, 10–12 May 2016; pp. 1–26.
8. Rhys-Tyler, G.; Legassick, W.; Bell, M. The significance of vehicle emissions standards for levels of exhaust pollution from light vehicles in an urban area. *Atmos. Environ.* **2011**, *45*, 3286–3293. [CrossRef]
9. Chen, Y.; Borken-Kleefeld, J. Real-driving emissions from cars and light commercial vehicles – Results from 13 years remote sensing at Zurich/CH. *Atmos. Environ.* **2014**, *88*, 157–164. [CrossRef]
10. Muncrief, R. *NOx Emissions from Heavy-Duty and Light-Duty Diesel Vehicles in the EU: Comparison of Real-World Performance and Current Type-Approval Requirements*; International Council on Clean Transportation: Washington, DC, USA, 2016.
11. Clima, D.G. *Revision of Regulation (EU) No 443/2009 and Regulation (EU) No 510/2011 Regulating CO2 Emissions from Light Duty Vehicles*; Brussels, Belgium, 2016. Available online: https://ec.europa.eu/clima/sites/clima/files/transport/vehicles/docs/evaluation_ldv_co2_regs_en.pdf (accessed on 21 June 2018).
12. Díaz, S.; Tietge, U.; Mock, P. *CO$_2$ Emissions from New Passenger Cars in the EU: Car Manufacturers' Performance in 2015*; International Council on Clean Transportation: Berlin, Germany, 2015.
13. Massiani, J. Cost-Benefit Analysis of policies for the development of electric vehicles in Germany: Methods and results. *Transp. Policy* **2015**, *38*, 19–26. [CrossRef]
14. Report of the Conference of the Parties on its twenty-first session, held in Paris from 30 November to 13 December 2015 Addendum Part two: Action taken by the Conference of the Parties at its twenty-first session. In Proceedings of the 2015 United Nations Climate Change Conference, Paris, France, 30 November–12 December 2015.
15. Rockström, J.; Schellnhuber, H.J.; Hoskins, B.; Ramanathan, V.; Schlosser, P.; Brasseur, G.P.; Gaffney, O.; Nobre, C.; Meinshausen, M.; Rogelj, J.; et al. The world's biggest gamble. *Earth's Future* **2016**, *4*, 465–470. [CrossRef]
16. European Commission. *Roadmap to a Single European Transport Area—Towards a Competitive and Resource Efficient Transport System*; European Commission: Brussels, Belgium, 2011.
17. Messagie, M. *Life Cycle Analysis of the Climate Impact of Electric Vehicles*; Brussels, Belgium, 2017. Available online: https://evobservatory.iit.comillas.edu/publicaciones/life-cycle-analysis-of-the-climate-impact-of-electric-vehicles (accessed on 21 June 2018).
18. Sanfélix, J.; de la Rúa, C.; Schmidt, J.; Messagie, M.; van Mierlo, J. Environmental and economic performance of an Li-Ion battery pack: A multiregional input-output approach. *Energies* **2016**, *9*, 584. [CrossRef]
19. Umicore raises $1.1 bn to invest in cathode business. *Focus Catal.* **2018**, *2018*, 4.
20. European Commission. *Report on Raw Materials for Battery Applications*; SWD (2018) 245 Final; European Commission: Brussels, Belgium, 2018.

21. Witkamp, B.; van Gijlswijk, R.; Bolech, M.; Coosemans, T.; Hooftman, N. The Transition to a Zero Emission Vehicles Fleet for Cars in the EU by 2050; Brussels, Belgium, 2017. Available online: https://cris.vub.be/files/35220288/The_Transition_to_a_ZEV_car_fleet_EU_2050_an_EAFO_study.pdf (accessed on 21 June 2018).

22. Pearre, N.S.; Kempton, W.; Guensler, R.L.; Elango, V.V. Electric vehicles: How much range is required for a day's driving? *Transp. Res. Part C Emerg. Technol.* **2011**, *19*, 1171–1184. [CrossRef]

23. Needell, Z.A.; McNerney, J.; Chang, M.T.; Trancik, J.E. Potential for widespread electrification of personal vehicle travel in the United States. *Nat. Energy* **2016**, *1*, 16112. [CrossRef]

24. Pasaoglu, G.; Fiorello, D.; Martino, A.; Scarcella, G.; Alemanno, A.; Zubaryeva, A.; Thiel, C. *Driving and Parking Patterns of European Car Drivers: A Mobility Survey*; European Commission: Sevilla, Spain, 2012.

25. Corchero, C.; Gonzalez-Villafranca, S.; Sanmarti, M. European electric vehicle fleet: Driving and charging data analysis. In Proceedings of the 2014 IEEE International Electric Vehicle Conference (IEVC), Florence, Italy, 17–19 December 2014; pp. 1–6.

26. Khan, M.; Kockelman, K.M. Predicting the market potential of plug-in electric vehicles using multiday GPS data. *Energy Policy* **2012**, *46*, 225–233. [CrossRef]

27. Gonder, J.; Markel, T.; Thornton, M.; Simpson, A. Using Global Positioning System Travel Data to Assess Real-World Energy Use of Plug-In Hybrid Electric Vehicles. *Transp. Res. Rec. J. Transp. Res. Board* **2007**, *2017*, 26–32. [CrossRef]

28. Redelbach, M.; Özdemir, E.D.; Friedrich, H.E. Optimizing battery sizes of plug-in hybrid and extended range electric vehicles for different user types. *Energy Policy* **2014**, *73*, 158–168. [CrossRef]

29. Ligterink, N.; Smokers, R. *Monitoring Van Plug-In Hybride Voertuigen (PHEVs) April 2012 t/m Maart 2016*; TNO: Den Haag, The Netherlands, 2016.

30. Tietge, U.; Díaz, S.; Yang, Z.; Mock, P. *From Laboratory to Road International: A Comparison of Official and Real-World Fuel Consumption and CO2 Values for Passenger Cars in Europe, the United States, China, and Japan*; International Council on Clean Transportation: Berlin, Germany, 2017.

31. Björnsson, L.H.; Karlsson, S. Plug-in hybrid electric vehicles: How individual movement patterns affect battery requirements, the potential to replace conventional fuels, and economic viability. *Appl. Energy* **2015**, *143*, 336–347. [CrossRef]

32. Neubauer, J.; Brooker, A.; Wood, E. Sensitivity of plug-in hybrid electric vehicle economics to drive patterns, electric range, energy management, and charge strategies. *J. Power Sources* **2013**, *236*, 357–364. [CrossRef]

33. Norman Shiau, C.-S.; Samaras, C.; Hauffe, R.; Michalek, J.J. Impact of battery weight and charging patterns on the economic and environmental benefits of plug-in hybrid vehicles. *Energy Policy* **2009**, *37*, 2653–2663. [CrossRef]

34. Michalek, J.J.; Chester, M.; Jaramillo, P.; Samaras, C.; Shiau, C.-S.N.; Lave, L.B. Valuation of plug-in vehicle life-cycle air emissions and oil displacement benefits. *Proc. Natl. Acad. Sci. USA* **2011**, *108*, 16554–16558. [CrossRef] [PubMed]

35. Nordelöf, A.; Messagie, M.; Tillman, A.-M.; Söderman, M.L.; van Mierlo, J. Environmental impacts of hybrid, plug-in hybrid, and battery electric vehicles—What can we learn from life cycle assessment? *Int. J. Life Cycle Assess.* **2014**, *19*, 1866–1890. [CrossRef]

36. Rusich, A.; Danielis, R. Total cost of ownership, social lifecycle cost and energy consumption of various automotive technologies in Italy. *Res. Transp. Econ.* **2015**, *50*, 3–16. [CrossRef]

37. Al-Alawi, B.; Bradley, T. Total cost of ownership, payback, and consumer preference modeling of plug-in hybrid electric vehicles. *Appl. Energy* **2013**, *103*, 488–506. [CrossRef]

38. Propfe, B.; Redelbach, M.; Santini, D.J.; Friedrich, H. Cost analysis of Plug-in Hybrid Electric Vehicles including Maintenance & Repair Costs and Resale Values. In Proceedings of the EVS26 International Battery, Hybrid and Fuel Cell Electric Vehicle Symposium, Los Angeles, CA, USA, 6–9 May 2012.

39. Wu, G.; Inderbitzin, A.; Bening, C. Total cost of ownership of electric vehicles compared to conventional vehicles: A probabilistic analysis and projection across market segments. *Energy Policy* **2015**, *80*, 196–214. [CrossRef]

40. ERTRAC. *Future Light and Heavy Duty ICE Powertrain Technologies*; ERTRAC: Brussels, Belgium, 2016.

41. Hooftman, N.; Oliveira, L.; Messagie, M.; Coosemans, T.; Van Mierlo, J. Environmental analysis of petrol, diesel and electric passenger cars in a Belgian urban setting. *Energies* **2016**, *9*, 84. [CrossRef]

42. Deb, S.; Tammi, K.; Kalita, K.; Mahanta, P. Impact of Electric Vehicle Charging Station Load on Distribution Network. *Energies* **2018**, *11*, 178. [CrossRef]

43. Meyer, D.; Wang, J. Integrating Ultra-Fast Charging Stations within the Power Grids of Smart Cities: A Review. *IET Smart Grid* **2018**, *1*, 3–10. [CrossRef]

44. European Comission for Standardization. *ISO 14040:2009—Environmental Management—Life Cycle Assessment—Principles and Framework*; European Comission for Standardization: Geneva, Switzerland, 2009.

45. International Organization for Standardization. *ISO 14044:2006 Environmental Management—Life Cycle Assessment—Requirements and Guidelines*; International Organization for Standardization: Gevena, Switzerland, 2006.

46. Messagie, M. *Environmental Performance of Electric Vehicles, a Life Cycle System Approach*; Vrije Universiteit Brussel: Brussels, Belgium, 2013.

47. US Department of Energy. Fuel Economy of New All-Electric Vehicles. Available online: http://www.fueleconomy.gov/feg/PowerSearch.do?action=alts&path=3&year1=2016&year2=2017&vtype=Electric&srchtyp=newAfv (accessed on 14 June 2017).

48. Goedkoop, M.; Heijungs, R.; Huijbregts, M.; de Schryver, A.; Struijs, J.; van Zelm, R. *A Life Cycle Impact Assessment Method Which Comprises Harmonised Category Indicators at the Midpoint and the Endpoint Level*; Amersfoort, The Netherlands, 2013. Available online: https://www.researchgate.net/publication/302559709_ReCiPE_2008_A_life_cycle_impact_assessment_method_which_comprises_harmonised_category_indicators_at_the_midpoint_and_the_endpoint_level (accessed on 21 June 2018).

49. Moro, A.; Lonza, L. Electricity carbon intensity in European Member States: Impacts on GHG emissions of electric vehicles. *Transport. Res. Part D Transp. Environ.* **2017**. [CrossRef]

50. Oliveira, L.; Messagie, M.; Rangaraju, S.; Sanfelix, J.; Rivas, M.H.; van Mierlo, J. Key issues of lithium-ion batteries—From resource depletion to environmental performance indicators. *J. Clean. Prod.* **2015**, *108*, 354–362. [CrossRef]

51. Messagie, M.; Boureima, F.-S.; Coosemans, T.; Macharis, C.; Mierlo, J. A Range-Based Vehicle Life Cycle Assessment Incorporating Variability in the Environmental Assessment of Different Vehicle Technologies and Fuels. *Energies* **2014**, *7*, 1467–1482. [CrossRef]

52. Rangaraju, S.; de Vroey, L.; Messagie, M.; Mertens, J.; van Mierlo, J. Impacts of electricity mix, charging profile, and driving behavior on the emissions performance of battery electric vehicles: A Belgian case study. *Appl. Energy* **2015**, *148*, 496–505. [CrossRef]

53. CERAM. *Test Report N°16/10681*; CERAM: Montlhéry, France, 2017.

Article

Comparisons of Energy Management Methods for a Parallel Plug-In Hybrid Electric Vehicle between the Convex Optimization and Dynamic Programming

Renxin Xiao, Baoshuai Liu, Jiangwei Shen, Ningyuan Guo, Wensheng Yan and Zheng Chen *

Faculty of Transportation Engineering, Kunming University of Science and Technology, Kunming 650500, China; xrx1127@foxmail.com (R.X.); baosliu@163.com (B.L.); shenjiangwei6@163.com (J.S.); gnywin@163.com (N.G.); yanwensheng65@sina.com (W.Y.)
* Correspondence: chen@kmust.edu.cn; Tel.: +86-186-6908-3001

Received: 13 December 2017; Accepted: 28 January 2018; Published: 31 January 2018

Abstract: This paper proposes a comparison study of energy management methods for a parallel plug-in hybrid electric vehicle (PHEV). Based on detailed analysis of the vehicle driveline, quadratic convex functions are presented to describe the nonlinear relationship between engine fuel-rate and battery charging power at different vehicle speed and driveline power demand. The engine-on power threshold is estimated by the simulated annealing (SA) algorithm, and the battery power command is achieved by convex optimization with target of improving fuel economy, compared with the dynamic programming (DP) based method and the charging depleting–charging sustaining (CD/CS) method. In addition, the proposed control methods are discussed at different initial battery state of charge (SOC) values to extend the application. Simulation results validate that the proposed strategy based on convex optimization can save the fuel consumption and reduce the computation burden obviously.

Keywords: battery power; convex optimization; dynamic programming; engine-on power; plug-in hybrid electric vehicle; simulated annealing

---

1. Introduction

Nowadays, plug-in hybrid electric vehicles (PHEVs) representing a positive research direction due to combination of a certain all electric range (AER) and hybrid drive, exhibit apparent advantages in environmental protection and petroleum savings over traditional hybrid electric vehicles (HEVs). Compared with HEVs, PHEVs are equipped with higher capacity energy storage systems that can be directly charged from the power grid [1,2]. Currently, automotive manufacturers and research institutes are actively devoted to developing PHEVs and improving controlling performances. For PHEVs, an appropriate and effective energy management is critical to improve the vehicle's fuel economy and reduce emissions.

For the energy management strategy, the main destination is to optimize the fuel economy. Since there exists some uncertainty for driving cycles, driver's habits, and weather conditions that can influence the energy distribution in the PHEV, from this point, it can be said that the energy management is a stochastic optimization problem. Actually, popular control candidates can be divided into four types: (1) rule based control method [3–5]; (2) intelligent control methods, including artificial neural network (ANN) [6,7], fuzzy logic [8,9], model predictive control (MPC) [10,11], and machine learning algorithm [12,13]; (3) analytic methods [14,15]; and (4) optimization based control method, including deterministic dynamic programming (DP) [1,16–19], Pontryagin's Minimum Principle (PMP) [20,21], quadratic programming (QP) [22,23], and convex optimization [24–26]. These methods' purpose can include improving the fuel economy, reducing emissions [27,28], prolonging cycling life of the battery pack [2,29], minimizing the operation cost [30], etc.

Among these methods, rule-based control methods are simpler and easier to implement, which have been widely applied in practical application [4,31,32]. It is relatively easy to implement with fixed control parameters according to experience and prior knowledge. The prevalent rule-based method is the charge depleting/charge sustaining (CD/CS) method. During the CD mode, the vehicle is powered by the motor which absorbs the energy from the battery. The CD mode tries to use up the energy stored in the battery until its state of charge (SOC) decreases to an allowable minimum value. After that, the vehicle operating mode turns to the CS mode, which is also called the hybrid mode. In this mode, the vehicle is powered by the motor and engine together and meanwhile the battery SOC maintains in the vicinity of the low threshold. Due to the complex coupling characteristic of driveline system of PHEVs, rule-based method may not achieve the optimal power split between the engine and the motor, even if it is simple and stable.

The intelligent methods and the analytic method have been widely researched in the energy management of the PHEVs. In Ref. [9], a fuzzy logic energy management strategy of a series-PHEV is proposed to achieve the power split between the battery and the engine, based on the battery working state and vehicle power demand, and simultaneously to control the engine working in the economic region. Nonetheless, the fuzzy logic table and related rules should be defined with care. In Ref. [11], the power split of a PHEV which is equipped with a semi-active hybrid energy storage system and an assistance power unit, is regulated by the MPC method; however, it does not consider the engine ON/OFF power threshold and the trip length. In Ref. [12], a reinforcement learning-based method of a PHEV is raised, which takes the minimizing electricity consumption, real-time control and different conditions into account. In Ref. [14], based on modeling the electric driveline loss and applying the piecewise linear fuel consumption, an analytic method is applied to establish the energy management strategy to minimize the fuel consumption via finding the engine-on power and the optimal battery power commands.

Generally, optimization-based control methods include PMP, QP, DP, and convex optimization [6]. The PMP algorithm is applied to achieve the energy management adaptively for the GM Chevrolet Volt [33]. Nevertheless, a complex Hamilton function and the local optimum solution need to be solved and determined. In Ref. [22], the energy management strategy is optimized by the QP algorithm to reduce the engine fuel consumption, whereas this algorithm does not consider the influence of the initial SOC variation. In Ref. [17], the DP based energy management strategy is constructed, which considered the discretization resolution of the relevant variables and the boundary constraint of their feasible regions. Currently, convex optimization has been substantially applied to energy management optimization of traditional HEVs and PHEVs [24,34–36]. In Ref. [34], a convex programming-based power management strategy of a PHEV, which covers expenditures of electricity charged from the grid, fuel consumed during on-road driving, and battery aging. In Ref. [36], a novel convex modeling approach is presented which allows for battery sized and energy management of a plug-in hybrid powertrain based on a semidefinite program. However, some fixed control parameters, such as the engine-on power, cannot be estimated by the convex optimization [37]. In order to calculate these variables, randomized heuristic searching algorithm are applied, such as genetic algorithm (GA), simulated annealing (SA). Compared with GA algorithm, the SA algorithm is simpler and with higher efficiency, which can search the optimal solution more quickly. Due to these merits, the SA algorithm has been applied in the research of energy management for HEVs and PHEVs [22,38,39].

Compared with the optimization-based method, the optimal control parameters of analytic method and rule-based energy management strategy for PHEV can be obtained difficultly. Even the intelligent based method for PHEV has been widely applied, nevertheless, the intelligent based control method is complex and difficult to find the optimal solutions. The optimization-based method can obtain the global optimal solutions, improve the vehicle's control performance and engine fuel economy. Motivated by these, we plan to compare the performance among different methods. In this paper, there typical methods including the CD/CS method, the DP method, and an intelligent method, i.e., SA combing with the convex optimization are employed to compare the fuel economy for a parallel

PHEV. As shown in Figure 1, the parallel PHEV can be powered by the engine and motor together. Since the gearbox can select the gear ratio according to the power and speed demand from the driveline, there exist two degrees of freedom in the vehicle, which increases certain complexity to decouple the powertrain dynamics. In order to simplify the problem, an automated gear shift rule strategy is adopted considering the vehicle speed. By this manner, the degrees of freedom changes from two to one and the problem becomes easier to solve. In this paper, the optimization purpose is to minimize the fuel consumption in a certain trip. Based on analysis of the energy flow and the vehicle working modes, the vehicle driveline system is simplified and translated into a series of quadratic equations, which can effectively express the relationship between the engine fuel rate and the battery power at different power demand and velocity. The SA algorithm is implemented to search the optimal engine-on power, in which the optimal sequence of battery power is optimized simultaneously by CP. With considering the referred discussions, the optimal engine-on power is quickly searched by the SA algorithm and then the battery power command is calculated by convex optimization algorithm based on the interior point method. Highway Fuel Economy Driving Schedule (HWFET), New European Driving Cycle (NEDC), and Urban Dynamometer Driving Schedule (UDDS) are selected as test cycles to verify the performance of the proposed algorithm. The CD/CS method and DP are adopted as the benchmark for comparison, and the extended study regarding different initial SOC is also proposed. It is necessary to mention that the proposed algorithm can be easily extended to series connected PHEVs and power-split PHEVs. Thus, it can potentially become a universal control algorithm solution for PHEVs.

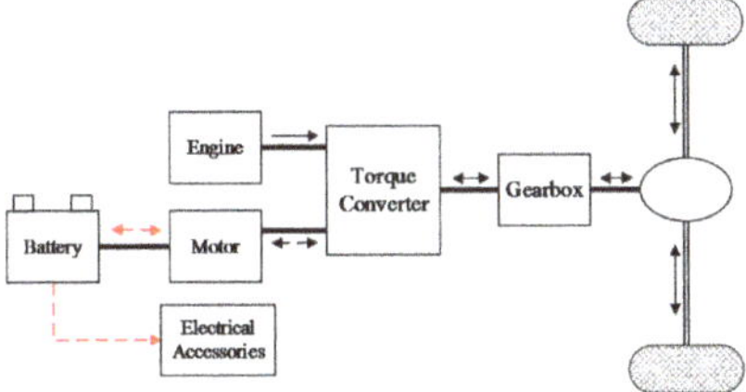

Figure 1. Powertrain structure of a parallel plug-in hybrid electric vehicle.

The remainder of this paper is structured as follows: Section 2 describes the vehicle model and the model simplification process; Section 3 presents the theory of convex optimization algorithm and the optimization method; Section 4 compares the proposed method with the CD/CS strategy and the DP strategy; and Section 5 concludes the proposed method in the paper.

2. Vehicle Model Analysis and Simplification

As shown in Figure 1, the vehicle driveline system consists of an engine, a battery pack, a clutch, an electric motor, a five-speed gearbox, etc. Compared with the traditional vehicles, two degrees of freedom exist in the PHEV [2]. The main parameters of the PHEV are shown in Table 1. The total vehicle mass is 1720 kg and the maximum engine power and motor power are 65 kW and 70 kW, respectively.

Table 1. Vehicle parameters.

Type	Parallel PHEV Value
Vehicle mass	1720 kg
Drive type	Front wheel drive
Engine	Maximum power 65 kW Maximum speed 6000 rpm
Motor	Rated power 30 kW Peak power 70 kW

2.1. Engine Model

As shown in Figure 1, the engine fuel consumption of the parallel PHEV as the target function of this paper can be calculated,

$$F = \int_0^{t_{total}} m_f dt \tag{1}$$

where F is the engine fuel consumption, t_{total} is the PHEV running time in a certain driving trip, m_f is the engine fuel rate. By proper assumption and simplification, m_f can be determined,

$$m_f = f(T_{eng}, w_{eng}, e_{on}) \tag{2}$$

where T_{eng} and w_{eng} are the engine torque and engine speed. e_{on} is the engine on command. Here we introduce an engine on/off threshold $P_{eng\_threshold}$, when the driveline power is more than $P_{eng\_threshold}$, the engine will be turned on, or else the engine will be turned off [26],

$$\begin{cases} e_{on} = 1 & P_0 \geq P_{eng\_threshold} \\ e_{on} = 0 & P_0 < P_{eng\_threshold} \end{cases} \tag{3}$$

where $e_{on} = 1$ represents the engine state is ON, and $e_{on} = 0$ means it is OFF. Under a certain drive cycle, the control sequences of the engine state need to be gained and imposed into the control system.

2.2. Electric Machine and Gearbox Unit

A five-speed gearbox is equipped in the parallel PHEV, and the gear ratios equal with 2.563, 1.522, 1.022, 0.727 and 0.52, respectively. In order to simplify the problem, an automated shift rule strategy is adopted considering the vehicle speed and the acceleration. The detailed gear shifting algorithm and the related rules are shown in Figure 2. The gear number up-shifting and down-shifting can be determined based on the shifting speed, chassis acceleration and the current gear number.

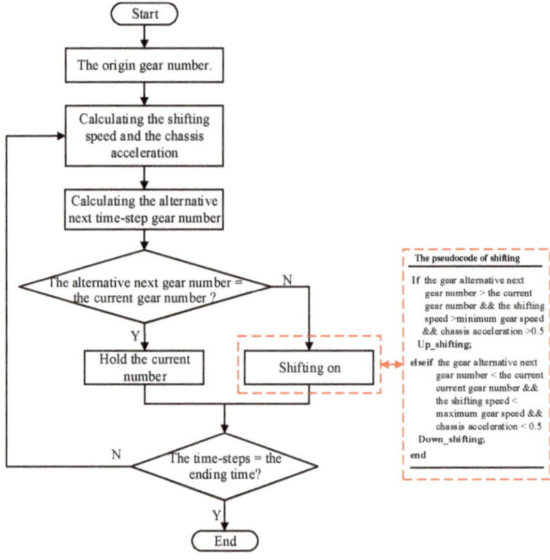

Figure 2. Gear shifting strategy.

Based on the vehicle speed w_0, and the vehicle acceleration and the driveline power demand P_0, the power P_{gb_in} and speed of gearbox input $w_{gb\_in}^k$ can be correspondingly determined,

$$P_{gb\_in} = P_0 / \eta(w^k_{gb\_in}, T_{gb}) \tag{4}$$

$$w^k_{gb\_in} = \frac{w_0 \times f_{fd\_ratio}}{r\_wheel} \times f^k_{gb\_ratio} \tag{5}$$

where $\eta(w_{gb}, T_{gb})$ and r_wheel state the efficiency of the gearbox and the radius of the wheel, f_{fd_ratio} is the final driveline ratio equaling with 4.4380, k denotes the numbers of the five-speed gearbox, and $f^k_{gb\_ratio}$ denotes the gear ratio.

As shown in Figure 1, the torque between the engine and the motor is distributed by a torque converter. Due to the fact that the battery and the motor are connected together, the motor torque T_{mot}, the motor speed w_{mot} and P_{eng} can be calculated,

$$w_{mot} = w^k_{gb\_in} \times f_{mot\_ratio} \tag{6}$$

$$T_{mot} = f_{mot}(P_{bat}, w_{mot}) \tag{7}$$

$$P_{eng} = (P_{mot} + P_{gb\_in}) \cdot \eta_{conv} \tag{8}$$

where η_{conv} denotes the efficiency of the torque convertor, f_{mot_ratio} is the motor drive ratio, P_{mot} denotes the motor power and equal the product of T_{mot} and w_{mot}. In this paper, the accessory power, equaling with 200 W, is considered in P_{bat}. The constraint conditions boundary of motor torque can be shown as,

$$T_{mot\_min} \leq T_{mot} \leq T_{mot\_max} \tag{9}$$

where T_{mot_min} and T_{mot_max} denote the minimum and maximum values of the motor torque. As shown in Figure 1, w_{eng} and T_{eng} can be determined,

$$w_{eng} = w_{mot}e_{on} \tag{10}$$

$$T_{eng} = f_{eng}(P_{eng}, w_{eng}) \tag{11}$$

From the above descriptions, we can conclude that m_f can be calculated by w_0, P_0, e_{on} and P_{bat},

$$m_f = f(T_{eng}, w_{eng}, e_{on}) = f(w_0, P_0, P_{bat}, e_{on}) \tag{12}$$

2.3. Battery Pack Model

A simplified battery model, shown in Figure 3, contains an internal resistor, an open circuit voltage (OCV) source connected in series topology to characterize the battery dynamic and static performance. The simplified model has been widely adopted in developing the energy management strategy without influencing the model precision [2]. The detailed battery parameters are listed in Table 2. It can be found that the capacity is 37 Ah and the nominal voltage is 259.2 V.

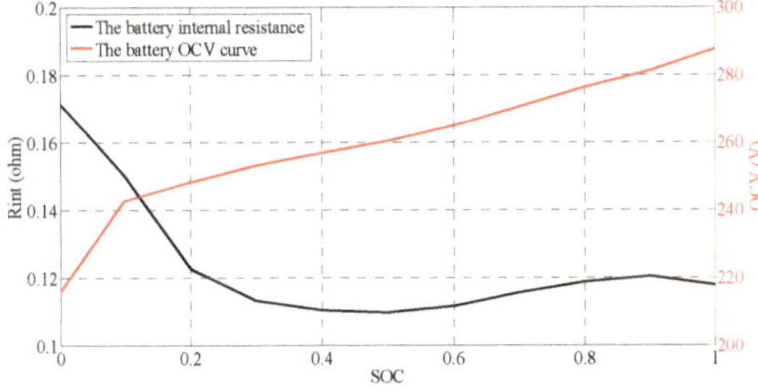

Figure 3. The battery internal resistance and the open circuit voltage (OCV) curve with respect to state of charge (SOC).

Table 2. Battery parameters. SOC: state of charge.

Type	Parameter
Battery type	Lithium-ion battery
Parallel number	1
Serial number	72
Minimum SOC	0.2
Maximum SOC	1
Initial SOC	0.9
Termination SOC	0.3
Capacity	37 Ah
Nominal voltage	259.2 V

The OCV and the internal resistance with respect to the battery SOC are shown in Figure 3. It can be observed that the OCV ranges from 216 V to 288 V. According to the acquired parameters and Figure 4 [40,41], the battery power P_{bat} can be calculated,

$$P_{bat} = U_{ocv}i - i^2 R_0 \tag{13}$$

where U_{ocv} and R_0 state the battery OCV and internal resistance, respectively. Now, the battery current i and SOC can be calculated,

$$i = \frac{U_{ocv} - \sqrt{U_{ocv}^2 - 4R_0 P_{bat}}}{2R_0} \tag{14}$$

$$SOC(t+1) = SOC(t) - \frac{i \times \Delta t}{C_{bat}} \tag{15}$$

where SOC_0 denotes the initial battery SOC, Δt and C_{bat} are the time interval and the battery capacity, respectively. It is noteworthy that, the above-mentioned specifications regarding battery and gearbox are derived from an existing vehicle model in the simulation software Autonomie.

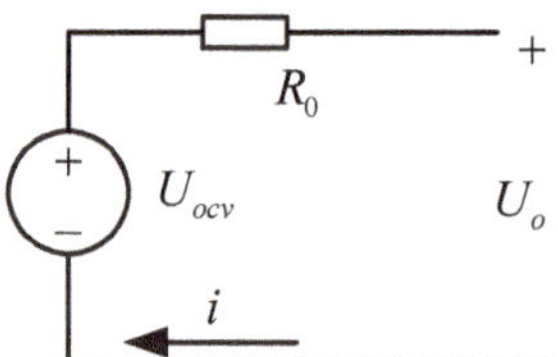

Figure 4. Simplified model of the battery pack.

2.4. Quadratic Static Equation

Based on the above discussion, P_{bat} can finally determine m_f with knowing w_0 and P_0. In addition, according to (13)–(15), the battery SOC can be calculated based on P_{bat}. Thus, P_{bat} can be treated as the control variable and the connection bridge to realize the power split between the engine and the battery. Due to the complex structure and coupling characteristics, the fuel rate can be simplified and can be herein considered as a series of quadratic equations with respect to battery power P_{bat}, as

$$m_f = \begin{cases} a_2(w_0, P_0) \cdot P_{bat}^2 + a_1(w_0, P_0) \cdot P_{bat} + a_0(w_0, P_0) & P_0 \geq P_{eng\_threshold} \\ 0 & P_0 < P_{eng\_threshold} \end{cases} \tag{16}$$

where $a_2(w_0, P_0)$, $a_1(w_0, P_0)$ and $a_0(w_0, P_0)$ are fitting coefficients, which can be determined by w_0 and P_0.

Figure 5 compares the engine fuel rate calculated by the quadratic equations with that looked up in the engine map, proving the proposed method can accurately describe the fuel rate variation. Hence, the fuel rate can be alternatively described by the quadratic equations efficiently. In terms of knowing the driving cycle, the vehicle speed and driveline power demand can be determined. Thus from (16), we can find that two control variables, i.e., P_{bat} and e_{on}, need to be calculated to achieve the energy management. As shown in Figure 6, $a_2(w_0, P_0)$ is always more than zero. Then, the proposed method can be translated into a typical convex problem when the engine is on, which can be solved via the interior point method. Additionally, as shown in (16), the engine fuel rate can be also influenced by the engine on/off command. Here, an engine on/off threshold $P_{eng\_threshold}$ stated in (3) needs to be determined properly to generate the proper engine on/off command and the simulated annealing algorithm is employed to estimate it.

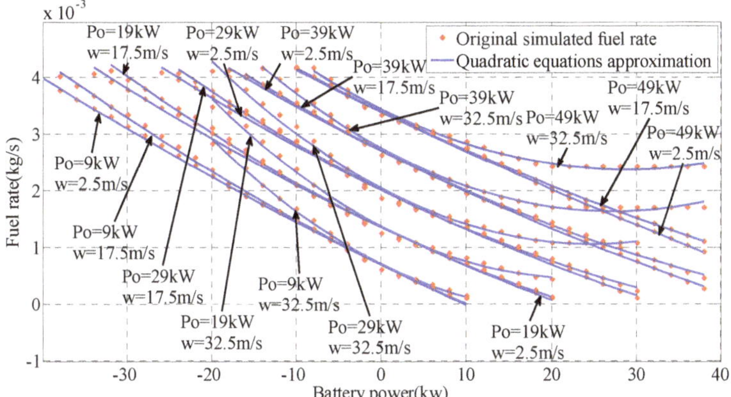

Figure 5. Fuel-rate validation and approximation.

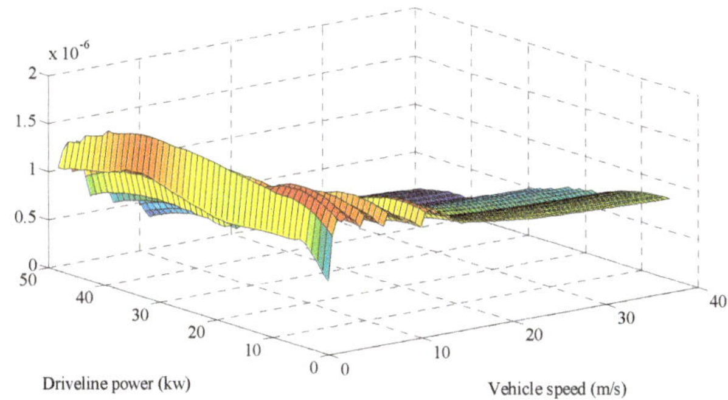

Figure 6. $a_2(w_0, P_0)$.

Next step, the interior point method and simulated annealing method are applied together to find the effective solutions for the battery power and engine ON/OFF commands, respectively.

3. Optimization Methods

As presented in (16), the total cost function of this paper can be expressed as,

$$\min \quad F = \int_0^{t_{total}} \left(a_2(w_0, P_o) \cdot P_{bat}^2(t) + a_1(w_0, P_o) \cdot P_{bat}(t) + a_0(w_0, P_o) \right) e_{on}(t) dt \tag{17}$$

Based on analysis of the vehicle model, related boundary constraints can be summarized,

$$\begin{cases} P_{eng\_on\_min} < P_{eng\_on} \leq P_{eng\_on\_max} \\ P_{bat\_min}(t) \leq P_{bat}(t) \leq P_{bat\_max}(t) \\ 0 \leq \Delta SOC \leq 0.7 \\ T_{mot\_min} \leq T_{mot} \leq T_{mot\_max} \end{cases} \tag{18}$$

where $P_{eng\_on\_max}$ is the upper limit of engine-on power threshold, which is equal to the maximum output power of engine, $P_{eng\_on\_min}$ is the lower limit of engine-on power threshold, P_{bat_min} and P_{bat_max} denote the minimum and maximum values of the battery power commands. ΔSOC denotes the range of the battery SOC variation, which is also called the depth of discharge (DOD). In this paper, the ending SOC is set to 0.3, the minimum SOC is 0.2 and thus ΔSOC belongs to $[0, 0.7]$.

In premise of knowing the driveline power requirements, the engine-on power threshold is solved by the SA algorithm, and the battery power command is calculated by the convex optimization algorithm when the engine is on, as shown in Figure 7. Compared with other optimal algorithms, the SA algorithm is faster and more efficient in solving such global optimization problems with boundary constraints. The calculation process can be described as follows. The first step is to acquire the information of the vehicle including speed, driving range and driveline power demand. Based on the sum of the whole power demand and the maximum battery supplied energy, a decision can be made that if the driving range is less than the AER, the vehicle is under pure EV mode, or else the vehicle is under the HEV mode. If the vehicle works under the HEV mode, an initial engine-on power can be supplied by the SA algorithm. Then, based on the setting constraints, the interior point method will be applied to calculate the battery power thereby realizing the power split between the engine and the battery. Now the fuel consumption based on (17) can be calculated. After that, the SA will be iterated to generate a new engine-on power, and accordingly the battery powers and the total fuel consumption will be updated and compared with the previous calculations. The algorithms will continue to iterate until reaching the termination conditions. Finally, the algorithm outputs the engine-on power and the corresponding battery power commands.

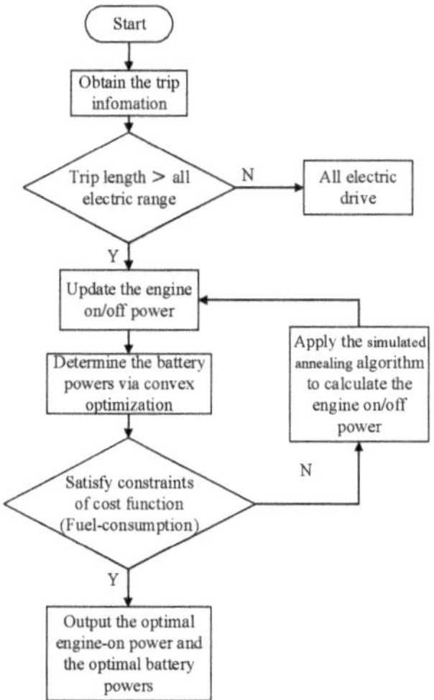

Figure 7. The whole optimization calculation process. SA: simulated annealing.

In (18), the lower bound and upper bound on the engine ON/OFF threshold are set to $P_{eng\_on\_min}$ and $P_{eng\_on\_max}$. Since the SA algorithm is insensitive to the initial value, the initial engine-on power threshold $P_{eng\_on\_initial}$ is determined randomly as the initial point between 10 kW and 15 kW and a sufficiently large initial temperature T is randomly selected based on the initial point. After each iteration, an updated state $P^*_{eng\_on\_update}$ is generated, and the temperature increment ΔT is calculated comparing with the initial iteration.

$$\Delta T = F(P^*_{eng\_on\_update}) - F(P_{eng\_on\_initial}) \tag{19}$$

where F denotes the cost function. If $\Delta T < 0$, $P^*_{eng\_on\_update}$ is accepted as the updated current solution. If $\Delta T > 0$, $P^*_{eng\_on\_initial}$ is accepted as the initial current solution with the probability $\exp(-\Delta T/T)$. Then, the battery power is programmed by the convex optimization algorithm combined with $P^*_{eng\_on\_update}$, and the cost consumption function is calculated. The temperature value decreases gradually until reaching the termination condition. Finally, the current solution can be achieved. In this paper, the SA algorithm is performed by MATLAB (2014a, Mathworks, Natick, MA, USA) [22]. In this paper, the number of iterations of the SA algorithm is set to 40, the terminated tolerance is 0.001, and the simulation step is 1 s.

As shown in Figure 8, the calculation process of SA algorithm under nine NEDC cycles is presented. It can be observed that the SA algorithm converges to a minimum value of the cost function after 14th iterations, and stops after 40 generations. The global optimal result can be found after 15 iterations. Next step, simulations will be conducted to verify the fuel savings of the proposed algorithm.

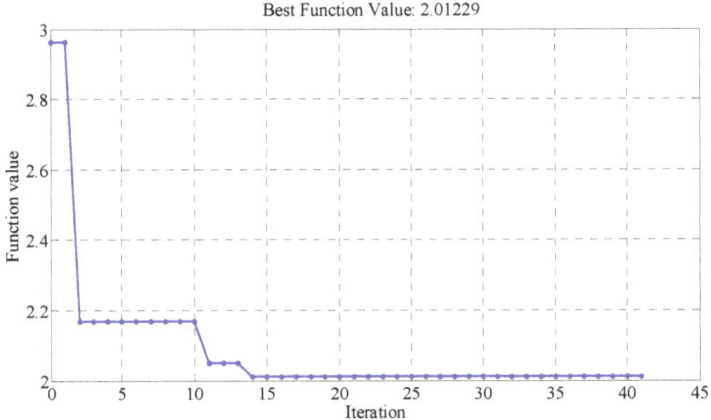

Figure 8. The calculation process of simulated annealing (SA) algorithm method.

4. Simulation Validation and Results Analysis

In this paper, the whole simulation is carried out under MATLAB and Autonomie. MATLAB is a mathematical calculation software for numerical analysis, matrix operation, algorithm development, etc. Autonomie, developed by the Argonne National Laboratory, is an intelligent vehicle simulation software based on MATLAB and Simulink [22,23].

In order to compare the optimization results, the default CD/CS strategy and the DP algorithm are applied as the benchmark. In the CD mode, the parallel PHEV is driven by the battery when the engine is off. In the CS mode, the engine will be turned on when the driveline power demand is more than the engine-on power threshold and the parallel PHEV is driven by the engine and the battery together. The engine will be turned on when the battery maximum power cannot satisfy the driveline power demand [1,6]. When the engine is on, the engine power should also consider the battery balance control, which means that if the battery SOC is lower than the pre-set value, the battery needs to be charged. The detailed CD/CS strategy can be formulated,

$$P_{bat} = \begin{cases} P_o & SOC \geq 36\% \\ \min(25317.8, P_o) & 33\% \leq SOC < 36\% \\ \min(25317.8 \cdot (SOC - 0.3)/0.03, P_o) & 30\% \leq SOC < 33\% \\ \max(-30717.3 \cdot (SOC - 0.3)/0.03, P_o) & P_o < 0, 27\% \leq SOC < 30\% \\ \max(-30717.3 \cdot (SOC - 0.3)/0.03, P_o - P_{eng\_max}) & P_o > 0, 27\% \leq SOC < 30\% \\ \max(-30717.3, P_o) & P_o < 0, SOC < 27\% \\ \max(-30717.3, P_o - P_{eng\_max}) & P_o > 0, SOC < 27\% \end{cases} \quad (20)$$

where P_{eng_max} indexes the maximum power of engine. Based on (20), the battery power is calculated according to the current SOC. In the CD mode, the vehicle is only powered by the battery if the battery SOC is more than 0.36. In the CS mode, the engine is turned on and the battery is charged to hold the SOC near 0.3. In order to verify the control performances more widely, different initial SOC values are considered. We select the initial SOC of 0.9, 0.8 and 0.7 to verify the controlling performances.

4.1. Simulation with Initial SOC of 0.9

We selected two standard drive cycles, i.e., NEDC and HWFET, to validate the performances of the proposed algorithm, DP based method and rule-based method. Figures 9 and 10 show their speed profile and the driveline power, from which we can find that their maximum speeds and maximum driveline power demand are 120 km/h, 96.40 km/h, 38.3 kW and 33.40 kW, respectively.

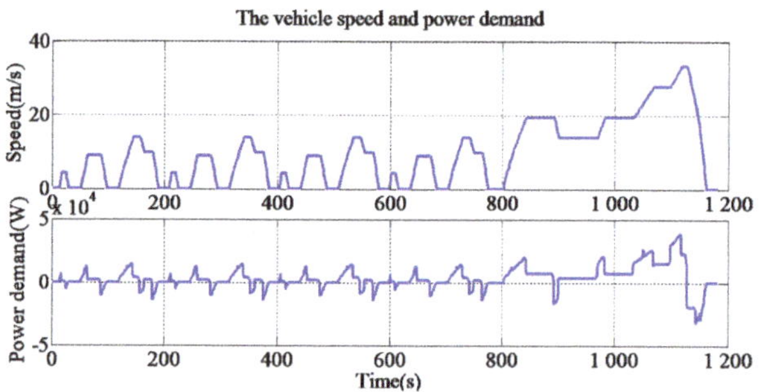

Figure 9. Speed and driveline power demand for the New European Driving Cycle (NEDC).

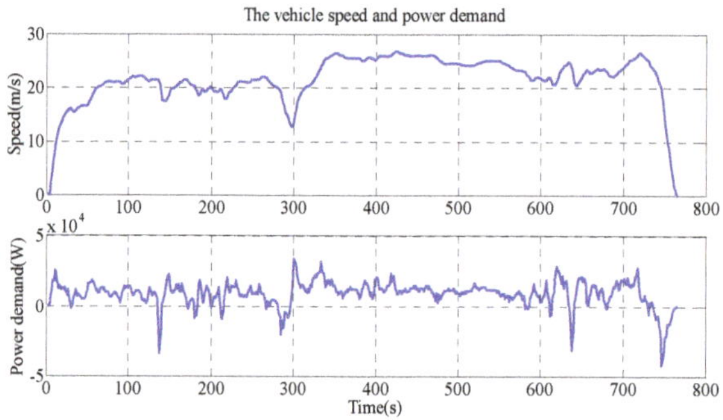

Figure 10. Speed and driveline power demand for the Highway Fuel Economy Driving Schedule (HWFET).

Based on the proposed algorithm, the optimal engine-on power threshold can be calculated by the SA algorithm. After calculation, the optimal engine-on power thresholds are 15.54 kW, 15.15 kW, and 13.52 kW when seven to nine standard NEDC drive cycles are simulated, and 13.25 kW, 13.05 kW, 12.65 kW and 12.50 kW when six to nine standard HWFET drive cycles are simulated.

Figure 11 shows the battery power commands, the engine output power, the driveline power demand and the engine on/off commands based on the proposed method when eight NEDC cycles are applied. It is clearly observed that the convex optimization-based method can start the engine more frequently, since it considers the global optimization of the battery and engine efficiency. Therefore, the DP based method is optimal and the convex optimization-based method is sub-optimal. For the existence of electrical accessories and energy loss, the battery power is higher than the driveline power demand when the engine is off. Figure 12 compares the SOC trajectory when different control methods are applied. Obviously, the battery SOC decreases more slowly when the proposed algorithm is applied due to its capability of the global optimization. Figure 13 compares the comparison of engine-efficiency points by DP and the proposed method, respectively, from which we can observe that the work efficiency by DP is superior than that by the proposed method and the CD/CS method. In addition, there are also less low efficiency points by the proposed method compared to those by CD/CS method. Actually, the purpose of the proposed algorithm is to increase the opportunity of the engine working in the high efficiency region and thus to improve the fuel economy. As shown in Figure 13, the convex optimization-based method increases the operating chances of engine working near the best operating line.

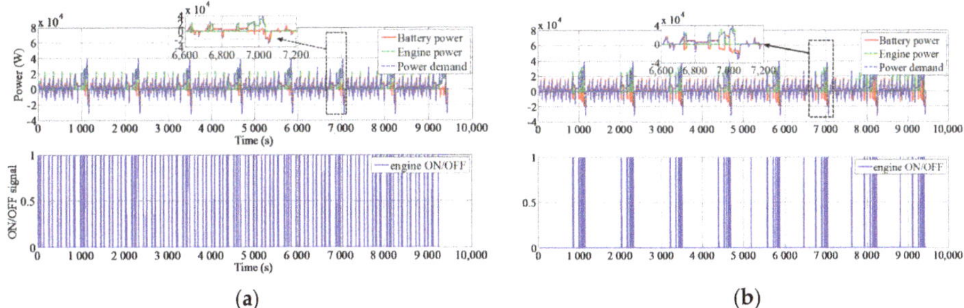

Figure 11. (**a**) The result based on dynamic programming (DP) when eight NEDC cycles are simulated; (**b**) The result based on convex optimization and SA when eight NEDC cycles are simulated.

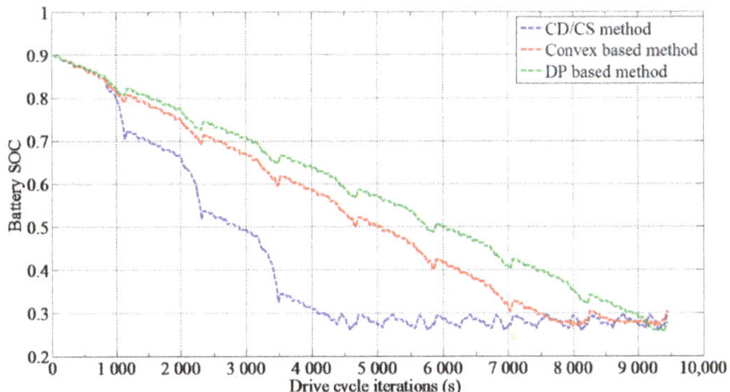

Figure 12. Battery SOC comparison.

In order to compare the fuel consumption, a linearly corrected method is applied to ensure the ending SOC with the same value when applying different strategies [2]. Table 3 list the final fuel consumption F, the terminal SOC and the rate of reducing fuel consumption at different cycles. It can be found that the proposed strategy can reduce the fuel consumption by 9.31%, 8.26%, 8.49% when seven to nine NEDC cycles are simulated, and 8.40%, 7.10%, 6.83%, 6.45% when six to nine HWFET cycles are simulated. As shown in Table 3, the fuel savings achieved with the convex optimization-based method are approximate to the DP based method for the HWFET and NEDC driving conditions. In addition, the fuel savings based on the proposed method are currently less than that based on the DP method. As shown in Table 4, the computation time of the DP based method is obviously longer than that of convex optimization-based method, based on a laptop computer of an i7 processor and 4 Gigabyte RAM. Thus, it justifies the effectiveness of reducing the fuel consumption based on the proposed method.

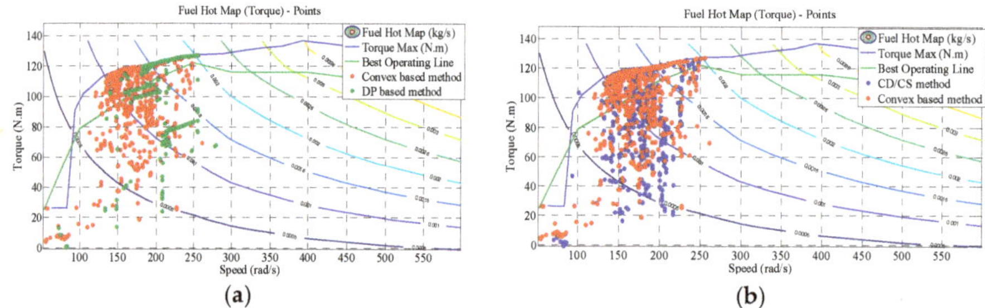

Figure 13. (**a**) Engine efficiency comparison between the convex programming-based strategy and the strategy based on DP; (**b**) Engine efficiency comparison between charging depleting–charging sustaining (CD/CS) strategy and the convex programming-based method.

Table 3. Results comparison with standard initial SOC. CD/CS: charging depleting–charging sustaining; DP: dynamic programming; HWFET: Highway Fuel Economy Driving Schedule; NEDC: New European Driving Cycle.

Drive Cycle	CD/CS Algorithm		DP Algorithm			Convex Algorithm		
	F (kg)	Ending SOC	F (kg)	Ending SOC	Savings (%)	F (kg)	Ending SOC	Savings (%)
9 HWFET	3.7004	0.2767	3.4980	0.3031	6.82	3.5030	0.2986	6.45
8 HWFET	3.1666	0.2767	2.9817	0.3027	7.39	2.9934	0.2995	6.83
7 HWFET	2.6328	0.2767	2.4719	0.3022	7.94	2.4768	0.2930	7.10
6 HWFET	2.0990	0.2767	1.9655	0.3017	8.62	1.9656	0.2994	8.40
9 NEDC	1.9803	0.2923	1.8449	0.3115	8.67	1.8486	0.3116	8.49
8 NEDC	1.6325	0.2923	1.4687	0.2772	8.29	1.5187	0.3034	8.26
7 NEDC	1.2847	0.2923	1.2059	0.3103	9.91	1.1910	0.3060	9.31

Table 4. Computation time comparison with standard initial SOC.

Drive Cycle	CPU Time (s)	
	DP Algorithm	Convex Algorithm
9 HWFET	170.5	3.1
8 HWFET	151.9	2.8
7 HWFET	133.1	2.5
6 HWFET	114.5	2.5
9 NEDC	227.9	7.0
8 NEDC	189.9	5.5
7 NEDC	176.7	4.2

4.2. Simulation with Different Initial SOCs

In actual application, it cannot guarantee the battery is always fully charged when the trip begins. Here, the proposed method is extended to consider different initial SOC values. The UDDS cycle, of which the speed and driveline power are shown in Figure 14, is chosen to verify the proposed method with different initial SOC values. The maximum speed and maximum driveline power demand for the UDDS cycle are 91.25 km/h and 41.92 kW, respectively.

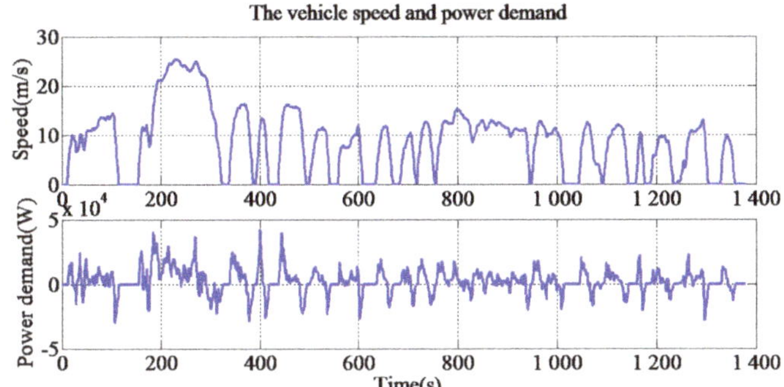

Figure 14. Speed and driveline power demand for Urban Dynamometer Driving Schedule (UDDS) drive cycle.

Figure 15 compares the SOC trajectories based on different energy management strategies when the battery initial SOC is 0.7. It shows that the battery is discharged more slowly when the proposed algorithm is applied than that when the CD/CS strategy is applied. Table 5 compares the final results, which show that the proposed method can save fuel consumption by 10.06%, 9.19% when the initial SOC are 0.7 and 0.8 with the SOC correction. Table 6 compares the CPU operation time based on different methods with respect to different initial SOC values. Obviously, the computation time based on convex optimization is obviously less than the DP based method. As shown in Figure 15 and Table 5, the solution based on the convex optimization method can be still acceptable and thus proving its feasibility [39].

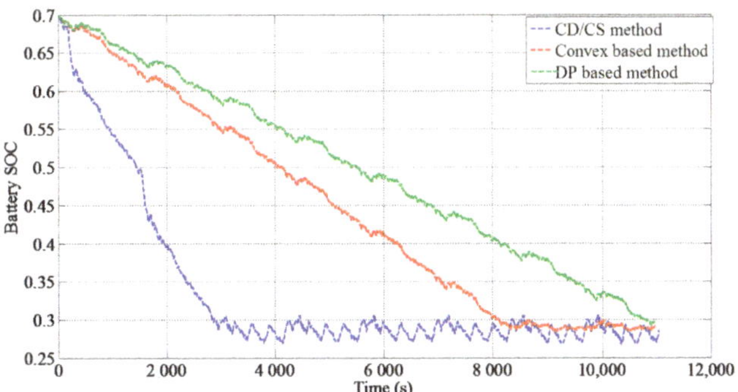

Figure 15. Battery SOC comparisons when the initial SOC is 0.7.

Table 5. Results comparison with different initial SOC.

Initial SOC	CD/CS Algorithm		DP Algorithm			Convex Algorithm		
	F (kg)	Ending SOC	F (kg)	Ending SOC	Savings (%)	F (kg)	Ending SOC	Savings (%)
0.7	2.1871	0.2859	1.9729	0.2969	10.75	1.9758	0.2905	10.06
0.8	1.9988	0.2859	1.7842	0.2986	11.93	1.8108	0.2836	9.19

Table 6. Results comparison with different initial SOC.

Initial SOC	Drive Cycle	CPU-Time (s)	
		DP Algorithm	Convex Algorithm Optimization
0.8	8 UDDS	269.6	2.7
0.7	8 UDDS	270.1	4.9

5. Conclusions

In order to improve the fuel economy and engine work efficiency, a time efficient energy management strategy is established for the parallel PHEV based on the convex optimization and the SA algorithm. By analyzing the dynamics of the driveline system, the convex quadratic function is built between the engine fuel-rate and the battery power considering requirements of the driveline power and speed. The fuel optimization problem is transformed and solved by the convex optimization algorithm based on the interior point method. In order to extend the proposed method, the convex function is solved at different initial battery SOCs. Compared with the DP based method and the CD/CS method, the proposed method can calculate the engine-on power and the battery power command efficiently, bringing improvement of the engine working efficiency and reduction of the fuel consumption.

Next step work can focus on the hardware-in-loop validation to verify the feasibility of the proposed algorithm and consideration of the battery performance variation under low temperature and degradation.

Acknowledgments: The work is supported by the project of National Science Foundation of China (Grant No. 51567012, 61763021) in part, Research and Development of Test Cycles for Chinese New Energy Vehicles—Data Acquisition in Kunming (Grant No. CF2016-0163) in part, the High-Level Overseas Talents Program of Yunnan Province (Grant No. 10978196) in part, the Innovation Team Program of Kunming University of Science and Technology (Grant No. 14078368) in part, and the Scientific Research Start-up Funding of Kunming University of Science and Technology (Grant No. 14078337) in part. The authors would like to thank them for their support and help. The authors would also like to thank the reviewers for their corrections and helpful suggestions.

Author Contributions: Renxin Xiao and Baoshuai Liu conceived the paper, discussed the convex optimization algorithm and carried out the simulation validation. Zheng Chen built the simplified vehicle model and revised the paper. Jiangwei Shen and Ningyuan Guo analyzed the data and conducted the figures drawing. Wensheng Yan provided some valuable suggestions.

Conflicts of Interest: The authors declare no conflict of interest.

References

1. Chen, Z.; Mi, C.C. An adaptive online energy management controller for power-split HEV based on dynamic programming and fuzzy logic. In Proceedings of the 2009 IEEE Vehicle Power and Propulsion Conference, 7–11 September 2009; pp. 300–304.
2. Zhang, X.; Mi, C.C. Vehicle Power Management: Modeling, Control and Optimization. *Power Syst.* **2011**, *14*, 91–118.
3. Lin, C.C.; Peng, H.; Grizzle, J.W.; Kang, J.M. Power management strategy for a parallel hybrid electric truck. *IEEE Trans. Control Syst. Technol.* **2003**, *11*, 839–849.
4. Sorrentino, M.; Rizzo, G.; Arsie, I. Analysis of a rule-based control strategy for on-board energy management of series hybrid vehicles. *Control Eng. Pract.* **2011**, *19*, 1433–1441. [CrossRef]
5. Peng, J.; He, H.; Xiong, R. Rule based energy management strategy for a series–parallel plug-in hybrid electric bus optimized by dynamic programming. *Appl. Energy* **2017**, *185 Pt 2*, 1633–1643. [CrossRef]
6. Chen, Z.; Mi, C.C.; Xu, J.; Gong, X.; You, C. Energy management for a power-split plug-in hybrid electric vehicle based on dynamic programming and neural networks. *IEEE Trans. Veh. Technol.* **2014**, *63*, 1567–1580. [CrossRef]
7. Park, J.; Chen, Z.H.; Murphey, Y.L. Intelligent vehicle power management through neural learning. In Proceedings of the International Joint Conference on Neural Networks, Barcelona, Spain, 18–23 July 2010.

8. Ahmadi, S.; Bathaee, S.M.T. Multi-objective genetic optimization of the fuel cell hybrid vehicle supervisory system: Fuzzy logic and operating mode control strategies. *Int. J. Hydrogen Energy* **2015**, *40*, 12512–12521. [CrossRef]

9. Li, S.G.; Sharkh, S.M.; Walsh, F.C.; Zhang, C.N. Energy and battery management of a plug-in series hybrid electric vehicle using fuzzy logic. *IEEE Trans. Veh. Technol.* **2011**, *60*, 3571–3585. [CrossRef]

10. Kermani, S.; Delprat, S.; Guerra, T.M.; Trigui, R.; Jeanneret, B. Predictive energy management for hybrid vehicle. *Control Eng. Pract.* **2012**, *20*, 408–420. [CrossRef]

11. Zhang, S.; Xiong, R.; Sun, F. Model predictive control for power management in a plug-in hybrid electric vehicle with a hybrid energy storage system. *Appl. Energy* **2017**, *185 Pt 2*, 1654–1662. [CrossRef]

12. Xiong, R.; Cao, J.; Yu, Q. Reinforcement learning-based real-time power management for hybrid energy storage system in the plug-in hybrid electric vehicle. *Appl. Energy* **2018**, *211*, 538–548. [CrossRef]

13. Chen, Z.; Masrur, M.A.; Murphey, Y.L. In Intelligent vehicle power management using machine learning and fuzzy logic. In Proceedings of the IEEE International Conference on Fuzzy Systems, Hong Kong, China, 1–6 June 2008; pp. 2351–2358.

14. Zhang, M.Y.; Yang, Y.; Mi, C.C. Analytical approach for the power management of blended-mode plug-in hybrid electric vehicles. *IEEE Trans. Veh. Technol.* **2012**, *61*, 1554–1566. [CrossRef]

15. Zhang, X.; Mi, C. Analytical approach for the power management of blended mode phev. In *Vehicle Power Management: Modeling, Control and Optimization*; Springer: London, UK, 2011; pp. 107–139.

16. Ansarey, M.; Panahi, M.S.; Ziarati, H.; Mahjoob, M. Optimal energy management in a dual-storage fuel-cell hybrid vehicle using multi-dimensional dynamic programming. *J. Power Sources* **2014**, *250*, 359–371. [CrossRef]

17. Wang, X.; He, H.; Sun, F.; Zhang, J. Application study on the dynamic programming algorithm for energy management of plug-in hybrid electric vehicles. *Energies* **2015**, *8*, 3225–3244. [CrossRef]

18. Patil, R.M.; Filipi, Z.; Fathy, H.K. Comparison of supervisory control strategies for series plug-in hybrid electric vehicle powertrains through dynamic programming. *IEEE Trans. Control Syst. Technol.* **2014**, *22*, 502–509. [CrossRef]

19. Wang, X.; Liang, Q. Energy management strategy for plug-in hybrid electric vehicles via bidirectional vehicle-to-grid. *IEEE Syst. J.* **2017**, *11*, 1789–1798. [CrossRef]

20. Li, L.; Yan, B.; Yang, C.; Zhang, Y.; Chen, Z.; Jiang, G. Application-oriented stochastic energy management for plug-in hybrid electric bus with amt. *IEEE Trans. Veh. Technol.* **2016**, *65*, 4459–4470. [CrossRef]

21. Zhang, S.; Xiong, R.; Zhang, C. Pontryagin's minimum principle-based power management of a dual-motor-driven electric bus. *Appl. Energy* **2015**, *159*, 370–380. [CrossRef]

22. Chen, Z.; Xia, B.; You, C.; Mi, C.C. A novel energy management method for series plug-in hybrid electric vehicles. *Appl. Energy* **2015**, *145*, 172–179. [CrossRef]

23. Chen, Z.; Mi, C.C.; Xiong, R.; Xu, J.; You, C.W. Energy management of a power-split plug-in hybrid electric vehicle based on genetic algorithm and quadratic programming. *J. Power Sources* **2014**, *248*, 416–426. [CrossRef]

24. Hadj-Said, S.; Colin, G.; Ketfi-Cherif, A.; Chamaillard, Y. Convex optimization for energy management of parallel hybrid electric vehicles. *IFAC Papersonline* **2016**, *49*, 271–276. [CrossRef]

25. Hu, X.S.; Murgovski, N.; Johannesson, L.M.; Egardt, B. Optimal dimensioning and power management of a fuel cell/battery hybrid bus via convex programming. *IEEE-ASME Trans. Mechatron.* **2015**, *20*, 457–468. [CrossRef]

26. Elbert, P.; Nusch, T.; Ritter, A.; Murgovski, N.; Guzzella, L. Engine on/off control for the energy management of a serial hybrid electric bus via convex optimization. *IEEE Trans. Veh. Technol.* **2014**, *63*, 3549–3559. [CrossRef]

27. Hu, X.; Moura, S.J.; Murgovski, N.; Egardt, B.; Cao, D. Integrated optimization of battery sizing, charging, and power management in plug-in hybrid electric vehicles. *IEEE Trans. Control Syst. Technol.* **2016**, *24*, 1036–1043. [CrossRef]

28. Bandeira, J.; Almeida, T.G.; Khattak, A.J.; Rouphail, N.M.; Coelho, M.C. Generating emissions information for route selection: Experimental monitoring and routes characterization. *J. Intell. Transp. Syst.* **2013**, *17*, 3–17. [CrossRef]

29. Zhang, Y.; Liu, H.P.; Guo, Q. Varying-domain optimal management strategy for parallel hybrid electric vehicles. *IEEE Trans. Veh. Technol.* **2014**, *63*, 603–616. [CrossRef]

30. Tie, S.F.; Tan, C.W. A review of energy sources and energy management system in electric vehicles. *Renew. Sustain. Energy Rev.* **2013**, *20*, 82–102. [CrossRef]
31. Zhang, B.Z.; Mi, C.C.; Zhang, M.Y. Charge-depleting control strategies and fuel optimization of blended-mode plug-in hybrid electric vehicles. *IEEE Trans. Veh. Technol.* **2011**, *60*, 1516–1525. [CrossRef]
32. Kelouwani, S.; Henao, N.; Agbossou, K.; Dube, Y.; Boulon, L. Two-layer energy-management architecture for a fuel cell hev using road trip information. *IEEE Trans. Veh. Technol.* **2012**, *61*, 3851–3864. [CrossRef]
33. Onori, S.; Tribioli, L. Adaptive pontryagin's minimum principle supervisory controller design for the plug-in hybrid gm chevrolet volt. *Appl. Energy* **2015**, *147*, 224–234. [CrossRef]
34. Hu, X.; Martinez, C.M.; Yang, Y. Charging, power management, and battery degradation mitigation in plug-in hybrid electric vehicles: A unified cost-optimal approach. *Mech. Syst. Signal Process.* **2017**, *87*, 4–16. [CrossRef]
35. Johannesson, L.; Murgovski, N.; Jonasson, E.; Hellgren, J.; Egardt, B. Predictive energy management of hybrid long-haul trucks. *Control Eng. Pract.* **2015**, *41*, 83–97. [CrossRef]
36. Murgovski, N.; Johannesson, L.; Sjöberg, J.; Egardt, B. Component sizing of a plug-in hybrid electric powertrain via convex optimization. *Mechatronics* **2012**, *22*, 106–120. [CrossRef]
37. Nüesch, T.; Elbert, P.; Flankl, M.; Onder, C.; Guzzella, L. Convex optimization for the energy management of hybrid electric vehicles considering engine start and gearshift costs. *Energies* **2014**, *7*, 834–856. [CrossRef]
38. Chen, Z.; Mi, C.C.; Xia, B.; You, C. Energy management of power-split plug-in hybrid electric vehicles based on simulated annealing and pontryagin's minimum principle. *J. Power Sources* **2014**, *272*, 160–168. [CrossRef]
39. Wang, Z.; Huang, B.; Xu, Y.; Li, W. In Optimization of series hybrid electric vehicle operational parameters by simulated annealing algorithm. In Proceedings of the IEEE International Conference on Control and Automation, Athens, Greece, 30 May 2007–1 June 2007; pp. 1536–1541.
40. Van Mierlo, J.; Van den Bossche, P.; Maggetto, G. Models of energy sources for ev and hev: Fuel cells, batteries, ultracapacitors, flywheels and engine-generators. *J. Power Sources* **2004**, *128*, 76–89. [CrossRef]
41. He, H.; Xiong, R.; Guo, H.; Li, S. Comparison study on the battery models used for the energy management of batteries in electric vehicles. *Energy Convers. Manag.* **2012**, *64*, 113–121. [CrossRef]

 *applied sciences*

Review

Electricity Generation in LCA of Electric Vehicles: A Review

Benedetta Marmiroli [1,2,*], **Maarten Messagie** [2], **Giovanni Dotelli** [1] and **Joeri Van Mierlo** [2,3]

[1] Mat4En2—Materials for Energy and Environment, Dipartimento di Chimica, Materiali e Ingegneria Chimica "Giulio Natta", Politecnico di Milano, Piazza Leonardo da Vinci 32, 20133 Milano, Italy; giovanni.dotelli@polimi.it

[2] MOBI—Mobility, Logistics and Automotive Technology Research Centre, Department of Electric Engineering and Energy Technology, Vrije Universiteit Brussel, Pleinlaan 2, 1050 Brussels, Belgium; maarten.messagie@vub.be (M.M.); joeri.van.mierlo@vub.be (J.V.M.)

[3] Flanders Make, 3001 Heverlee, Belgium

* Correspondence: benedetta.marmiroli@polimi.it; Tel.: +39-02-2399-3232

Received: 25 July 2018; Accepted: 13 August 2018; Published: 16 August 2018

Abstract: Life Cycle assessments (LCAs) on electric mobility are providing a plethora of diverging results. 44 articles, published from 2008 to 2018 have been investigated in this review, in order to find the extent and the reason behind this deviation. The first hurdle can be found in the goal definition, followed by the modelling choice, as both are generally incomplete and inconsistent. These gaps influence the choices made in the Life Cycle Inventory (LCI) stage, particularly in regards to the selection of the electricity mix. A statistical regression is made with results available in the literature. It emerges that, despite the wide-ranging scopes and the numerous variables present in the assessments, the electricity mix's carbon intensity can explain 70% of the variability of the results. This encourages a shared framework to drive practitioners in the execution of the assessment and policy makers in the interpretation of the results.

Keywords: LCA; Well-to-Wheel; electric vehicle; plug-in hybrid; electricity mix; consequential; attributional; marginal; system modelling; energy system; meta-analysis

---

1. Introduction

Electric mobility is gaining momentum as a promising technology for decarbonisation of the transport sector and lots of scientific papers assessing environmental impacts of electric vehicles (EVs) are being produced. However, as the literature grows, so do the number of conflicting results.

A few reviews have tried to find a pattern in the Life Cycle Assessment (LCA) results: Hawkins et al. [1] identify the lack of a transparent and complete Life Cycle Inventory (LCI) as one of the main gaps in LCA. On the other hand, a more recent review by Nordelöf et al. [2] argues that the absence of a complete goal definition is the main hurdle to correctly interpret results and find trends in the literature.

Since the goal dictates the line for the subsequent scope and defines the applications of the study, omitting this phase leaves the study as a missive without address in the scientific community, and thus potentially ineffective.

Hawkins et al. [1] focused on the necessity to find consensus on the inventory. At the time of Hawkins' publication, it was found that most studies limited their attention to a well-to-wheel analysis—since the use phase was seen to dominate the life cycle of vehicles, or to the battery production. The author's purpose was then to investigate all those aspects that were not sufficiently addressed, providing the practitioner with a standardised system boundary and a set of relevant sub-components that could be distinguishable in the production phase.

The main causes of divergence in the literature, which make it difficult to compare studies are identified as follows by Hawkins et al. [1]: Different system boundaries, different level of detail and quality in the datasets, different lifetimes, different vehicles' typologies and masses, battery technologies, vehicles performances, and then the electricity mix.

Nordelöf et al. [2] present an exhaustive analysis, and performs various meta-analyses from the findings of LCAs. They also widen the discussion to impact categories like resource depletion and toxicity, while the Hawkins's focus remained on climate change.

Both reviews identified electricity production as the most impactful phase when it comes to climate change, and agreed on the need to find consensus on the appropriate electricity mix.

Since these two seminal reviews have been published, a variety of papers appeared in the literature, paving the way to new and interesting discussions as well as diverging results and methods.

A significant change in the way to account for electricity in the use phase of electric vehicles has occurred in the last few years: from an overgeneralisation of the inventory, using generic datasets (from EcoInvent, Emissions & Generation Resource Integrated Database [eGRID], Greenhouse gases Regulated Emissions and Energy use in Transportation [GREET], etc. ...), the trend in the most recent LCAs has been a high detail of temporal and spatial variability, following the work by Graff Zivin et al. [3].

A blossoming of different methods to account for the "correct" electricity mix has led to lots of different, and sometime conflicting, results.

Lack of consensus in LCI data selection and lack of clear goal definitions are still the key factors to explain the difficult path of providing policy makers with robust and clear results.

This review analyses the methodological choices made by the scientific papers when assessing EVs. Since its relevance has been proved, a special focus is placed on electricity generation. The aim is to find a trend in accounting for the electricity production method, thus helping users navigate the conflicting results found in the literature, providing guidelines for future EVs studies and informing policy makers on the right method to use when assessing political choices.

2. Method

2.1. Articles Selection

The review analyses 44 LCA studies available in Scopus and Web of Science databases, published between 2008 and 2018, in which at least one of the analysed vehicles has an electric powertrain. This way both LCAs focusing only on electric vehicles (EVs) or hybrid electric vehicles (HEVs) and both comparative studies between traditional cars and electric vehicles are included.

The review includes both complete LCAs and Well-to-Wheel analysis, following the classification accepted in the literature [2]. A well-to-wheel (WtW) analysis is a partial LCA, which limits its system boundary to the cycle of the energy carrier used to propel the vehicle, such as liquid fuel or electricity. This type of studies had a large diffusion in the first attempt to assess and compare different powertrain options [4], owing to known relevance of the operation phase in the total life cycle. Conversely, a complete LCA also includes the stages related to vehicle production, maintenance and dismantling (see Figure 1).

A separate chapter in the field of electric mobility assessment is taken up by LCAs of batteries used for electromotive applications. Batteries have had a place of honour in the literature on electromobility. When the production phase of vehicles gained attention, lots of studies focused only on battery production [5,6]. Due to the relevance of battery production and performance in the assessment of EVs, several LCAs focused on this device and its performance during the use phase rather than on the entire vehicle [7]. Since the matter of which electricity mix to use in the evaluation is the same for studies analysing the entire vehicle and studies focusing on batteries, also the LCAs of batteries that includes the use phase in their system boundaries have been analysed.

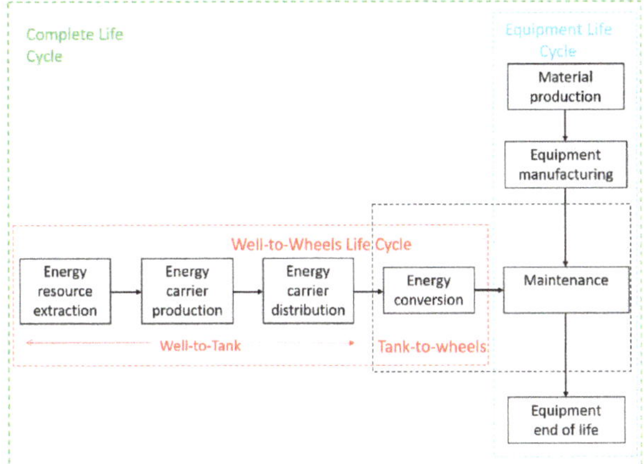

Figure 1. Simplified view of the well-to-wheels (WtW) and equipment flows [2]. (Reproduced with permission from [2], Springer, 2014).

Studies on energy production and LCA of energy systems have been referred to, in order to better understand the modelling choices used in EVs' LCAs. Many LCAs of EVs, which propose a new way to account for the electricity used to charge the vehicles, directly reflect the methodology described in studies on energy systems.

LCAs of energy intensive products (buildings, aluminium products, chemicals produced via electrolysis) have also been consulted, since they provide an interesting insight into the modelling of electricity production and the methodological choices.

National and intergovernmental reports and energy scenarios has been investigated for the data regarding the energy mix and their emissions, as well as documentations from the main databases used in the LCA studies (EcoInvent, GREET, eGRID, etc.).

2.2. Review Approach

After a selection process, the articles complying with the requirements have been reviewed and their consistency between goal, scope, modelling choices, selected inventory and recommendation provided to the audience has been investigated.

In order to find a trend between the goal and scope, the modelling choice and the electricity mix accounted, the articles have been analysed looking for the following information:

- Methodological characteristics: Goal, intended audience and applications (both explicit or inferred), and modelling choice (attributional versus consequential, whenever the study cohere to this distinction);
- Descriptive characteristics of the assessed electricity mix: Regional boundaries, time horizon, calculation methods, technology involved (average versus marginal suppliers), and data source.

3. Literature Results

3.1. Goal and Scope

The lack of clear goals in the literature had originally been highlighted by Norderlöf et al. in 2014 [2] and no improvement has been noticed so far. Among the revised studies only 4 abided by all the requirements from ISO 14040 [8–11]. According to ISO 14044 "In defining the goal of an LCA, the following items shall be unambiguously stated: The intended application; the reasons for carrying out the study; the intended audience, i.e., to whom the results of the study are intended to be communicated; whether the results are intended to be used in comparative assertions intended to be disclosed to the public". Most of the remaining studies presents detailed information about the objective of the study but omits application and intended audience. However, in many cases this information can be inferred a posteriori in the conclusion and discussion sections; to a different extent almost all the articles are found to inform policy or decision makers explicitly or implicitly (only the study by Garcia et al., 2017 [12] is presented as a pure eco-design study, thus implying first an optimisation chain application rather than decision making orientation).

This is done through a wide range of scopes, ranging from WtW comparative analysis, complete LCA analysis, battery LCA, analysis of the infrastructure, comparison of different vehicle segments and technologies, analysis of a single vehicle or of an entire fleet (see Table 1).

The characterisation of the object of the study presents different levels of detail in the literature. Its description ranges from 'average passenger vehicle' [13] to more detailed vehicle segment (mid-size, compact, sport-utility, etc. ...), to identification of archetypes (Nissan Leaf as a paradigm of small size EV, Toyota Prius for HEV, etc.) to comparison within specific models with different powertrains: Piaggio Porter [14], Iveco Daily [15], Smart [16,17], GM Chevrolet Malibu [18].

Passenger vehicles represent most of the analysed vehicles (41 out of 44) with only three exceptions:

1.   Giordano et al. [15] and Bartolozzi et al. [14], evaluating light duty vehicles for goods delivery;
2.   Lee et al. [10] evaluating medium duty trucks.

Studies can be classified based on their scale, between:

1.   Vehicle based LCA;
2.   Fleet based scenarios.

Vehicle based LCA evaluates the performance of a single vehicle technology, or compare it with another vehicle equipped with a different powertrain (generally electric versus conventional internal combustion engine vehicle [ICEV], but also HEV versus Battery Electric Vehicle [BEV], BEV versus Fuel Cell Electric Vehicle [FCEV] [14] etc.).

However, in more than one situation the result obtained from vehicle-based comparison has been extended to larger deployment consideration, multiplying the value obtained for a supposed penetration rate, thus implying a linear relation between the single vehicle purchase and nationwide policies. This implication is questionable and is not confirmed in LCA standards, which propose different approaches depending on the scale of the analysis, see Table 2 [19]. On the other hand, fleet-based analyses suggest that some scale and time dependent aspects cannot be detected from a single product analysis. Only two studies belong to this category [20,21].

Garcia et al., 2015 [21] developed a dynamic fleet-based life cycle model, able to include the effects of technology turnover and other time related parameters, such as ICEVs fuel consumption reduction and electricity mix impacts, fleet penetration scenarios, fleet and distance travelled growth rates, and changes in vehicle weight and composition and battery technologies over time.

Bohnes et al., 2017 [20] developed a fleet based LCA under different deployment scenarios in order to meet the urban transport demand of a specific city for a given time period. As a proof-of-concept they applied it to the Copenhagen urban area from 2016 to 2030.

Table 1. Synoptic view of the literature review.

Authors	Region	Time Horizon	Penetration Rate	Vehicle	Electricity mix				BEV Impact (GHG)
					Average	Marginal	Source	GHG Intensity	
				Complete LCA					
Archsmith et al., 2017 [22]	US NERC regions	2011–2012	-	ICEV, EV		a	EPA's Continuous Emissions Monitoring Systems (CEMS); GREETnet	513–1258 g CO_2-eq/kWh	4.22–6.11 ton CO_{2eq}/vehicle/year (rebound effect included) (average 5.22)
		2030 2040			x		EIA forecasts	-	-
Bartolozzi et al., 2013 [14]	Italy	-	-	BEV, ICHE, FC LCV (Piaggio Porter)	x	-	EcoInvent (unspecified version)	-	178.25 g CO_2-eq/km
Bauer et al., 2015 [23]	Switzerland (EU 27 mix in use phase)	2012	-	ICEV, HEV, Plug-in Hybrid (PHEV), BEV, FCV, FCHEV	x		EcoInvent v. 2.2	-	214 g CO_2-eq/km EU electricity mix (BEV mid size, average efficiency 0.77 mj/km)
		2030			x		Kannan and Turton [24] and ESU-services/IFEU 2008	-	117 g CO_2-eq/km EU mix (BEV mid size, average efficiency 0.64 mj/km)
Bohnes et al., 2013 [20]	Denmark	From 2016 to 2030	-	ICV, BEV, HEV, PHEV, nPHEV, REEV, FCEV fleet analysis		d	EcoInvent 3.1 Substitution, consequential updated with EU and DK projections	-	From 292 g CO_2-eq/km in 2016 to 202 g CO_2-eq/km in 2030
Crossin and Doherty 2016 [25]	Australia	2015	-	PHEV, ICEV	x		Australian Energy market operator	1006 g CO_2-eq/kWh	246 g CO_2-eq/km
						a		794 g CO_2-eq/kWh	204 g CO_2-eq/km
Faria et al., 2013 [26]	Poland			gasoline and diesel ICEVs, PHEVs, BEVs, FCEVs	x		European Environment Aagency (EEA) 2011	979 g CO_2-eq/kWh	210 g CO_2-eq/km (compact)" 175 g CO_2-eq/km (subcompact peugeot iOn)"
	Portugal	2011	-					376 g CO_2-eq/kWh	125 g CO_2-eq/km (compact)" 95 g CO_2-eq/km (subcompact peugeot iOn)"
	France							103 g CO_2-eq/kWh	75 g CO_2-eq/km (compact)" 60 g CO_2-eq/km (subcompact peugeot iOn)"
Freire and Marques 2012 [27]	Portugal	2004	-	BEV, PHEV, gasoline, Diesel (compact and subcompact passenger vehicles)	x		REN	666 g CO_2-eq/kWh	170 g CO_2-eq/km (compact) 110 g CO_2-eq/km (subcompact)
		2009						560 g CO_2-eq/kWh	150 g CO_2-eq/km (compact) 100 g CO_2-eq/km (subcompact)
		2010						390 g CO_2-eq/kWh	110 g CO_2-eq/km (compact) 80 g CO_2-eq/km (subcompact)
Garcia and Freire 2016 [28]	Portugal	2015–2017	35–59 GWh in 2017	BEV	x		REN	352 g CO_2-eq/kWh	352 g CO_2-eq/km
						a			723 g CO_2-eq/km

166

Table 1. *Cont.*

Authors	Region	Time Horizon	Penetration Rate	Vehicle	Electricity mix				BEV Impact (GHG)
					Average	Marginal	Source	GHG Intensity	
Girardi et al., 2015 [9]	Italy	2013 (data from 2012)	"few EVs"	EV, ICEV		a	TRENA		155 g CO_2-eq/km (vehicle efficiency 0.19 kWh/km)
		2030	+17.5 TWh			c(b)	Lanati et al. [29]	480 g CO_2-eq/kWh	148.88 g CO_2-eq/km (vehicle efficiency 0.19 kWh/km)
Hawkins et al., 2012 [1]	EU	-	-	EV Li-NCM, EV LifePO4, ICEV diesel, ICEV gasoline	x		EcoInvent v. 2.2	562 g CO_2-eq/kWh EU mix*	196.79 g CO_2-eq/km (0.713 kWh/km)
Helmers et al., 2017 [16]	Germany	Present (data 2004)	-	BEV, ICEV, FCV new ICEV Smart, new e-Smart, Smart converted from combustion engine to electric; BEV, ICEV, FCV	x		EcoInvent v. 2.2	719.5 g CO_2-eq/kWh	-
		2013					IEA	707.4 g CO_2-eq/kWh	180 g CO_2-eq/km
		Generic future based on renewables					Nitsch et al., 2012 [30]	130.6 g CO_2-eq/kWh	97 g CO_2-eq/km
Helmers and Marx 2012 [17]	Germany	2010	-	BEV, ICEV, FCV smart: ICEV, e-SMART; BEV, ICEV, FCV	x		German federal environmental agency	536 g CO_2-eq/kWh	140 g CO_2-eq/km
Lee et al., 2017 [10]	US states	2014	-	medium duty TRUCK diesel, biodiesel, biodiesel hybrid, CNG, CNG hybrid electric	x		EPA's CEM hourly data and NEI database	-	-
						a		-	-
Lucas et al., 2012 [31]	Portugal	2010	-	gasoline, diesel, FCHEV, FCPHEV, EV	x		REN	-	61 g CO_2-eq/km
Ma et al., 2012 [32]	UK	2015+ (data from 2009–2010)	-	ICEV, HEV, BEV Mid size in UK	x		BM report	518.4 g CO_2-eq/kWh	130.6 g CO_2-eq/km
						a		799.2 g CO_2-eq/kWh	178.7 g CO_2-eq/km
	California	2015+ (data from 2010)		SUV in California;	x		McCarthy and Yang 2009 [33]	-	220.4 g CO_2-eq/km
						c(b)		-	326.6 g CO_2-eq/km
McCarthy and Yang 2009 [33]	California	2010	1% of VMT	ICEV, HEV, PHEV, BEV, FCV	x		eGRID v.1.1 (2007)	250 g CO_2-eq/kWh	118.32 g CO_2-eq/km (WTW) 107.44 g CO_2-eq/km (WTW)
						b		626 g CO_2-eq/kWh (off peak marginal) 571 g CO_2-eq/kWh (load level marginal)	
Marshall et al., 2013 [34]	Michigan	2009	10% infiltration, additional 2.41×10^{10} MJ	mid size PHEV		b	-	-	547–611 g CO_2-eq/km (PHEV, efficiency 0.45 kWh/km)
Noshadravan et al., 2015 [35]	US	2009	-	EV, ICEV				227.1–894.2 g CO_2-eq/kWh 560.65 g CO_2-eq/kWh	190 g CO_2-eq/km"
Messagie et al., 2015 [36]	Belgium	2012 2013 2017	-		x		ELIA (Belgian DSO)	-	-

Table 1. *Cont.*

Authors	Region	Time Horizon	Penetration Rate	Vehicle	Electricity mix				BEV Impact (GHG)
					Average	Marginal	Source	GHG Intensity	
Onat et al., 2015 [37]	US states	2009	-	mid size BEV, ICEV, PHEV, HEV	x		eGrid	663.4 g CO_2-eq/kWh	187.72 g CO_2-eq/km
		2020				c (b)	Hadley and Tsvetkova [38]	range 644–911 g CO_2-eq/kWh	
Stephan and Sullivan 2008 [39]	US regions	2002	"a significant number"	PHEV	x		EPA	608.4 g CO_2-eq/kWh'	177 g CO_2/km (PHEV, efficiency 0.92 MJ/km) (WTW)
		2002				a			
		2030				d			157 g CO_2-eq/km
Van Mierlo et al., 2017 [40]	Belgium	2011	-	BEV, Compressed Natural Gas (CNG), Liquid Petrol Gas (LPG), Biogas (BG), PHEV, HEV	x		Messagie et al., 2014 [41]	190 g CO_2-eq/kWh	31–24 g CO_2-eq/km (wtw)
Weis et al., 2016 [42]	US PJM	2010 2018	-	EV, HEV and conventional gasoline		b	UCED model (data from NEEDS database and EPA projections)	-	-
Yuksel et al., 2016 [43]	US states	2011	-	2013 Nissan Leaf BEV; 2013 Chevrolet Volt PHEV; 2013 Toyota Prius PHEV; Toyota Prius HEV; the Mazda 3		c (a)	Siler-Evans et al. [44]	430–932 kg CO_2-eq/MWh	-
Giordano et al., 2017 [15]	UK	2015	-	light duty vehicle (IVECO daily) BEV, diesel	x		EcoInvent 3.0 updated with data for 2015 from Entso-e	688 g CO_2-eq/kWh	263 g CO_2-eq/km, ni-nacl2 battery 2015 bev
	Germany							579 g CO_2-eq/kWh	260 g CO_2-eq/km, ni-nacl2 battery 2015 bev
	Portugal							553 g CO_2-eq/kWh	250 g CO_2-eq/km, ni-nacl2 battery 2015 bev
	Italy							512 g CO_2-eq/kWh	235 g CO_2-eq/km, ni-nacl2 battery 2015 bev
	France							98 g CO_2-eq/kWh	81 g CO_2-eq/km, ni-nacl2 battery 2015 bev
	Norway							36. g CO_2-eq/kWh	58 g CO_2-eq/km, ni-nacl2 battery 2015 bev
Lombardi et al., 2017 [18]	Italy	Present (data 2004)	-	gasoline ICEV, EV, gasoline PHEV, PHFCV	x		EcoInvent v. 2.2	640.8 g CO_2-eq/kWh	226 g CO_2-eq/km
	USA							770.4 g CO_2-eq/kWh	-
	France							93.6 g CO_2-eq/kWh	-

Table 1. *Cont.*

Authors	Region	Time Horizon	Penetration Rate	Vehicle	Electricity mix				BEV Impact (GHG)
					Average	Marginal	Source	GHG Intensity	
Tagliaferri et al., 2016 [45]	EU	2012	-	BEV (Nissan Leaf), ICEV (Toyota Yaris); PHEV	x		EcoInvent v. 2.2	-	120 g CO_2-eq/km (Nissan leaf EVI 0.50 MJ/km)111 g CO_2-eq/km (Nissan leaf EVII 0.50 MJ/km)
		2050					Behrens et al., 2013 (2050)	-	-
Tamayao et al., 2015 [46]	US regions	2009	-	PHEV (Chevrolet Volt), HEV (Toyota Prius), BEV (Nissan Leaf)	x		ANL	-	-
						a	Siler-Evans et al., 2012 [44] Graff Zivin et al., 2014 [3]	-	-
W-t-W Analysis									
Thomas 2012 [47]	US electrical power regions	2020	-	HEV, PHEV, BEV		c (b)	Hadley and Tsvetkova [38]	-	-
Dallinger et al., 2012 [48]	Germany	2030	12 milion EV	BEV		b	own calculation Elgowainy et al., 2010 [4]	247.26 construction and dispatch of RES to serve EVs 245.42 construction and dispatch of RES to serve EVs614 long term high RES mix (2030) 558.21 long term high RES mix (2030)	23.4 g CO_2-eq/km last trip charging; 10.7 g CO_2-eq/km DSM charging; 122.84 g CO_2-eq/km DSM charging; least marginal cost dispatch 111.64 g CO_2-eq/km last trip charging; least marginal cost dispatch
Van Vliet et al., 2011 [49]	Nederland	2015	-	SHEV, BEV, PHEV, gasoline, diesel		b	Van den Broek et al. [50]	-	62 g CO_2-eq/km (uncoordinated charging) 57 g CO_2-eq/km (off-peak charging)
Faria et al., 2012 [51]	EU	Portugal 2009		ICE, HEV, PHEV, BEV, future BEV	x	-	Eea Eurostat	365 g CO_2-eq/kWh	65.3 g CO_2-eq/km′
		France	-					78 g CO_2-eq/kWh	14 g CO_2-eq/km′
		2009						378 g CO_2-eq/kWh	68 g CO_2-eq/km′
		2010						360 g CO_2-eq/kWh	67.3 g CO_2-eq/km′
		2020						230 g CO_2-eq/kWh	43.3 g CO_2-eq/km′
		2030						160 g CO_2-eq/kWh	30 g CO_2-eq/km′
		2040						150 g CO_2-eq/kWh	28 g CO_2-eq/km′
		2050						135 g CO_2-eq/kWh	25.3 g CO_2-eq/km′
Woo et al. 2017 [52]	70 countries	2014	-	EV, ICEV	x	-	IEA (2015b), EIA (2015) World Bank (2016)	-	78.1 g CO_2-eq/km (world average)
Huo et al., 2015 [13]	US (3 regions), China (3 regions)	2012		EV, PHEV, ICEV	x			-	190–290 g CO_2-eq/km′ "(China), 110–225 (US) g CO_2-eq/km′ "
		2025					EIA (2015)	-	110–160 g CO_2-eq/km′ " (China), 60–150 (US) g CO_2-eq/km′ "
Gao and Winfeld 2012 [53]	US			CV (Toyota Corolla), HEV(Prius), PHEV (Prius Plug-in), EREV (GM Volt), EV (Nissan Leaf), FCV (Honda Clarity)	x		GREET	-	240 g CO_2-eq/km″ (efficiency 0.213 kWh/km)

Table 1. *Cont.*

Authors	Region	Time Horizon	Penetration Rate	Vehicle	Electricity mix				BEV Impact (GHG)
					Average	Marginal	Source	GHG Intensity	
					Battery LCA				
Ambrose et al., 2016 [54]	US states	see Archsmith	-			c(a)	Archsmith et al. [22]	-	-
Majeau-Bettez et al., 2011 [7]	EU		-		x		EcoInvent v. 2.2	-	-
Deng et al., 2017 [55]	US		-		x		Thinkstep (unspecified version)	-	-
Zackrisson et al., 2010 [56]	West Europe, Scandinavia, China		-		x			-	-
Garcia et al., 2017 [12]	EU-27 France		-		x			-	-
Oliveira et al., 2015 [57]	Belgium (EU mix)	2011 (data from 2004)	-		x			516 g CO_2-eq/kWh	-
Sanfelix et al., 2015 [58]	EU		-		x				33.9 g CO_2-eq/km (WtW 25.7 g CO_2-eq/km)
Notter et al., 2010 [6]	EU		-	BEV golf class	x		EcoInvent v. 2.01	592.61 g CO_2-eq/kWh	155 g CO_2-eq/km (vehicle efficiency 0.17 kWh/km including auxiliary energy consumption)
					Eco balance				
Noori et al., 2015 [11]	US electric regions	2030	-	ICEV, gasoline HEV, gasoline PHEV, gasoline EREV, BEV	b		EIA projections	-	-
					LCA of energy-demanding products				
Roux et al., 2017 [59]	France	reference year (averaging annual economic and meteorological variations)			x		RTE		61.4 to 84.9 g CO_{2eq}/kWh of heating in households
						b			765.1 to 928.7 g CO_{2eq}/kWh of heating in households
Alvarez Gaitan et al., 2014 [60]	Australia	2030	+1 Mg of NaClO and1Mg $FeCl_3$			b, d			
Colett et al. [61]	US (different boundary levels)	2010				nested approach			19.0 and 19.9 kg CO_2-eq/kg primary aluminium ingot

* Value obtained a posteriori looking for the EcoInvent dataset "Electricity production mix low voltage RER" (IPCC 2007 100a); "value normalised by the author"; value inferred from bar graph; a = marginal mix obtained through bottom up approach; b = marginal mix obtained through top down approach; c() = marginal mix obtained from the literature; d = long term marginal.

3.2. System Modelling and Inventory Choices

Since the International Workshop on Electricity Data for Life Cycle Inventories organised by the Environmental Protection Agency (EPA) and held in October 2001 at the Breidenbach Research Center in Cincinnati (Ohio) to discuss life cycle inventory data for electricity production, two system modelling approaches have been opposed in LCA [62]: Attributional (ALCA) and consequential (CLCA). ALCA methodology accounts for immediate physical flows (i.e., resources, material, energy, and emissions) involved across the life cycle of a product. ALCA typically utilises average data for each unit process within the life cycle. CLCA, on the other hand, aims to describe how physical flows can change as a consequence of an increase or decrease in demand for the product system under study. Unlike ALCA, CLCA includes unit processes inside and outside of the product's immediate system boundaries. It utilises economic data to measure physical flows of indirectly affected processes [63].

3.2.1. Consequential System Modelling

According to Weidema, a consequential approach is a "[s]ystem modelling approach in which activities in a product system are linked so that activities are included in the product system to the extent that they are expected to change as a consequence of a change in demand for the functional unit" [64].

A CLCA is basically concerned with identifying the cause and effect relationship between possible decisions and their environmental impacts. The cause and effect relationships are based on models of equilibrium between supply and demand, borrowed from neoclassical economics. In practice, this consist in the identification of the potential suppliers/technologies that will be affected by a change in demand (marginal suppliers/technologies).

Some studies manifested the need to include other mechanisms when assessing the consequences of a decision, such as rebound effect [63], learning curves and the so called positive feedback or third order consequences [65].

CLCA in Electric Mobility

In the electric mobility literature, the only field where marginal technology has been investigated is electricity generation. Despite the growing concern about resource scarcity of rare earths and precious metals involved in batteries and electric motors and the risk related to their supply [66,67], no study has attempted to asses it from a consequential point of view.

Consequential LCA should be more than the use of marginal mixes, as pointed out in the 62° LCA conference [68]. However, in the LCA of EV, it emerges that the difference between ALCA and CLCA is often identified with the difference between average and marginal electricity mixes, and the terms are sometimes used as synonyms.

The goal should define the methodological choices, and from the methodological choice should stem the inventory. This chain is often inverted in the literature, where the goal is missing or is not strong enough to justify methodological choices, and the inventory selection defines the study. In several studies it is the selection of the marginal mix in the inventory what defines the work as a consequential LCA.

In the following section all the studies adopting marginal mixes will be presented, even if in the authors' opinion they do not comply with the requirement of a CLCA. This allows for a more inclusive literature analysis.

Marginal Mixes Selection

Among the selected articles, 17 use marginal mix when assessing EVs from a life cycle point of view. 12 are complete LCAs, 4 are WtW analysis, and the last one is a battery-LCA accounting also for the use phase (see Figure 2).

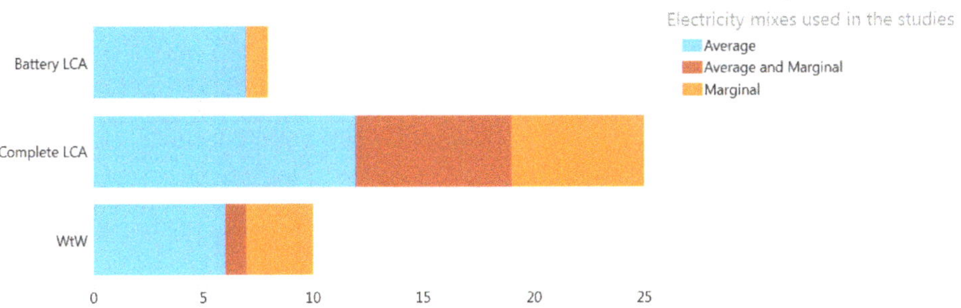

Figure 2. Number of LCAs sorted per study type and electricity mix selected.

The interesting aspect is that among these, seven studies use marginal mixes along with average mixes. Some of these consider alternatively one mix or the other as a matter of sensitivity analysis [22], others for testing results with the mix of higher GHG intensity [10], others use different mixes for different time horizon, due to difficulties in determining the marginal mix in future energy systems [22,37].

This aspect is crucial in demonstrating the detachment of the inventory from the methodological choice, as our work previously stated after highlighting a diffuse weakness in the definition of goal and scope phases.

Since the use of marginal mixes is no longer linked to the system modelling choice as it was expected, the reasons to adopt marginal mixes have been investigated. Among the literature, these reasons are:

- It is a way of testing the robustness of results [22];
- It is used as a sensitivity analysis [10,25,37,69];
- It is considered conceptually more appropriate [22,25,28,33,46];
- It is the result of a consequential modelling choice [20,42];
- It is suggested from previous studies [10];
- It is required from GHG protocol [47];
- Not specified [43,49];
- EVs are considered the marginal consumer [9].

The last point, made explicit by few studies, is actually the implicit assumption made by every study that uses a marginal mix when assessing EVs.

Some studies state that using the marginal mix is more appropriate, but their explanations barely offer any insight as to why. Some of them read as follows:

" . . . small changes in the composition of the vehicle stock, replacing an ICE with an EV will represent an increase on the margin of electricity generation . . . " [22]

"Marginal grid GHG intensity gives a more realistic measure of the GHG impact of the growth of electric vehicles than does average grid GHG intensity." [32,70]

" . . . assessing a technology that entails a change in electricity consumption require MEF . . . " [28]

"It better represents the effects of the of EV adoption in the near future . . . " [22]

"It is useful for short term forecasts of electricity demand . . . " [25]

The widespread lack of specification and justification of the modelling choices made in published LCA studies has been already pointed out by Weidema et al. [71] and it is confirmed by our selection of articles. Only two studies actively describe their choice in the framework of the CLCA methodology, either adopting it [20], or refusing it [9].

The study by Bohnes et al. [20] explains the reason for assessing a consequential LCA. It estimates the impacts of transitioning to an electric fleet in the Copenhagen urban area from 2016 to 2030. The decision context is ascribed to the macro level decision support or case B according to the ILCD handbook (see Table 2) and the consequential approach is chosen [19]. Therefore the consequential EcoInvent dataset is selected and the "medium-term" marginal mix for Denmark is used. This is derived from EcoInvent database 3.1, and adapted with projections from Danish government [72] and goals set by the European Union [73].

Table 2. Combination of two main aspects of the decision-context: Decision orientation and kind of consequences in background system or other systems [19]. (Reproduced with permission from [19], Publications Office of the European Union, 2014)

Decision support?		Kind of Process-Changes in Background System/Other Systems	
		None or small-scale	Large-scale
	Yes	**Situation A** **"Micro-level decision support"**	**Situation B** **"Meso/macro-level decision support"**
	No	**Situation C** **"Accounting"** **(with C1: Including with other systems, C2: Excluding interactions with other systems)**	

On the other hand, Girardi et al. [9] do not define their study as a CLCA, since their system boundary does not include "processes to the extent of their expected change caused by a demand (affected processes) and do not solve multifunctionality through system expansion". Instead, they adopt the approach by Zamagni et al. [74]. According to the authors, CLCA is not strictly a methodology but an approach to "deepen LCA" taking into account market mechanisms, rather than a modelling principle with defined rules.

Time Horizon

When assessing the changes due to the additional demand of the selected functional unit, the definition of the temporal boundary is a key issue. The literature distinguishes between short term and long term effects of a change [75]. This simplification, derived from economy, defines short term effects as the ones affecting the existing production capacity, while in the long term production capacity is allowed to adapt to the changes in demand, and a second simplification set "the utilization*[sic]* of this capacity to be constant".

Even though CLCA practitioners agree in defining the long term changes as the relevant ones [76], most of the studies on EVs focus on short term changes (see Table 1).

When referring to electricity generation, the difference between short term and long term is usually identified with operation versus new installed capacity; "[l]ong-term marginal supply is sometimes also referred to as the 'build-marginal', i.e., the technology of the capacity to be installed next" [77]. Soimakallio refers to them as "operational margin" and "build margin" respectively [78]. Short term will affect only existing production capacity whereas, in the longer term, it will require the installation of new capacity, which reflects the most common and persistent effect [79].

The concept of 'marginal' has different meaning in LCA and in energy systems modelling, leading to some confusion in the literature, especially when it comes to long term consequences. The long term marginal can be referred to as the marginal power plant running in a projected energy system in the future when an additional unit is required to the system, or as the new capacity installed from the time being to the year of interest (usually at least five years after the decision to install new capacity [77]), due to investment in new power plants pushed by the increased electricity demand.

As mentioned before, a second simplification implies "the utilization*[sic]* of the new capacity to be constant" [75].

In the field of energy systems, this means that the dynamics in the operation of the marginal capacities are ignored, so the marginal supply will be fully produced at such capacity.

According to Lund et al. [80], this marginal change in capacity does not reflect the change in marginal electricity supply. He suggests to simulate the new installed capacity integrated in the pre-existing system via ESA simulation, in order to obtain the latter.

This paved the way to many LCAs simulating how EVs charged on the margin will behave in a projected energy system. However, as explained by Schmidt et al. [81], sticking to the main definition of long term consequences, this represents a short term marginal (albeit of a future energy system), since the additional demand is met by the existing capacity, even though it is a projected capacity.

Adhering to the definition proposed by Weidema, this study distinguished the studies assessing short term marginal (both in present energy systems and in projected energy systems) from those appraising long term marginal (see Table 1).

Despise the suggestion to consider long term consequences, only three studies address long term marginal mixes. The study by Stephan and Sullivan [39] endorses this allocation method ante litteram. It states that "if the utilities add more base capacity (beyond that projected by the U.S. Energy Information Administration [EIA]) as a result of the growth of PHEVs, then the correct CO_2 emission allocation should be based on that new base capacity". Assuming that the extra capacity has the same repartition as the EIA projection, the authors find the impacts of PHEV to be 157 g CO_2/km.

A clear application of the Weidema's heuristic approach (see Figure 3) to define the long term marginal technology has been found in Alvarez-Gaitan et al. [60], whose study is worth mentioning, even if not related to EVs. Alvarez-Gaitan et al. [60] set the analysis of chlor-alkali chemicals at 2030. Following the guidelines from Weidema et al. [79], the analysis focuses on the effects on the long term. The emphasis is on the long term marginal supply of electricity because of its relevance in the production of the chlor-alkali. Other inputs are considered less relevant, so the marginal suppliers are assumed to remain unchanged. The new installed capacity, assumed to be installed as a consequence of the increased demand, is identified using the 'five-step Weidema approach' [82] (see Figure 3). In the long term all the generation technologies are supposed to be unconstrained (with the exception of political and resource restriction). The new installed capacity is then found to be supercritical pulverised black coal and wind on shore and solar photovoltaic, based on levelised cost of energy (LCOE) projections by the Australian Government. Afterwards, they compare these technologies, defined as the simple marginal, echoing Mathiesen [83], with the one identified with a partial equilibrium model (complex marginal technology).

All the other studies were found to assess what in the authors' opinion is a short term marginal consequence, regardless of whether it is run in present or future energy systems.

Short Term Marginal Mix—Calculation Method

Even when the methodological choice and the time horizon are set, defining the marginal mix is not straightforward. The results can differ due to the various calculation methods available to define the emissions, to the method selected to identify the eligible marginal technologies, to temporal granularity, geographical boundaries etc. In the following section, a brief description of all the methods available in the literature is reported.

Some studies already tried to develop a framework for different allocation methods, to be populated with the studies available in the literature. The first one, to the authors' best knowledge, was published by Yang in 2013 [84]. Yang identifies three key parameters in the classification of the methods for assigning greenhouse gas emissions from electricity generation (beside the choice of average and marginal):

1. Temporal resolution or granularity (aggregate versus temporally explicit);
2. Time frame (retrospective versus prospective);

3.   Spatial boundary.

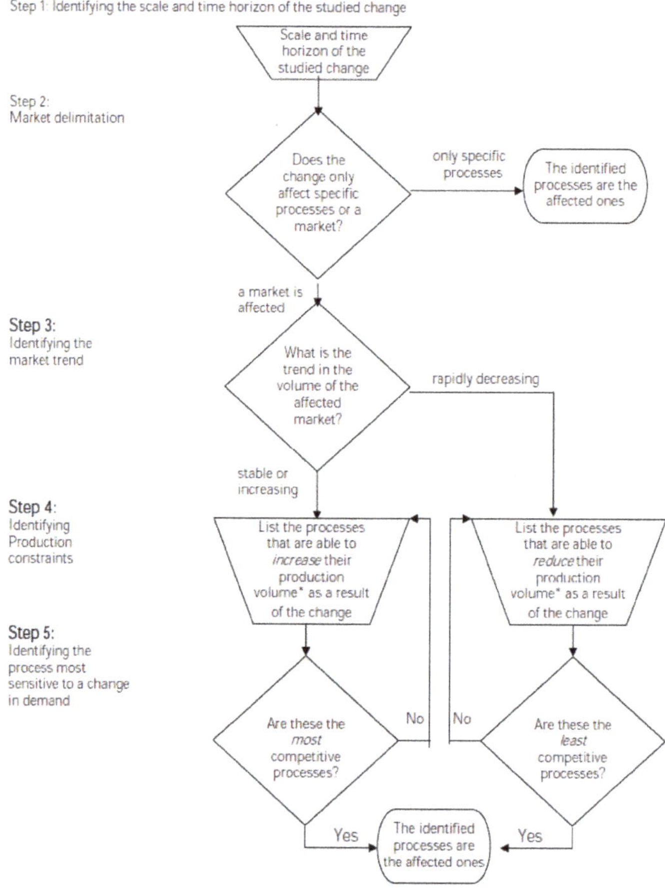

Figure 3. Decision tree outlining the five-step procedure for identifying the processes affected by a change in demand for a specific intermediate product [82]. (Reproduced with permission from [82], 2.-0 LCA consultants, 2011)

At the simplest extreme of this classification are aggregate methods, which use data of the total energy production from one year to define the emission from the grid (both average and marginal, but mainly average). These studies rely on robustness and simplicity, and need few data available from many reliable sources (national databases, Transmission System Operators [TSO], Distribution System Operators [DSO]). They can be both retrospective and prospective, but of course the prospective version has higher uncertainties and arbitrariness, since the definition of future scenarios is always uncertain and subject to estimation errors.

On the other side, temporally explicit studies capture coincidence between variations in supply and demand. They need more data than aggregate methods and usually require models to simulate the operation of the power system. This approach is accompanied mainly, but not exclusively, by prospective and marginal approaches. The author identifies the division between "short run approach" and "system response approach" as another distinction in the temporal definition in these cases. The former accounting only for operational changes, where the existing grid is responding to a change in demand, the latter including also structural changes that could occur as a result of the addition of EV charging, such as changes in grid composition.

The problem of identifying the correct emission factor is common to many studies, that do not necessarily refer to EVs as product system. Some studies analyse it from an energetic point of view [80,83,85], while others are interested in the topic because the energy use is a relevant part in their product system—such as, the building sector [59], chemicals produced via chlor-alkali electrolysis [60], and aluminium production [61]. Another famous review is the one by Ryan et al. [85]. The review by Ryan is not directly addressed to EV charging, but instead wants to orientate the evaluation for every type of electrical load, even though a special mention is set aside for EV fields as the worst case of inconsistency in methods' selection.

The methods found in the literature are divided into "empirical data and relationship models"—either simple average emission factors multiplied by the load of interest or statistical correlation between demand and emissions, but all relying on historical data—and "power system optimisation models". The latter are used for evaluating projected electricity generation operations bounds by physical laws of power generation and economic optimisation.

Ryan's main operational distinction [85] is based on time perspective. On one side, there are empirical and relationship models, which rely on historical data to define emissions of past loads, or small changes of future loads While on the other side, power system optimisation models allow for evaluation of projected electricity generation, which is constrained by physical laws of power generation and economic optimisation.

Ryan [85] states that a correct method on the whole does not exist. It is rather a matter of defining the correct method for the objective of the study on a case by case basis. The match is defined in terms of load characteristic; and the selection of time granularity is function of load and energy system. Yearly average emission factors can apply to industries operating at constant load, while hourly emission factors suit best to loads with strong diurnal variations, or to energy system with a consistent share of wind and hydro, which can cause variability in the supply. Empirical data and relationship models (relying on historical data) are applicable only on loads consistent with recent historical perspective, while power system optimisation models fit when analysing policies, processes, or products with a multiyear forward-looking view. The adoption of average emission factors in the case of EVs could be justified by the low rate penetration of EVs, while the use of marginal could be explained by the high variation in load [85].

In the end it turns out to be again a matter of equity, wondering whether it is fair or not to separate future and current consumption, since existing consumption dictates the emissions from new consumption. This consideration is defended by the incremental approach proposed by Messagie et al. [36], and endorsed by Soimakallio et al. [78].

Tamayao et al. [46], comparing the effect of different methods on carbon footprint of PHEVs, BEVs and HEVs, distinguish the methods to calculate marginal emission factors into: (1) Bottom up, and (2) top down methods. The distinction is close to the classification made by Ryan et al. [85] between "empirical data and relationship models" and "power system optimization*[sic]* models" presented in this chapter. In fact, in the category "bottom up approach" Tamayao et al. include all the models that defines how a system will respond to a load profile due to normative, operational and economic constraints; within "top down approach" they includes regression models relying on observed data.

Entering the details of each model is behind the scope of this study. Characteristics and suggested applications of these two methods have been collected and the relevance of each statement has been assessed based on its statistical occurrence in the energy system literature. The application of these methods in the selected articles has been then discussed in the following chapters.

Top down approach

Top down approach applies regression models using observed data to assess how generation and/or emissions change as a function of changes in the load [46]. The relevant characteristics of regression models are:

- They require few data and calculation time compared to models [85];
- They are retrospective in nature [85];
- They can describe only small changes in generation load [42];
- They suffer from error in counterfactual analysis because correlations in past data do not necessarily imply causality [42,46];
- They are not suitable to capture significant changes (infrastructural changes, fuel price, energy policies, changes to the grid due to economic conditions or other factors, generator or transmission additions) [43,85].

Despite their limitations, regression methods have been widely applied in the literature. Archsmith et al. [22] and Tamayao et al. [46] used it to identify the effect of different charging time and regional variability in Marginal Emission Factors (MEFs) in the U.S. regions. The results have been used to suggest a better cohesion between regional MEF and federal subsidies for EVs household purchase.

Garcia and Freire [28] use regression to assess the introduction of BEVs in Portuguese fleet from 2015 to 2017 (displacing ICEV). Fewer than 20,000 vehicles (causing an additional electric demand of 60 GWh in the most energy demanding scenario) are assumed to displace either new gasoline or new diesel cars in varying percentages according to the scenarios.

Girardi et al. [9] apply a probabilistic approach to determine the marginal technology for a 2013 scenario with "few EVs". Even though the marginal mix is still mainly based on fossil fuels, EVs are found to perform better than ICEV because of a 60% of efficient combined cycle gas turbine power plants on the margin.

Ma et al. [32] use a regression method which employs historical data from between 2009–2010 for the UK and an estimation for 2010 for California, to assess the expected EV market in 2015 and beyond. They obtain that BEV perform worse than ICEV in the UK market and worse than HEV in both the analysed markets.

The studies by Yuksel et al. [43] and Ambrose et al. [54] rely on regression results respectively from Siler-Evans et al. [44] and Archsmith et al. [22]. They do not specify the time horizon of their study, but they rely on regression data from 2011 and 2012 respectively.

Bottom up approach

As seen in the aforementioned reviews, there are plenty of system models, with various pro and cons; their level of complexity can range from simple dispatch curves to detailed simulation or optimisation models [46]. The common aspect is that they define how a system is bound to respond to a load profile, due to normative, operational or economic constraints (e.g., ramp rates, plant availability, emissions and transmission constraints, etc.).

In order to gain a better understanding of these methods' application for environmental assessments, this study tried to highlight common characteristics influencing the limitations and range of application of the electricity mix obtained, rather than provide details for each method, as this was beside the scope of this analysis.

Some relevant aspects of these methods for LCA application are the following:

- They are suitable to model future power plant scenarios and large load changes [42];
- They have limited scalability [42];
- They could require large number of inputs and their complexity represents a significant hurdle for incorporation in LCA [85];
- The results depend heavily on the input data and on assumptions made by the user [85].

Girardi et al. [9] rely on a previous study by Lanati et al. [29], which uses a model to determine the optimal long term evolution of the generation set, with and without EV demand. They obtained no significant changes in the two scenarios, because of the limited impact of EV demand, found to be less than 5% of the total 2030 end-use demand, with an EV penetration rate of 25%.

A second model has then been used to model the operation of the previously defined generation set, obtaining that all the electricity supplied to EVs will be produced using fossil fuels.

McCarthy and Yang [33] use a simple dispatch model to assess the introduction of EV fleet in 2010; if 1% of Vehicle Miles Travelled is driven with EVs, the increase demand of electricity would be 0.1–0.3%. Their model predicts that marginal mix used to charge EV will be mainly provided with relatively inefficient NGCT plants. Marginal electricity in California is more carbon-intensive than gasoline, but in most cases, the improved efficiency of electric-drive trains outweighs the difference in fuel carbon intensity, and the vehicles considered here reduce GHG emissions compared to HEVs.

Both Onat et al. [37] and Thomas [47] rely on the marginal mix, determined by Hadley and Tsvetkova [38], for the introduction of PHEV in the 2020 energy system across all the states of the U.S., under a certain penetration rate and charging condition, and apply it to their own scenarios that include BEVs introduction in the transport system, under unspecified penetration rates.

Weis et al. [42] employ the UCED model, a model used to optimise the operation of energy systems. They use data from EPA's NEEDs database to model the 2010 US energy system, while for the future US scenario (2018) they include retirement of power plants predicted by the EPA and a 3% wind penetration. A third scenario includes the power plant retirements predicted by EPA, and 20% wind penetration.

Dallinger et al. [48] use an agent-based electricity market equilibrium model to estimate variable electricity prices and power plant utilisation in a 2030 German scenario.

Considerations

Despite three exceptions, all the marginal mixes identified in the literature are found to be short term marginal mixes, according to the definition by Weidema. The identification of marginal is usually detached from the penetration scenarios (in Table 1 can be seen that only a few studies using marginal mixes provide penetration scenarios); this tendency is confirmed by the use of MEF from other studies, which are applied to other circumstances (different technologies and different energy demand in the case of Thomas [47] and Onat et al. [37] using the results from the study by Hadley and Tsvetkova [38]).

What is common to every study applying marginal mixes is the identification of the EVs as the marginal consumers, both in present and in future energy system. The result is that the effects of using present or future energy systems convey similar results, because technologies on the margin tend to remain the same also in projected scenarios. Therefore, EVs do not benefit from the general decarbonisation in the energy system that is happening at present and that will tend to continue in the future.

The identification of EVs as marginal consumers can be meaningful if a big amount of EVs are inserted in the transport system before the energy system can react to it. However, this is an unlikely scenario, due to the low penetration rate the transport sector is experiencing.

The use of short term marginal can address some specific questions on the increase of demand due to incentives to EVs and can provide answers on how to optimise their charging time; however, limiting the evaluation of EVs in the short run represents a burden shifting in time. Similarly, considering EVs as the short-run marginal consumers in projected scenarios is not coherent when the goal is to inform policy makers on the introduction of EVs in the transport system. As stated by Soimakallio et al. [78], if 'new consumption' is adequately anticipated before it occurs, there is no unambiguous reason to assign short term marginal production to this particular consumption. Most future energy scenarios already include the supply required for electric vehicles in the projected electricity demand.

3.2.2. Attributional System Modelling

While there is no biunivocal link between marginal mixes and consequential studies, ALCA studies are more straightforward. All the attributional studies apply the average national mix, and they mainly model it as background processes using renowned databases (EcoInvet, GREET, eGRID . . .). Other sources are Transmission System Operators (TSO), Distribution System Operators (DSO), national and regional statistics etc.

No author provides explanations for the modelling choice when performing ALCAs, probably because it is perceived as the norm in scientific literature, due to its overwhelming majority among LCA papers [86].

The identification of the average mix is methodologically less equivocal, since it represents the total generation allocated evenly to the total load. However, various sources of variability can be listed:

- Data quality: Transparent and up-to-date electricity data are not the norm. Especially when relying on background databases, studies tend to overlook the data quality of the selected dataset.
- Regional boundary: The region of production of electricity and that of consumption do not always coincide. Selecting the adequate regional boundary is a trade-off between representativeness and the problem of modelling cross boundary flows.
- Time: Generally temporal granularity in attributional LCA is one year. However also for average electricity mixes, there can be significant differences from one year to another. To overcome this obstacle, some studies average the emissions over a longer time span.
- Future scenarios' definition and stylised states adoption. (A 'stylised state' is denoted an extreme state (e.g. a state where all electricity and heat is produced from coal) that is unlikely to materialise but that could illustrate important technology differences in a clear way [87]).

Data Quality

Data on electricity mix are provided from DSO and TSO, national and international agencies (EPA, German federal environmental agency, IEA) or are derived from databases. The former generally provide more updated data (usually from the previous year) than the latter; also, studies gathering data on electricity from energy operators are generally aware of the relevance of these data in the overall results.

When relying on data from databases, studies do not provide the emission intensity of the selected mix. Obtaining it a posteriori is either time demanding or impossible because database versions are not always made explicit [14] and the selected dataset is rarely expounded (an exception is the study by Helmers et al. [16]).

Ignoring the database version and the dataset selected for the electricity mix, besides contravening the basic principle of reproducibility of LCA, makes it difficult to understand the results and the role played by electricity. Selecting one dataset or another can lead to relevant differences. In Figure 4, the g CO_2-eq/kWh are reported according to different datasets and versions of EcoInvent for the main countries analysed in the literature regarding EVs. This is to show how much datasets from the same database can differ. EcoInvent database has been selected for exemplification sake because of its widespread use in LCA of EVs.

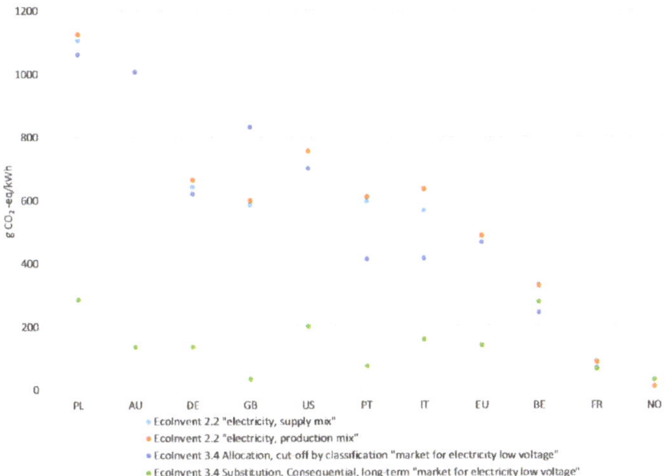

Figure 4. Climate change of 1 kWh of different electricity mixes provided at EcoInvent database. Characterisation model: IPCC 2007; Characterisation factor: GWP 100a. Time period: 2004 for EcoInvent 2.2 series, 2014 for EcoInvent 3.4 series.

When relying on databases, studies overlook the quality of the data and their time period. Seven of the revised papers adopt EcoInvent as a background database; and among those, five rely on the version 2.2. Data on electricity supply mix from EcoInvent v.2.2 were updated to 2004, meaning that some papers published in 2017 still rely on electricity mix data from that year [16,18].

Geographical Boundary

Most of the studies use mixes which have been aggregated at national level or at regional and subcontinental level, trying to follow the electricity infrastructure and trade (PJM regions in the US, ENTSO-E in Europe, NORDEL in Scandinavia, etc.). The definition of the regional boundary has relevant influence on the final result: Tamayao et al. found the use of state boundaries versus NERC region boundaries leads to estimates that differ by as much as 120% for the same location (using average emission factors) [46]. Choosing the adequate regions for the accounting of electricity emissions is a trade-off between including spatial heterogeneity and modelling cross-boundary electricity flows. This is naturally linked to the problem of defining the consumption mix of a country, rather than using its production mix, as highlighted in [37,88].

Production and Supply

In EcoInvent, the supply is created as the sum of the production and the import, which is modelled as a percentage of the production of the importing country. However, this simplification can represent a problem in countries where the trades are a significant percentage of the domestic demand, such as Switzerland, where the traded electricity volume is about 85% of the domestic demand [89].

Modelling electricity trades is complicated: A net importer could export part of the electricity, and since it is unknown which electrons have been produced by which plant, this generates a recursive problem.

Itten et al. [89] present a model to deal with this aspect: The electricity mix of the domestic supply is modelled according to the integration of the electricity declarations of all electric utilities in a country. The declaration includes a differentiation according to technology and whether or not the electricity is produced domestically or is imported. It usually includes a share of "unknown" electricity, which in that study is represented by the ENTSO-E electricity mix.

Temporal Boundary

Attributional studies rarely present the time horizon of their analysis (see Figure 5). The lack of time horizon in the scope definition has been already highlighted in Nordelöf et al. [2]. It makes it difficult to determine the temporal validity of both results and conclusions and, as far as electricity generation is concerned, also to determine whether the dataset used is suitable to the goal and scope.

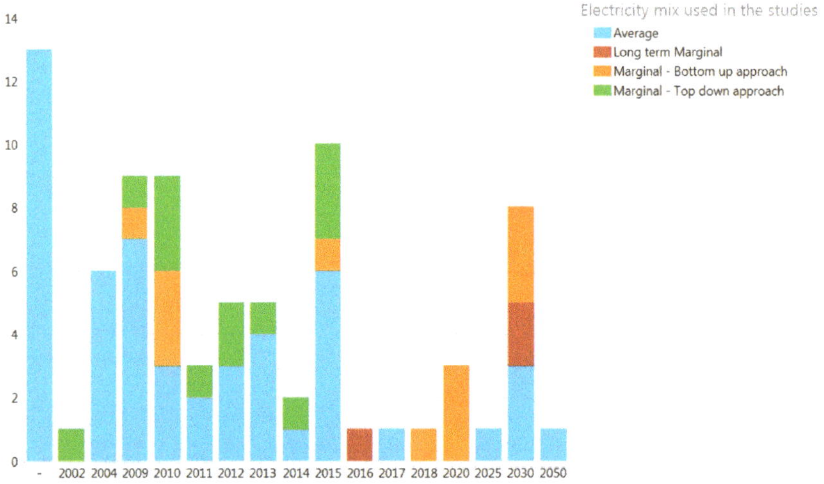

Figure 5. Number of analyses sorted by time horizon. Some studies do not make explicit the time horizon of their study. In this case the year of the electricity dataset used in the analysis has been considered, if available; otherwise they are accounted in the first column. Note that each study can have more than one time-horizon in its analysis.

A relevant exception is represented by the article by Bauer et al. [23]. In their parametric study on environmental performance of current and future mid-size passenger vehicles, every input varies with time. Vehicle mass, aerodynamic drag, tire rolling, performances of powertrain components are expected to change due to increasing know how and to compelling emission standards. However, despite the clear time scope (2012 for current fleet), the electricity mix used—from EcoInvent 2.2—is still based on data from 2004.

Despite the temporal granularity discussed in the paragraph "Short Term Marginal Mix—Calculation Method" (in Section 3.2.1), even temporal aggregated data can present some inconvenient. The composition of the country mix can differ a lot from one year to another, especially when a relevant share of the country generation capacity relies on renewable energy sources (RES).

Freire and Marques [27] detailed how GHG intensity of Portuguese mix reduced from 560 to 390 g CO_2-eq/kWh from 2009 to 2010, due to unusual meteorological condition in 2010: 1374 mm of rain compared to 950 mm of the previous year [90].

Roux et al. [59] reported that GHG of French mix varied from one year to another, due to economic and meteorological hazards, thus suggesting the need to define a reference year to mitigate these variations and to convey more representative average impacts. They also highlight that using EcoInvent v3.1 data for low voltage supply mix in order to evaluate the impact of the average mix for year 2014 leads to an overestimation of the carbon footprint by 25%.

Considering that the EVs represent an emerging technology, it is quite exemplar that no ALCA studies analyse future energy scenarios, while they are rather linked to a blurred present (not defined in the time scope) and rely on outdated data. Tagliaferri et al. [45] consider also a future electricity mix set in 2050 but the results for this time horizon are not made explicit, along with the mix composition.

It is instead common the use of scenarios where the electricity is assumed to be supplied entirely from a single technology, or a bunch of homogeneous technologies (e.g., all RES, all fossil fuels) also known as stylised states according to the definition by [87]. These technologies are generally at the extremes of the polluting bar, in a sort of stress test of the potential benefits (or disadvantages) of implementing such a technology. Sixteen among the review papers investigate how EVs perform when charged on a specific technology or when are charged solely by certified RES energy (see Figure 6 and Table 3).

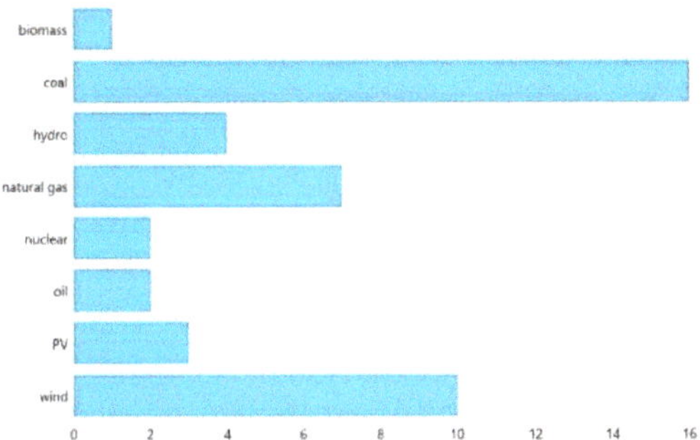

Figure 6. Number of stylised states presented in the literature sorted by energy source.

Table 3. Energy GHG intensity and GHG of BEV in stylised states across the literature.

	Stylised State	g CO$_2$-eq/kWh	g CO$_2$-eq/km
Bartolozzi et al. [14]	Biomass	-	110.35
Nordelöf et al., 2014 [2]	CNG	642	93.5 (WtW)
Van Mierlo et al., 2017 [40]	CNG	-	93.5 (WtW)
Giordano et al., 2017 [15]	Coal	1180	482
Nordelöf et al., 2014 [2]	Coal	1080	157 (WtW)
Stephan and Sullivan 2008 [39]	Coal	954	274 (WtW, PHEV)
Van Mierlo et al., 2017 [40]	Coal	-	157 (WtW)
Bauer et al., 2015 [23]	Coal (average efficiency in 2012)	-	371
Bauer et al., 2015 [23]	Coal (average efficiency in 2030)	-	308
Freire and Marques 2012 [27]	Coal (desulfurisation and denitrification	1050	225 (Compact BEV) 160 (Subcompact BEV)
Hawkins et al., 2012 [1]	Coal (EcoInvent Dataset)	1260	231
Hawkins et al., 2012 [1]	Coal IGCC	936	185
Hawkins et al., 2012 [1]	Hydro	6.12	48
Stephan and Sullivan 2008 [39]	Hydro	0	0 (WtW, PHEV)
Bauer et al., 2015 [23]	Hydro (average efficiency in 2012)	-	66.9
Bauer et al., 2015 [23]	Hydro (average efficiency in 2030)	-	55.8
Dallinger et al. [48]	Marginal mix in a scenario with dedicated RES for additional EV loads	247.26 (last trip charging) 245.42 (DSM charging)	23.4 (WtW) 10.7 (WtW)
Stephan and Sullivan 2008 [39]	Natural gas	-	184 (WtW, PHEV)
Bauer et al., 2015 [23]	Natural gas (average efficiency in 2012)	-	186
Bauer et al., 2015 [23]	Natural gas (average efficiency in 2030)	-	155
Hawkins et al., 2012 [1]	NGCC	504	120
Van Vliet 2011 [49]	NGCC	430	55 (WtW)
Bauer et al., 2015 [23]	Nuclear (average efficiency in 2012)	-	68
Bauer et al., 2015 [23]	Nuclear (average efficiency in 2030)	-	56.7
Nordelöf et al., 2014 [2]	Oil	885	128.5 (WtW)
Stephan and Sullivan 2008 [39]	Oil	892.8	262 (WtW, PHEV)
Oliveira et al., 2015 [57]	PV	89	-
Bauer et al., 2015 [23]	PV (average efficiency in 2012)	-	87.2
Bauer et al., 2015 [23]	PV (average efficiency in 2030)	-	72.6
Bartolozzi et al. [14]	Wind	-	121.2
Freire and Marques 2012 [27]	Wind	23	50 (Compact BEV) 40 (Subcompact BEV)

Table 3. *Cont.*

	Stylised State	g CO_2-eq/kWh	g CO_2-eq/km
Nordelöf et al., 2014 [2]	Wind	11	1.50 (WtW)
Oliveira et al., 2015 [57]	Wind	11.2	-
Van Mierlo et al., 2017 [40]	Wind	-	1.50 (WtW)
Crossin and Doherty 2016 [25]	Wind	24	76
Bauer et al., 2015 [23]	Wind (average efficiency in 2012)	-	70.4
Bauer et al., 2015 [23]	Wind (average efficiency in 2030)	-	58.7

However, even for a single fuel, a huge variability in GHG emission is noticed, due to technology differences (CC plants, CHP plants, etc.). Soimakallio et al. [78] present how the impact of a single technology varies along with the allocation method when energy production has to solve problem of multifunctionality. For the same coal fired CHP plant, the results obtained vary from 400 g CO_2-eq/kWh if an allocation on the base of energy content is applied to the two outputs (heat and electricity), to 1200 g CO_2-eq/kWh if all the emission are allocated on the production of electricity.

3.3. Data Quality

Data quality depends on the type of source providing the country's generation data and its elaboration (in case of marginal mixes). Articles rely on several sources. In this article these sources have been grouped into fewer categories on the basis of common characteristics of the data provided ((1) LCA databases; (2) Grid databases; (3) the existing literature; and (4) electricity mix forecasts). The links between the articles reviewed and the source of the electricity mix are presented in Figure 7 Studies which employ more than one database (to cover different time horizons, different areas or different modelling approaches, etc.) are split and linked to both.

3.3.1. LCA Databases

LCA databases allow for transparent and replicable results. They have accessible and verifiable data and simplify calculations, since most of their datasets are supplemented with characterisation methods. They also include impacts form upstream processes, contrary to other sources which only provide combustion emissions.

The most used database is EcoInvent, with 8 studies relying on it for the selection of the electricity mix. Other studies depend on this database for other data, but only those using the EcoInvent mixes are represented in Figure 7. Bohnes et al. [20] and Giordano et al. [15] source the generation datasets from EcoInvent, but their mixes are updated with data from IEA and Entso-E, respectively.

EcoInvent is a non-profit association founded by institutes of the ETH Domain and the Swiss Federal Offices, whose data are updated every time a new version is released. Except from one study, whose version is not specified, all the articles published so far rely on EcoInvent version 2.2.

Electricity country mix from this version dates to 2004, for all the countries considered in the literature. It is noteworthy that studies published in 2017 still rely on data from 2004.

Gabi Database is created by Thinkstep and updated every year [91]. In the literature, two studies employ this database: Deng et al. [55] do not specify the version, and Tagliaferri et al. [45] use the version from 2015 to assess Italian electricity mix.

GREET is a life cycle assessment model, initially born as an Excel datasheet, now available also in the more appealing graphic user interface GREET.NET. It consists of two sub-models, the GREET fuel cycle model and GREET vehicle cycle. For the electricity generation it offers complete life cycle inventory of different production pathways. It has been widely used in the literature, especially for the evaluation of upstream emission of fossil-fuel generated energy [46,47].

3.3.2. Grid Databases

National grid databases offer updated information on electricity generation. However, they are not intended primarily for LCA applications. They generally provide up-to-date and detailed information on electricity generation throughout the region they cover and updated and transparent stack emissions. Furthermore, depending on the dataset, they can provide also aggregated indicators (CO_2-eq, etc.), but they do not provide upstream information about the electricity produced.

Articles that rely on this kind of database, thus, either integrate the analysis with other databases providing upstream emissions for each generation technology (e.g., Tamayao et al. [46] rely on eGRID data (see later on) and integrate them with GREET model [92]), or neglect upstream emissions, focusing only on combustion emissions.

eGRID is the most used database by articles assessing EVs introduction in the American grid. For every power plant in the United States, eGRID supplies detailed emissions profiles and Generation resource mix.

It also provides updated average generation emissions [35,37,46], but due to its level of detail on power plants information it is the source for studies that require input for their model to detect the marginal technologies affected by BEVs introduction [33,34].

EPA's Continuous Emissions Monitoring Systems (CEMS) dataset provides hourly gross generation load, consumed fuel, and CO_2 emissions from direct combustion for all grid-connected electricity generating units with a capacity of 50 MWh or more in the continental United States from 1997 through the present. For this reason, it has been widely used by American studies applying regression models to identify the short term marginal mixes.

Other studies look at national TSO and DSO in order to gather the most updated data available on national electricity mix (Giordano et al., 2017 [15] use data from Entso-e for European countries, Messagie from ELIA for Belgium [41], Freire and Marques from REN for Portugal [27], etc.) or to obtain non aggregated data (usually 1 h time resolution) to elaborate linear regression in order to determine marginal emission factors [9,25,28].

Faria et al. [51] draw the EU electricity mix for 2009 from European Environment Agency (EEA) and the upstream emissions are filled from the literature [93–96]

3.3.3. Literature Sources

Some studies rely on electricity mixes published in other works, particularly for the marginal mixes, due to the difficulties encountered in determining them and to the large amount of data required. Aside from the use of not updated data due to articles publication time, the issue of scalability of the results from one application to another is questioned.

Onat et al. [37] and Thomas [47] consider the mix determined by Hadley and Tsvetkova [38] for the introduction of PHEV in the 2020 energy system, under certain penetration rates and charging conditions, and apply it to their own scenarios that include BEVs introduction in the transport system, under unspecified penetration rates.

3.3.4. Forecasts

Studies investigating future scenarios are subject to higher uncertainties linked to future generation mixes as well as future plant/technology emissions and efficiency. Energy generation forecast from Energy Information Administration [97], and International Energy Agency [98] has been used in the literature.

Some databases are available to characterise power plants in future scenarios. One of these is the NEEDS Life Cycle Inventory Database, developed by the homonym EU project [99].

It is difficult to obtain the value of the energy emission intensity of the mix used in the studies. Even when studies focus on the role of electricity production for charging the vehicle, it is not always clear whether this represents the stack emission or also includes upstream emissions. As a rule of thumb, studies relying on grid databases only consider stack emission unless otherwise explicitly specified.

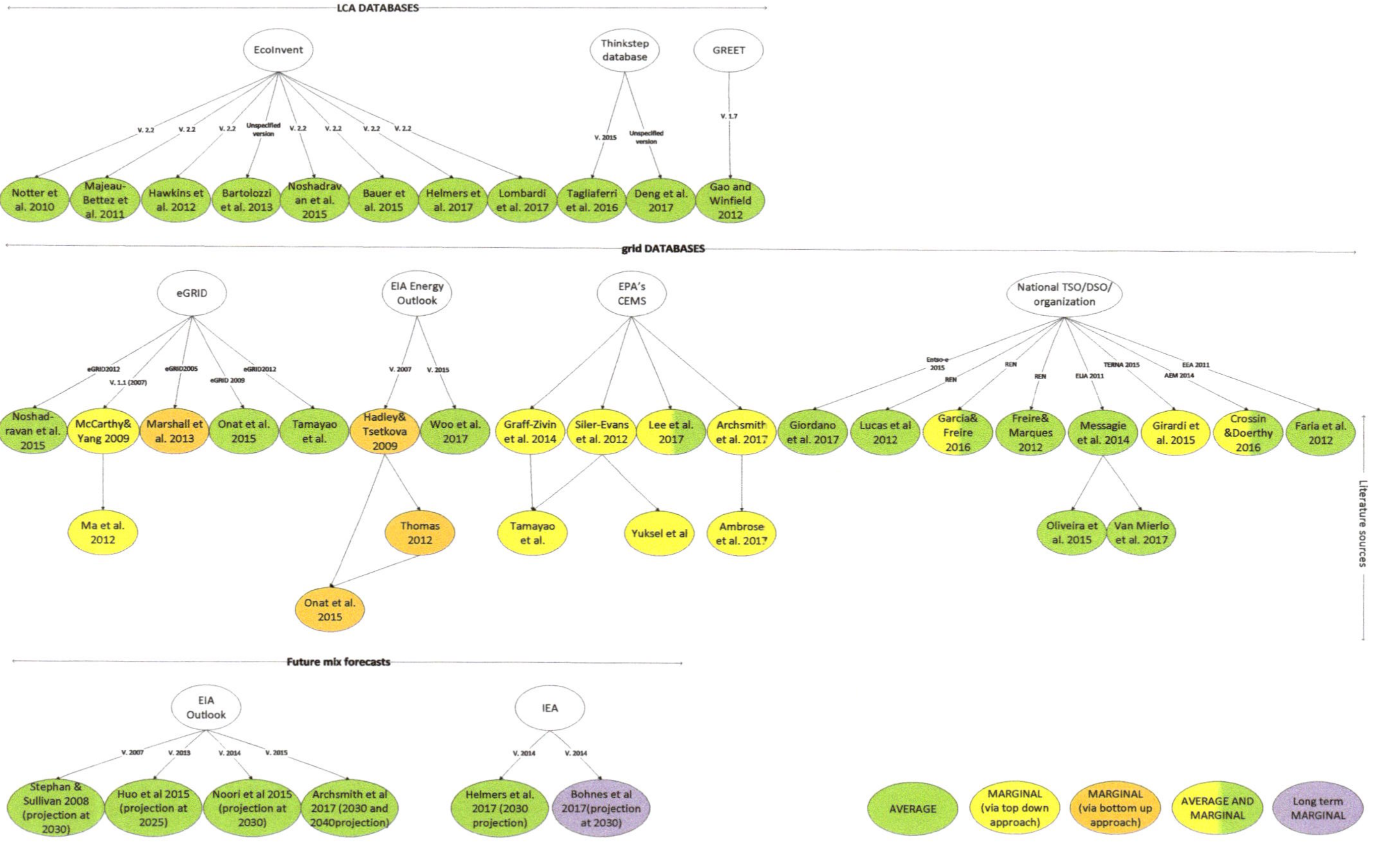

Figure 7. Sources of electricity mix's data used in the reviewed papers.

3.4. Results

3.4.1. Quantitative Results

Current quantitative results have a wider variability than older ones: Frischknecht and Flury were the first to pinpoint this aspect; they found that the results of the impact of electric vehicles—expressed in the most common indicator g CO_2-eq/km—ranged from 95 to 240 g CO_2-eq/km [100].

In the present review results are even more spread. By excluding results obtained from stylised states, whose role is exactly to denote an extreme state rather than a situation that is likely to happen, the aforementioned indicator spans from 27.5 g CO_2-eq/km, obtained by Van Mierlo et al. [40] presenting the WtW results of an EV in the Belgium environment (data from 2011), to 326 g CO_2-eq/km, obtained by Ma et al. [32] when assessing EV introduction in the UK market in 2015, using short term marginal electricity mix (see Figure 8).

In Figure 8, the GHG of electric vehicles found in the literature are listed. Since results in the literature are expressed according to different functional units, in this review they have been transformed in the most common functional unit (g CO_2-eq/kWh) when enough information allows for the conversion. If more than a single value is provided, bars represent the average value, while the error bars delimit the upper and lower value (see Figure S1 for the value with the associated electricity mix carbon intensity). For the numeric values see Table 1.

The reason behind this spread is due to wider scopes and more detailed LCI and does not necessarily entail weaker studies. However, such is the variety of study types and scopes that the use of a framework and a rigid application of the ISO guidelines should be applied more than ever. A tentative framework for the main applications found in literature can be find in Table S2.

Influence of Electricity Mixes in the Results

Within this plethora of results, the focus of this literature review was to understand the role of a specific data, the electricity mix, because its influence has been highlighted by many authors [101] but never quantified. Among studies that present a level of detail fair enough to obtain homogeneous results (expressed in the most diffuse functional unit CO_2-eq/km) and CO_2-eq emissions for kWh, correlation has been investigated. These articles are represented in Figure S2 along with the aforementioned values. Some studies present results for more than one time-horizon and for more than one country, others present results according to different stylised states. Thus, in Figure S2 studies are labelled with their peculiarity: Country mix, time-horizon (or the year of the data of the electricity mix) or stylised state, depending whether the information was available.

In Figure 9 a linear regression between g CO_2-eq/km and g CO_2-eq/kWh shows the correlation between the selection of the electricity mix and final result. The pool includes all the data available in the literature, and no normalisation has been applied to them. This means that in the graph are reported values from WtW analysis, and complete life cycle assessments, with different system boundaries and modelling approaches, with different vehicle lifetimes and energy consumptions per km (PHEV results have been excluded from the regression, since their use phase relies also on other fuels in addition to electricity).

The purpose of this data collection is not to fully harmonise the studies and create comparable results, but rather aims at showing how influential the electricity mix selection is, despite the extent of the scopes available in the literature.

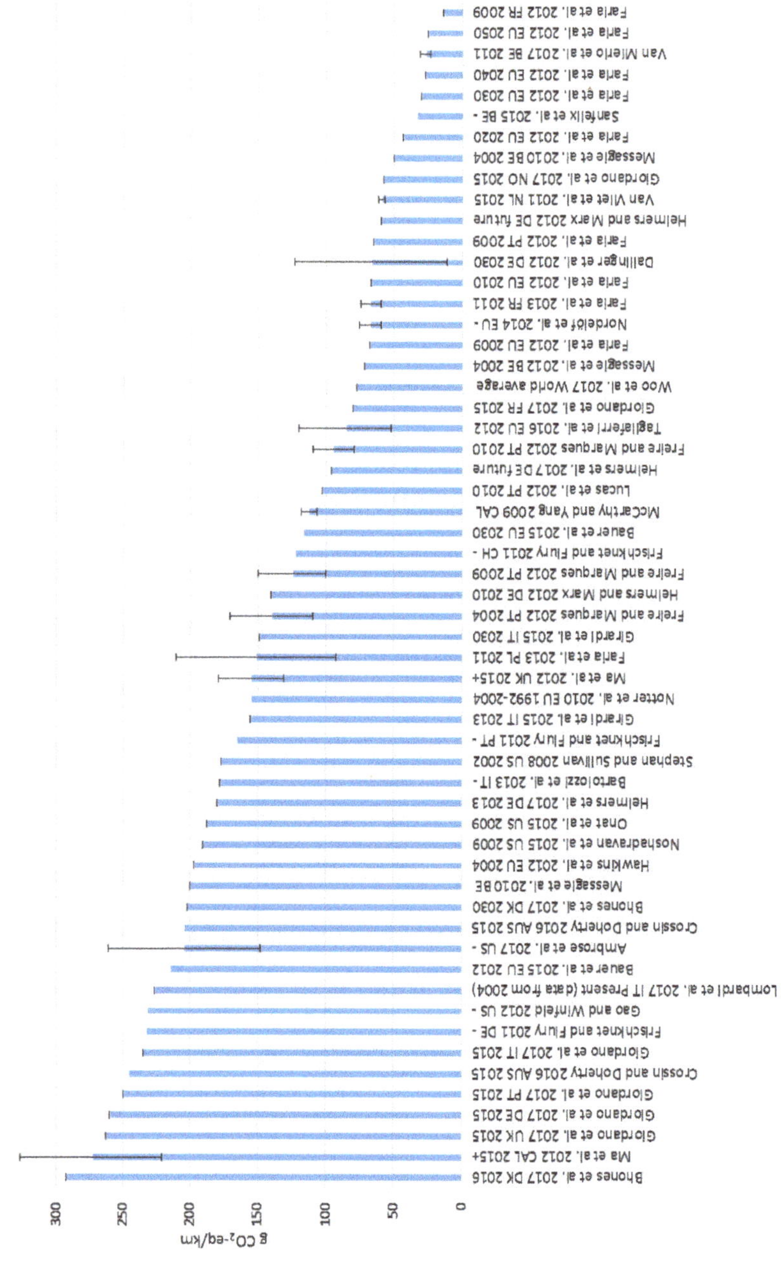

Figure 8. Literature results normalised at the most common indicator "g CO_2-eq/km" (blue bars). If a study presents more than one value, the average value has been reported (blue bar) with the associated range.

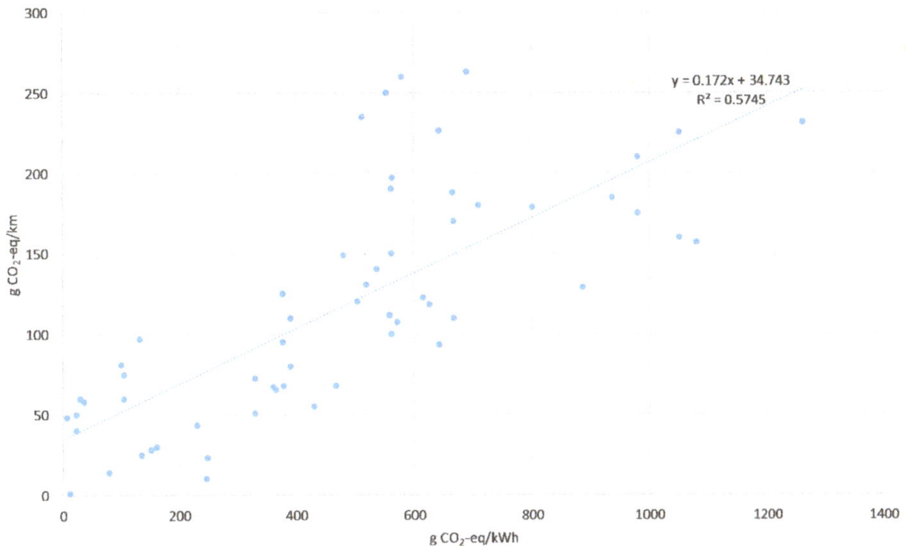

Figure 9. Linear regression between EVs climate change and carbon intensity of the electricity mix used to recharge them.

The coefficient of determination R^2 is 0.57. Considering the vast number of variables taking place in the LCAs and the heterogeneity of the scopes, it is noticeable that the carbon intensity of the electricity mix alone can explain almost the 60% of the variability of the results available in the literature.

In Figure 10 the regression has been applied to more homogeneous groups. In the first place, only passenger cars have been considered, excluding light duty vehicles. Then results have been divided into WtW analyses and complete LCA and two separate linear regressions are performed.

The coefficients of determination R^2 increase significantly: for WtW analyses it is 0.91; for complete LCA it is 0.73. What is worth noticing is that the regressions present almost the same slopes. As expected the intercept of the WtW analyses is almost null.

Factors Influencing the Mix

The carbon intensity of the electricity mixes used in the literature to assess EVs varies with country, time horizon, modelling choice, scenarios definition and data quality.

In Figure 11 the carbon intensities used in the literature are reported, in order to highlight possible correlation with selected parameters (country, time horizon, modelling choice, scenarios definition, and data quality).

Values find in the literature are represented in the graph sorted by region; values with homogenous characteristic are marked with the same label. Average and marginal electricity mixes are identified with different labels; electricity mixes of projected energy system are identified with another label. For clarity sake the year of the mix is not presented in the graph and all the electricity mixes identified for future scenarios are grouped under the label "future electricity mix". For detailed information regarding the time horizon of the electricity mixes refer to Table 1. Since EcoInvent is the most used database in the literature, also the value from its country mix datasets are reported for version 2.2 and for the new release 3.4 (datasets name "market for electricity low voltage"). For the equivalence between the datasets across the two EcoInvent versions see [102].

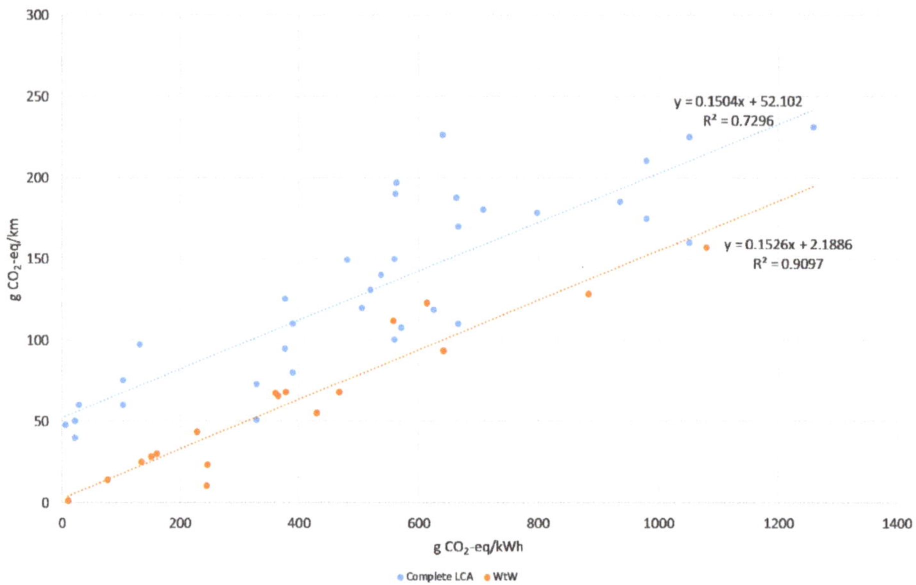

Figure 10. Linear regression between passenger-EVs climate change and GHG emission intensity of the electricity mix used to recharge them, divided between WtW analysis and complete LCA studies.

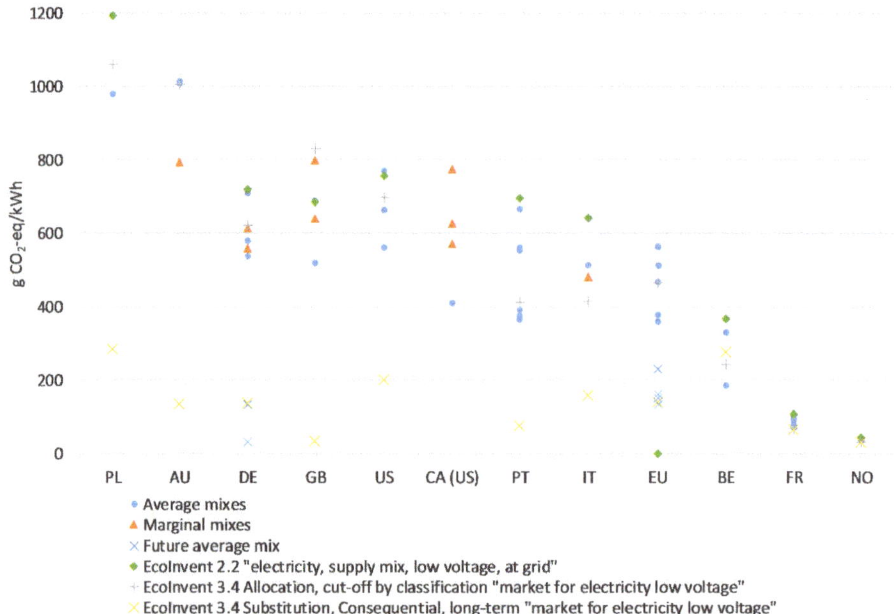

Figure 11. Spread of the carbon intensity of electricity mixes used in the articles reviewed and in the main database.

The first evidence is the results' spread, from country to country and with time horizon (in order to improve the graph's clarity data are distinguished only between present and future scenarios).

Figures range from the very high values of the country relying on coal, such as Poland, Australia and the US, to the very low ones of France and Norway. These countries at the extreme of the polluting bar present the less dispersed data, mainly because they rely mainly on a single source (coal for Poland and Australia, nuclear for France, hydro for Norway).

Short term marginal mixes are generally higher than the average, with the relevant exceptions of Australia (where the average mix is dominated by coal, while the cleaner but more expensive natural gas is mainly on the margin) and Italy.

It has to be noted that average and marginal results cannot be directly compared, because they originate from different studies relying on different sources and years; the representation is more aimed at capturing trends rather than allowing for a fully harmonised comparison.

As mentioned before, only one study uses the long term marginal mix. To present the effect that this selection would have on the studies, the EcoInvent 3.4 Substitution, consequential, long term "market for electricity low voltage" for the various country have been reported in Figure 11 (time validity of EI 3.4 consequential 2016–2030).

It emerges that in the long run variability among countries tends to lower.

3.4.2. Recommendations for Policy Makers in the Literature

For a practitioner or a policy maker who approaches literature, the indication provided can be quite contradictory; among studies comparing EVs and ICEVs we found:

- 7 studies stating that EVs are not decreasing GHG (6 using marginal mixes);
- 4 studies being cautious on the adoption of EVs (all using marginal mixes);
- 13 studies presenting EVs as an efficient decarbonising technology (5 using marginal mixes).

These results could be not regarded as conflicting if their goal was more detailed and explained the audience how they could be used.

The studies discouraging EVs focus on the short term introduction of EVs (both in present and future scenarios) where the energy system is not able to adjust to the increased demand. Aside some doubts regarding the methodology with which the short term marginal electricity mix of each system is calculated, we think that this option is not suitable to inform policy makers on wide-ranging/far-reaching policies, such as the paradigm shift in transport sector, while it can have applications in optimising the recharging time and balancing the grid load, since EVs are a more flexible load than others.

4. Conclusions and Recommendations

The reason for this review stemmed from the nebulous indications that a practitioner approaching the subject, or a decision maker, can encounter reading available literature results. To understand these inconsistencies, all the steps of a life cycle analysis have been analysed, as follows (for a synoptic view see Table S1).

Goal and modelling choice: The first hurdle is related to the goal definition, which is often missing or incomplete; the subsequent scope definition and in particular the modelling choice are not justified or inconsistent.

Thus, the selection of supplier/technology (the most influential of which is the electricity mix) is not straightforward, and sometimes, in a paradoxical inversion, it becomes the parameter that defines the modelling choice.

LCI and data quality: LCI is often inexplicit, thus making the identification of relevant information quite difficult, contravening the dogma of reproducibility of any LCA study.

The datasets which most of the background processes are based on are outdated. Considering the influence of the electricity mix, the use of 15-year-old data is found to be an unwise practice.

Time inconsistency: Despite the willingness of informing policy makers and thus influencing effect/developments in the long run, most of the scopes are focused on the here and now. Time and fleet prospective are investigated by only a handful of studies, while others apply the results from vehicle technology comparison and extend it to a wider level using a simple scaling factor, thus neglecting time and scale dependent aspects that cannot be detected from a single product analysis.

Most of the studies do not even include any penetration scenario in the analysis, instead providing policy makers with a simple vehicle to vehicle comparison. Studies are aimed at informing policy makers but most of the analyses lack a political dimension: No clear time frame, no clear and reliable future scenarios, inconsistency between variables in the scenarios.

In the issue of policy information, the selection of short term marginal electricity mixes has to be taken into account. Even though these mixes are useful for the modelling of short term effects of a rapid, albeit unlikely, introduction of EVs, in the authors' opinion they are not the correct instruments to inform policy makers, since they only offer a partial view: Focusing on short term effect is no more than a form of burden shifting in time.

Literature results: Range of quantitative results is widening compared to the past. GHG emissions of electric vehicles range from 27.5 g CO_2-eq/km [40] to 326 g CO_2-eq/km [32] in the literature. This is due also to the wide range of scopes investigated by scientific papers. The selected papers cover a wide range of scopes: Various vehicle typologies are investigated using different system boundaries and different modelling approaches at various scales. Even with so many variables involved, the selection of the electricity mix has been found to be a key issue. A linear regression between g CO_2-eq/km and the carbon intensity of the electricity mix considered in the analyses (g CO_2-eq/kWh). The GHG of the electricity mix was found to explain almost 70% of the variability in the LCA results. Even if the electricity mix has always been seen as an influencing factor, this represent the first attempt to quantify its role.

Recommendations from the literature: We found some articles warmly promoting the introduction of EVs as a way to reduce GHG in transport sector, some highly discouraging it and others recommending caution in their adoption. These results could be considered not conflicting if their goal and scope were more detailed and if they explained to the audience how the results should be used.

The studies discouraging EVs are works focusing on the short term introduction of EVs (both in present and future scenarios), where the energy system is not able to adjust to the increased demand of electricity. Besides some doubts regarding the methodology which the short term marginal electricity mix of each system has been calculated with, we think this is not the best option to inform policy makers on wide-ranging policies such as the paradigm shift in transport sector. However, it can have applications when it comes to optimising recharging time, since EVs are loads more flexible than others.

On the other hand, studies using average mix often suffer from the use of old data and do not present a wide-ranging situation either, focusing on the here and now.

In conclusion, this review confirms the weak trends pinpointed by previous reviews, which have not changed in the last years—missing goal definitions, and weak LCIs. Lack of consensus on LCI data selection and missing clear goal statements are still the key factors to explain the difficult path to inform policy makers with robust and clear results.

Another element this review has found missing is the link between the two. A consensus on the framework that would inform on the correct modelling choices, and from the defined goal would orient the practitioner to the right selection of the inventory data, in particular the most relevant one in this field—the electricity mix.

Supplementary Materials: The following are available online at http://www.mdpi.com/2076-3417/8/8/1384/s1, Figure S1: The literature results normalised at the most common indicator "g CO_2-eq/km" (blue bars, left axis). If a study presents more than one value, the average value has been reported with the associated range. For studies that explicit electricity mix intensity, they are represented with a red dot (right axis); Figure S2: Studies reporting g CO_2-eq/km and g CO_2-eq/kWh in descending order of electricity carbon intensity; Table S1: Literature results and recommendations for practitioners and policy makers; Table S2. Framework for the main applications found in the literature.

Author Contributions: B.M. and M.M. conceived the paper. B.M. carried out the papers selection and the extensive search of data in the literature and wrote the paper. M.M. helped finding the slant in the literature data and orienting the discussion and revised the first draft. G.D. provided valuable suggestions on how to elaborate data from the literature and proofread and edited the manuscript. J.v.M. provided supervision guidance to this research.

Funding: This research received no external funding.

Acknowledgments: We acknowledge Flanders Make for the support to our research group.

Conflicts of Interest: The authors declare no conflict of interest.

References

1. Hawkins, T.R.; Gausen, O.M.; Strømman, A.H. Environmental impacts of hybrid and electric vehicles-a review. *Int. J. Life Cycle Assess.* **2012**, *17*, 997–1014. [CrossRef]

2. Nordelöf, A.; Messagie, M.; Tillman, A.; Söderman, M.L.; Van Mierlo, J. Environmental impacts of hybrid, plug-in hybrid, and battery electric vehicles—What can we learn from life cycle assessment? *Int. J. Life Cycle Assess.* **2014**, *19*, 1866–1890. [CrossRef]

3. Graff Zivin, J.; Kotchen, M.; Mansur, E. Spatial and temporal heterogeneity of marginal emissions: Implications for electric cars and other electricity-shifting policies. *J. Econ. Behav. Organ.* **2014**, *107*, 248–268. [CrossRef]

4. Elgowainy, A.; Burnham, A.; Wang, M.; Molburg, J.; Rousseau, A. Well-To-Wheels Energy Use and Greenhouse Gas Emissions of Plug-in Hybrid Electric Vehicles. *SAE Int. J. Fuels Lubr.* **2009**, *2*, 627–644. [CrossRef]

5. Gaines, L.; Sullivan, J.; Burnham, A.; Belharouak, I. Life-cycle analysis of production and recycling of lithium ion batteries. *Transp. Res. Rec.* **2011**, *2252*, 57–65. [CrossRef]

6. Notter, D.; Gauch, M.; Widmer, R.; Wäger, P.; Stamp, A.; Zah, R.; Althaus, H. Contribution of Li-Ion Batteries to the Environmental Impact of Electric Vehicles. *Environ. Sci. Technol.* **2010**, *44*, 6550–6556. [CrossRef] [PubMed]

7. Majeau-Bettez, G.; Hawkins, T.R.; Strømman, A.H. Life cycle environmental assessment of lithium-ion and nickel metal hydride batteries for plug-in hybrid and battery electric vehicles. *Environ. Sci. Technol.* **2011**, *45*, 4548–4554. [CrossRef] [PubMed]

8. Hawkins, T.; Singh, B.; Majeau Bettez, G.; Stromman, A.; Strømman, A. Comparative Environmental Life Cycle Assessment of Conventional and Electric Vehicles. *J. Ind. Ecol.* **2013**, *17*, 53–64. [CrossRef]

9. Girardi, P.; Gargiulo, A.; Brambilla, P. A comparative LCA of an electric vehicle and an internal combustion engine vehicle using the appropriate power mix: The Italian case study. *Int. J. Life Cycle Assess.* **2015**, *20*, 1127–1142. [CrossRef]

10. Lee, D.; Thomas, V. Parametric modeling approach for economic and environmental life cycle assessment of medium-duty truck electrification. *J. Clean. Prod.* **2017**, *142*, 3300–3321. [CrossRef]

11. Noori, M.; Gardner, S.; Tatari, O. Electric vehicle cost, emissions, and water footprint in the United States: Development of a regional optimization model. *Energy* **2015**, *89*, 610–625. [CrossRef]

12. Garcia, J.; Millet, D.; Tonnelier, P.; Richet, S.; Chenouard, R. A novel approach for global environmental performance evaluation of electric batteries for hybrid vehicles. *J. Clean. Prod.* **2017**, *156*, 406–417. [CrossRef]

13. Huo, H.; Cai, H.; Zhang, Q.; Liu, F.; He, K. Life-cycle assessment of greenhouse gas and air emissions of electric vehicles: A comparison between China and the U.S. *Atmos. Environ.* **2015**, *108*, 107–116. [CrossRef]

14. Bartolozzi, I.; Rizzi, F.; Frey, M. Comparison between hydrogen and electric vehicles by life cycle assessment: A case study in Tuscany, Italy. *Appl. Energy* **2013**, *101*, 103–111. [CrossRef]

15. Giordano, A.; Fischbeck, P.; Matthews, H.S. Environmental and economic comparison of diesel and battery electric delivery vans to inform city logistics fleet replacement strategies. *Transp. Res. Part D* **2017**. [CrossRef]

16. Helmers, E.; Dietz, J.; Hartard, S. Electric car life cycle assessment based on real-world mileage and the electric conversion scenario. *Int. J. Life Cycle Assess.* **2017**, *22*, 15–30. [CrossRef]

17. Helmers, E.; Marx, P. Electric cars: Technical characteristics and environmental impacts. *Environ. Sci. Eur.* **2012**, *24*, 14. [CrossRef]

18. Lombardi, L.; Tribioli, L.; Cozzolino, R.; Bella, G. Comparative environmental assessment of conventional, electric, hybrid, and fuel cell powertrains based on LCA. *Int. J. Life Cycle Assess.* **2017**, *22*, 1989–2006. [CrossRef]

19. European Commission, Joint Research Centre, Institute for Environment and Sustainability (Ed.) *ILCD Handbook-General Guide for Life Cycle Assessment-Detailed Guidance*; Publications Office of the European Union: Luxembourg, 2010; 417p. [CrossRef]

20. Bohnes, F.; Gregg, J.; Laurent, A. Environmental Impacts of Future Urban Deployment of Electric Vehicles: Assessment Framework and Case Study of Copenhagen for 2016–2030. *Environ. Sci. Technol.* **2017**, *51*, 13995–14005. [CrossRef] [PubMed]

21. Garcia, R.; Gregory, J.; Freire, F. Dynamic fleet-based life-cycle greenhouse gas assessment of the introduction of electric vehicles in the Portuguese light-duty fleet. *Int. J. Life Cycle Assess.* **2015**, *20*, 1287–1299. [CrossRef]

22. Archsmith, J.; Kendall, A.; Rapson, D. From Cradle to Junkyard: Assessing the Life Cycle Greenhouse Gas Benefits of Electric Vehicles. *Res. Transp. Econ.* **2015**, *52*, 72–90. [CrossRef]

23. Bauer, C.; Hofer, J.; Althaus, H.; Del Duce, A.; Simons, A. The environmental performance of current and future passenger vehicles: Life cycle assessment based on a novel scenario analysis framework. *Appl. Energy* **2015**, *157*, 871–883. [CrossRef]

24. Kannan, R.; Turton, H. Cost of ad-hoc nuclear policy uncertainties in the evolution of the Swiss electricity system. *Energy Policy* **2012**, *50*, 391–406. [CrossRef]

25. Crossin, E.; Doherty, P.J.B. The effect of charging time on the comparative environmental performance of different vehicle types. *Appl. Energy* **2016**, *179*, 716–726. [CrossRef]

26. Faria, R.; Marques, P.; Moura, P.; Freire, F.; Delgado, J.; de Almeida, A. Impact of the electricity mix and use profile in the life-cycle assessment of electric vehicles. *Renew. Sust. Energy Rev.* **2013**, *24*, 271–287. [CrossRef]

27. Freire, F.; Marques, P. Electric vehicles in Portugal: An integrated energy, greenhouse gas and cost life-cycle analysis. In Proceedings of the 2012 IEEE International Symposium on Sustainable Systems and Technology (ISSST), Boston, MA, USA, 16–18 May 2012; IEEE: Piscataway, NJ, USA, 2012.

28. Garcia, R.; Freire, F. Marginal Life-Cycle Greenhouse Gas Emissions of Electricity Generation in Portugal and Implications for Electric Vehicles. *Resources* **2016**, *5*, 41. [CrossRef]

29. Lanati, F.; Benini, M.; Gelmini, A. Impact of the penetration of electric vehicles on the Italian power system: A 2030 scenario. In Proceedings of the 2011 IEEE Power and Energy Society General Meeting, Detroit, MI, USA, 24–28 July 2011; p. 3403.

30. Nitsch, J.; Pregger, T.; Naegler, T.; Heide, D.; Luca de Tena, D.; Trieb, F.; Scholz, Y.; Nienhaus, K.; Gerhardt, N.; Sterner, M. Langfristszenarien und Strategien für den Ausbau der erneuerbaren Energien in Deutschland bei Berücksichtigung der Entwicklung in Europa und global. 2012. Available online: https://www.researchgate.net/publication/259895385 (accessed on 14 August 2018).

31. Lucas, A.; Alexandra Silva, C.; Costa Neto, R. Life cycle analysis of energy supply infrastructure for conventional and electric vehicles. *Energy Policy* **2012**, *41*, 537–547. [CrossRef]

32. Ma, H.; Balthasar, F.; Tait, N.; Riera Palou, X.; Harrison, A. A new comparison between the life cycle greenhouse gas emissions of battery electric vehicles and internal combustion vehicles. *Energy Policy* **2012**, *44*, 160–173. [CrossRef]

33. McCarthy, R.; Yang, C. Determining marginal electricity for near-term plug-in and fuel cell vehicle demands in California: Impacts on vehicle greenhouse gas emissions. *J. Power Sources* **2010**, *195*, 2099–2109. [CrossRef]

34. Marshall, B.M.; Kelly, J.C.; Lee, T.; Keoleian, G.A.; Filipi, Z. Environmental assessment of plug-in hybrid electric vehicles using naturalistic drive cycles and vehicle travel patterns: A Michigan case study. *Energy Policy* **2013**, 358–370. [CrossRef]

35. Noshadravan, A.; Cheah, L.; Roth, R.; Freire, F.; Dias, L.; Gregory, J. Stochastic comparative assessment of life-cycle greenhouse gas emissions from conventional and electric vehicles. *Int. J. Life Cycle Assess.* **2015**, *20*, 854–864. [CrossRef]

36. Messagie, M.; Coosemans, T.; Van Mierlo, J. The influence of electricity allocation rules in environmental assessments of electric vehicles. In Proceedings of the 28th International Electric Vehicle Symposium and Exhibition, Goyang, Korea, 3–6 May 2015; Korean Society of Automotive Engineers: Seoul, Korea, 2015; p. 1252.

37. Onat, N.; Kucukvar, M.; Tatari, O. Conventional, hybrid, plug-in hybrid or electric vehicles? State-based comparative carbon and energy footprint analysis in the United States. *Appl. Energy* **2015**, *150*, 36–49. [CrossRef]

38. Hadley, S.; Tsvetkova, A. Potential Impacts of Plug-in Hybrid Electric Vehicles on Regional Power Generation. *Electr. J.* **2009**, *22*, 56–68. [CrossRef]

39. Stephan, C.; Sullivan, J. Environmental and Energy Implications of Plug-In Hybrid-Electric Vehicles. *Environ. Sci. Technol.* **2008**, *42*, 1185–1190. [CrossRef] [PubMed]

40. Van Mierlo, J.; Messagie, M.; Rangaraju, S. Comparative environmental assessment of alternative fueled vehicles using a life cycle assessment. *Transp. Res. Procedia* **2017**, *25*, 3435–3445. [CrossRef]

41. Messagie, M.; Boureima, F.; Coosemans, T.; Macharis, C.; Mierlo, J. A Range-Based Vehicle Life Cycle Assessment Incorporating Variability in the Environmental Assessment of Different Vehicle Technologies and Fuels. *Energies* **2014**, *7*, 1467–1482. [CrossRef]

42. Weis, A.; Jaramillo, P.; Michalek, J. Consequential life cycle air emissions externalities for plug-in electric vehicles in the PJM interconnection. *Environ. Res. Lett.* **2016**, *11*, 024009. [CrossRef]

43. Yuksel, T.; Tamayao, M.; Hendrickson, C.; Azevedo, I.M.L.; Michalek, J. Effect of regional grid mix, driving patterns and climate on the comparative carbon footprint of gasoline and plug-in electric vehicles in the United States. *Environ. Res. Lett.* **2016**, *11*, 044007. [CrossRef]

44. Siler Evans, K.; Azevedo, I.; Morgan, M.G. Marginal Emissions Factors for the U.S. Electricity System. *Environ. Sci. Technol.* **2012**, *46*, 4742–4748. [CrossRef] [PubMed]

45. Tagliaferri, C.; Evangelisti, S.; Acconcia, F.; Domenech, T.; Ekins, P.; Barletta, D.; Lettieri, P. Life cycle assessment of future electric and hybrid vehicles: A cradle-to-grave systems engineering approach. *Chem. Eng. Res. Des.* **2016**, *112*, 298–309. [CrossRef]

46. Tamayao, M.; Michalek, J.; Hendrickson, C.; Azevedo, I.M.L. Regional Variability and Uncertainty of Electric Vehicle Life Cycle CO_2 Emissions across the United States. *Environ. Sci. Technol.* **2015**, *49*, 8844–8855. [CrossRef] [PubMed]

47. Thomas, C.E. US marginal electricity grid mixes and EV greenhouse gas emissions. *Int. J. Hydrogen Energy* **2012**, *37*, 19231–19240. [CrossRef]

48. Dallinger, D.; Wietschel, M.; Santini, D.J. Effect of demand response on the marginal electricity used by plug-in electric vehicles. *World Electr. Veh. J.* **2012**, *4*, 2766–2774. [CrossRef]

49. Van Vliet, O.; Brouwer, A.; Kuramochi, T.; van den Broek, M.; Faaij, A. Energy use, cost and CO_2 emissions of electric cars. *J. Power Sources* **2011**, *196*, 2298–2310. [CrossRef]

50. Van den Broek, M.; Veenendaal, P.; Koutstaal, P.; Turkenburg, W.; Faaij, A. Impact of international climate policies on CO_2 capture and storage deployment. *Energy Policy* **2011**, *39*, 2000–2019. [CrossRef]

51. Faria, R.; Moura, P.; Delgado, J.; de Almeida, A.T. A sustainability assessment of electric vehicles as a personal mobility system. *Energy Convers. Manag.* **2012**, *61*, 19–30. [CrossRef]

52. Woo, J.; Choi, H.; Ahn, J. Well-to-wheel analysis of greenhouse gas emissions for electric vehicles based on electricity generation mix: A global perspective. *Transp. Res. Part D* **2017**, *51*, 340–350. [CrossRef]

53. Gao, L.; Winfield, Z.C. Life cycle assessment of environmental and economic impacts of advanced vehicles. *Energies* **2012**, *5*, 605–620. [CrossRef]

54. Ambrose, H.; Kendall, A. Effects of battery chemistry and performance on the life cycle greenhouse gas intensity of electric mobility. *Transp. Res. Part D* **2016**, *47*, 182–194. [CrossRef]

55. Deng, Y.; Li, J.; Li, T.; Gao, X.; Yuan, C. Life cycle assessment of lithium sulfur battery for electric vehicles. *J. Power Sources* **2017**, *343*, 284–295. [CrossRef]

56. Zackrisson, M.; Avellán, L.; Orlenius, J. Life cycle assessment of lithium-ion batteries for plug-in hybrid electric vehicles—Critical issues. *J. Clean. Prod.* **2010**, *18*, 1519–1529. [CrossRef]

57. Oliveira, L.; Messagie, M.; Rangaraju, S.; Sanfelix, J.; Hernandez Rivas, M.; Van Mierlo, J. Key issues of lithium-ion batteries—From resource depletion to environmental performance indicators. *J. Clean. Prod.* **2015**, *108*, 354–362. [CrossRef]

58. Sanfélix, J.; Messagie, M.; Omar, N.; Van Mierlo, J.; Hennige, V. Environmental performance of advanced hybrid energy storage systems for electric vehicle applications. *Appl. Energy* **2015**, *137*, 925–930. [CrossRef]

59. Roux, C.; Schalbart, P.; Peuportier, B. Development of an electricity system model allowing dynamic and marginal approaches in LCA—tested in the French context of space heating in buildings. *Int. J. Life Cycle Assess.* **2017**, *22*, 1177–1190. [CrossRef]

60. Alvarez Gaitan, J.; Short, M.; Peters, G.; MacGill, I.; Moore, S. Consequential cradle-to-gate carbon footprint of water treatment chemicals using simple and complex marginal technologies for electricity supply. *Int. J. Life Cycle Assess.* **2014**, *19*, 1974–1984. [CrossRef]

61. Colett, J.; Kelly, J.; Keoleian, G. Using Nested Average Electricity Allocation Protocols to Characterize Electrical Grids in Life Cycle Assessment. *J. Ind. Ecol.* **2016**, *20*, 29–41. [CrossRef]

62. Curran, M.; Mann, M.; Norris, G. The international workshop on electricity data for life cycle inventories. *J. Clean. Prod.* **2005**, *13*, 853–862. [CrossRef]

63. Earles, J.M.; Halog, A. Consequential life cycle assessment: A review. *Int. J. Life Cycle Assess.* **2011**, *16*, 445–453. [CrossRef]

64. Why and When? Available online: www.consequential-lca.org (accessed on 6 February 2018).

65. Sandén, B.; Karlström, M. Positive and negative feedback in consequential life-cycle assessment. *J. Clean. Prod.* **2007**, *15*, 1469–1481. [CrossRef]

66. Hernandez, M.; Messagie, M.; De Gennaro, M.; Van Mierlo, J. Resource depletion in an electric vehicle powertrain using different LCA impact methods. *Resour. Conserv. Recycl.* **2016**. [CrossRef]

67. Gemechu, E.; Sonnemann, G.; Young, S. Geopolitical-related supply risk assessment as a complement to environmental impact assessment: The case of electric vehicles. *Int. J. Life Cycle Assess.* **2017**, *22*, 31–39. [CrossRef]

68. Frischknecht, R.; Benetto, E.; Dandres, T.; Heijungs, R.; Roux, C.; Schrijvers, D.; Wernet, G.; Yang, Y.; Messmer, A.; Tschuemperlin, L. LCA and decision making: When and how to use consequential LCA, 62nd LCA Forum, Swiss Federal Institute of Technology, Zürich, 9 September 2016. *Int. J. Life Cycle Assess.* **2017**, *22*, 296–301. [CrossRef]

69. Onat, N.; Noori, M.; Kucukvar, M.; Zhao, Y.; Tatari, O.; Chester, M. Exploring the suitability of electric vehicles in the United States. *Energy* **2017**, *121*, 631–642. [CrossRef]

70. McCarthy, R.W. Impacts of electric-drive vehicles on California's energy system. In Proceedings of the 23rd International Battery, Hybrid and Fuel Cell Electric Vehicle Symposium and Exposition 2007—Sustainability: The Future of Transportation, EVS 2007, Anaheim, CA, USA, 2–5 December 2007; Volume 3, pp. 1771–1789.

71. Weidema, B.P. Attributional or consequential Life Cycle Assessment: A matter of social responsibility. *J. Clean. Prod.* **2018**, *174*, 305–314. [CrossRef]

72. Ministry of Climate and Energy (Ed.) *Energy Strategy 2050—from Coal, Oil and Gas to Green Energy*; Danish Government: Copenhagen, Denmark, 2011; 66p. Available online: http://www.danishwaterforum.dk/activities/Climate%20change/Dansk_Energistrategi_2050_febr.2011.pdf (accessed on 14 August 2018).

73. International Energy Agency. *Energy Policies of IEA countries—Denmark 2011 Review*; IEA Publications: Paris, France, 2011; p. 158.

74. Zamagni, A.; Guinée, J.; Heijungs, R.; Masoni, P.; Raggi, A. Lights and shadows in consequential LCA. *Int. J. Life Cycle Assess.* **2012**, *17*, 904–918. [CrossRef]

75. Ekvall, T.; Weidema, B.P. System boundaries and input data in consequential life cycle inventory analysis. *Int. J. Life Cycle Assess.* **2004**, *9*, 161–171. [CrossRef]

76. Example. Marginal Electricity in Denmark. Available online: www.consequential-lca.org (accessed on 30 January 2018).

77. Further Theory on the Special Case of Electricity—Forecasting and Time Horizon. Available online: www.consequential-lca.org (accessed on 30 January 2018).

78. Soimakallio, S.; Kiviluoma, J.; Saikku, L. The complexity and challenges of determining GHG (greenhouse gas) emissions from grid electricity consumption and conservation in LCA (life cycle assessment)—A methodological review. *Energy* **2011**, *36*, 6705–6713. [CrossRef]

79. Weidema, B.P.; Ekvall, T.; Heijungs, R. Guidelines for application of deepened and broadened LCA. Technical report for CALCAS Project. 2009. Available online: https://lca-net.com/publications/show/guidelines-applications-deepened-broadened-lca/ (accessed on 14 August 2018).

80. Lund, H.; Mathiesen, B.; Christensen, P.; Schmidt, J. Energy system analysis of marginal electricity supply in consequential LCA. *Int. J. Life Cycle Assess.* **2010**, *15*, 260–271. [CrossRef]

81. Schmidt, J.H.; Thrane, M.; Merciai, S.; Dalgaard, R. Inventory of Country Specific Electricity in LCA-Consequential and Attributional Scenarios. Methodology Report v2. 2011. Available online: https://lca-net.com/publications/show/inventory-country-specific-electricity-lca-consequential-attributional-scenarios-methodology-report-v2/ (accessed on 14 August 2018).

82. Weidema, B.P. Market Information in Life Cycle Assessment. 2003. Available online: http://citeseerx.ist.psu.edu/viewdoc/download?doi=10.1.1.197.5739&rep=rep1&type=pdf (accessed on 14 August 2018).

83. Mathiesen, B.; Münster, M.; Fruergaard, T. Uncertainties related to the identification of the marginal energy technology in consequential life cycle assessments. *J. Clean. Prod.* **2009**, *17*, 1331–1338. [CrossRef]

84. Yang, C. A framework for allocating greenhouse gas emissions from electricity generation to plug-in electric vehicle charging. *Energy Policy* **2013**, *60*, 722–732. [CrossRef]

85. Ryan, N.; Johnson, J.; Keoleian, G. Comparative Assessment of Models and Methods to Calculate Grid Electricity Emissions. *Environ. Sci. Technol.* **2016**, *50*, 8937–8953. [CrossRef] [PubMed]

86. Plevin, R.; Delucchi, M.; Creutzig, F. Using Attributional Life Cycle Assessment to Estimate Climate-Change Mitigation Benefits Misleads Policy Makers. *J. Ind. Ecol.* **2014**, *18*, 73–83. [CrossRef]

87. Hillman, K.M.; Sanden, B.A. Time and scale in Life Cycle Assessment: The case of fuel choice in the transport sector. *Int. J. Altern. Propul.* **2008**, *2*, 1–12. [CrossRef]

88. Marriott, J.; Matthews, H.S. Environmental Effects of Interstate Power Trading on Electricity Consumption Mixes. *Environ. Sci. Technol.* **2005**, *39*, 8584–8590. [CrossRef] [PubMed]

89. Itten, R.; Frischknecht, R.; Stucki, M. *Life Cycle Inventories of Electricity Mixes and Grid*; Treeze Ltd.: Uster, Switzerland, 2014. [CrossRef]

90. Climate Change Knowledge Portal. Available online: http://sdwebx.worldbank.org/climateportal/index.cfm?page=country_historical_climate&ThisCCode=PRT (accessed on 1 March 2018).

91. GaBi Databases. Available online: http://www.gabi-software.com/italy/databases/gabi-databases/ (accessed on 21 March 2018).

92. Argonne National Laboratory Greet 2016 Manual. 2016. Available online: https://greet.es.anl.gov/publication-greet-manual (accessed on 14 August 2018).

93. Lenzen, M. Life cycle energy and greenhouse gas emissions of nuclear energy: A review. *Energy Convers. Manag.* **2008**, *49*, 2178–2199. [CrossRef]

94. Sovacool, B. Valuing the greenhouse gas emissions from nuclear power: A critical survey. *Energy Policy* **2008**, *36*, 2950–2963. [CrossRef]

95. Fthenakis, V.; Kim, H.; Alsema, E. Emissions from Photovoltaic Life Cycles. *Environ. Sci. Technol.* **2008**, *42*, 2168–2174. [CrossRef] [PubMed]

96. Pehnt, M. Dynamic life cycle assessment (LCA) of renewable energy technologies. *Renew. Energy* **2006**, *31*, 55–71. [CrossRef]

97. U.S. Energy Information Administration. *Annual Energy Outlook 2013—With Projections to 2040*; U.S. Energy Information Administration: Washington, DC, USA, 2013; p. 233.

98. IEA. *Energy Policies of IEA Countries: European Union 2014 Review*; IEA Publications: Paris, France, 2014; p. 312.

99. NEEDS. Available online: http://www.needs-project.org/needswebdb/search.php (accessed on 16 February 2018).

100. Frischknecht, R.; Flury, K. Life cycle assessment of electric mobility: Answers and challenges—Zurich, April 6, 2011. *Int. J. Life Cycle Assess.* **2011**, *16*, 691–695. [CrossRef]

101. Doucette, R.; McCulloch, M. Modeling the CO_2 emissions from battery electric vehicles given the power generation mixes of different countries. *Energy Policy* **2011**, *39*, 803–811. [CrossRef]

102. Treyer, K. Datasets Related to Electricity Production and Supply in Ecoinvent Version 3—Short Overview. 2015. Available online: http://www.proyectaryproducir.com.ar/public_html/Seminarios_Posgrado/Material_de_referencia/ecoinvent-electricity%20datasets%20201305%20final.pdf (accessed on 14 August 2018).

Article

Power Sharing and Voltage Vector Distribution Model of a Dual Inverter Open-End Winding Motor Drive System for Electric Vehicles

Yi-Fan Jia, Liang Chu, Nan Xu *, Yu-Kuan Li, Di Zhao and Xin Tang

State Key Laboratory of Automotive Simulation and Control, Jilin University, Changchun 130022, China; jiayf16@mails.jlu.edu.cn (Y.-F.J.); chuliang@jlu.edu.cn (L.C.); liyk13@mails.jlu.edu.cn (Y.-K.L.); zhaod14@mails.jlu.edu.cn (D.Z.); tangxin16@mails.jlu.edu.cn (X.T.)
* Correspondence: nanxu@jlu.edu.cn; Tel.: +86-431-8509-5165

Received: 4 January 2018; Accepted: 5 February 2018; Published: 8 February 2018

Featured Application: control strategy of motor drive system for electric vehicles

Abstract: A drive system with an open-end winding permanent magnet synchronous motor (OW-PMSM) fed by a dual inverter and powered by two independent power sources is suitable for electric vehicles. By using an energy conversion device as primary power source and an energy storage element as secondary power source, this configuration can not only lower the DC-bus voltage and extend the driving range, but also handle the power sharing between two power sources without a DC/DC (direct current to direct current) converter. Based on a drive system model with voltage vector distribution, this paper proposes a desired power sharing calculation method and three different voltage vector distribution methods. By their selection strategy the optimal voltage vector distribution method can be selected according to the operating conditions. On the basis of the integral synthesizing of the desired voltage vector, the proposed voltage vector distribution method can reduce the inverter switching frequency while making the primary power source follow its desired output power. Simulation results confirm the validity of the proposed methods, which improve the primary power source's energy efficiency by regulating its output power and lessen inverter switching loss by reducing the switching frequency. This system also provides an approach to the energy management function of electric vehicles.

Keywords: electric vehicle; open-end winding; dual inverter; voltage vector distribution; power sharing; energy management

---

1. Introduction

Permanent Magnet Synchronous Motor (PMSM) has been widely used as drive motor on electric vehicles for its high power density, outstanding low-speed torque output and high efficiency [1–5]. However, the back EMF (electromotive force) increases rapidly along with motor speed due to the uncontrollable constant magnetic field of the permanent magnet, which means a higher DC-bus (direct current) voltage and a high-level flux-weakening control are required [6–8]. Furthermore, limited by the restricted energy of power battery with present technology, an extra energy source is often required to reach an acceptable driving range. Usually a DC/DC converter is needed to acquire power distribution and energy management function between the two power sources, which increases system complexity and brings additional losses [9,10].

The above problems can be perfectly solved by using an open-end winding permanent magnet Synchronous Motor (OW-PMSM) fed by two inverters, as shown in Figure 1. Each inverter is powered by an independent power source, and the two power sources are electrically isolated. This configuration inherently eliminates common-mode currents and can make the two sources virtually cascaded by proper control of dual inverter. Thus, the motor can obtain a higher voltage without increasing DC-bus voltage, resulting in an easier flux-weakening control and a higher top speed [11–14]. By using dual inverter, the actual number of motor phase voltage level is increased. For example, dual two-level inverter operates in three-level mode with equal DC-bus voltages, and operates in four-level mode when the dc voltages are in 2:1 ratio [15]. Moreover, the power flow between the two sources can be transferred through the motor controllably, making the system free of DC/DC converter [15,16].

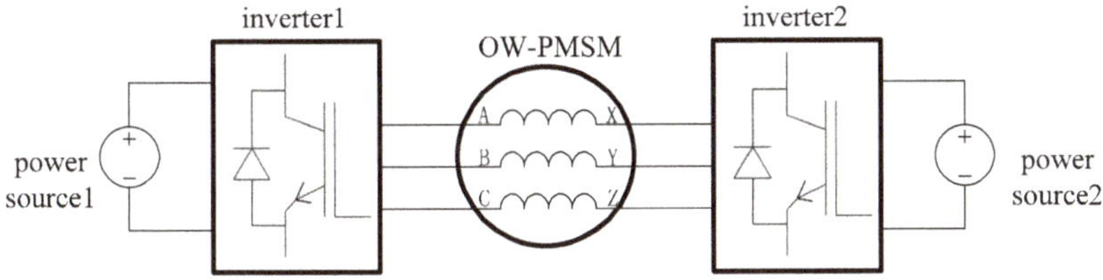

Figure 1. Structure of dual inverter open-end winding permanent magnet synchronous motor (OW-PMSM) drive system.

If we define power source1 as primary power source, power source2 as secondary power source, one of the most popular scheme is using a power battery as primary power source, and a floating capacitor as secondary power source. Due to the low internal resistance and high-rate discharge capability, the capacitor often acts as a power buffer, providing the reactive power consumed by the load, and smoothening power output of primary power source by absorbing instantaneous power fluctuation [16–20]. Joon Sung Park et al. proposed a dual inverter strategy for this scheme. By setting inverter2 to take care of reactive voltage and controlling active power flow, the torque and power level of high speed was enhanced, and the voltage of the capacitor was well regulated [20]. A hybrid PWM (pulse-width modulation) based flux-weakening control strategy for this scheme was proposed by Dan Sun et al. A double vector based PWM, utilizing one active vector and one optimal zero vector in a switching period, was applied to inverter1, resulting in the decreasing of switching frequency as well as switching loss. The flux-weakening control strategy also fully utilized the DC-bus voltage and widened motor operating range [21].

However, because of the low energy capacity, using a capacitor as secondary power source usually could not satisfy the requirements of most driving conditions for electric vehicle. Thus, the scheme including some type of energy conversion device, such as an engine generator or fuel cell as primary power source, some type of energy storage element, such as a battery or super capacitor as secondary power source comes into fashion [15,22]. Brian A. Welchko proposed three methods to achieve the combined motor control and energy management functions for this scheme, which are unity power factor control, voltage quadrature control, and optimum inverter utilization control [15]. By choosing the proper method, the energy flow in secondary energy source, the available system voltage to load when secondary energy source is not outputting active power, or the power flow to load can be maximized, respectively. A modulation strategy was proposed by Domenico Casadei et al., able to regulate the power sharing between the power sources by means of a special switching sequence for dual inverter. In addition, the limit of power sharing ratio was determined as a function of the modulation index [23]. However, the proposed strategy could only split the voltage vector linearly so that the modulation ranges of two inverters are underutilized.

In our previous work, we proposed a multi-level current hysteresis modulation algorithm, able to set major power source and switch it at any moment, which could accomplish power distribution between two power sources [24]. However, the proposed methods could only distribute the power between two power sources in a fixed ratio according to the major power source at each working point of the motor. To control the power flow appropriately, the major power source must be switched many times in a working cycle, just like a PWM controller.

Aiming at electric vehicles with an OW-PMSM driving and powered by two power sources, in this paper, a voltage vector-synthesizing model based on the midpoint voltage, and the power sharing principle are discussed. Then a power sharing calculation method is proposed to acquire the present desired power of primary power source. After that, three voltage vector distribution methods are introduced. The selection strategy of these three methods is then proposed to select the optimal method. On the basis of the integral synthesizing of the desired voltage vector, lowering the inverter switching frequency while making the output power of primary power source follow the desired value with specified accuracy can be achieved. At last a complete simulation of the drive system is executed to verify the control methods proposed.

In particular, the coordination transformations among static three-phase coordinate, static two-phase coordinate and rotary two-phase coordinate in this paper are equivalent power conversions.

2. Operating Principle

The control methods proposed by this paper is based on voltage vector generated by dual inverter, so in this section the mathematical system model is built, which provides theoretical support for the voltage vector distribution methods. In addition, the principle of the power sharing and power flow between two power sources are illuminated to support the proposed control methods.

2.1. System Modeling

First, we will introduce the concept of mid-point voltage. By equally dividing primary power source and secondary power source into two parts according to their voltages, we can acquire virtual mid-points m and n. Voltage difference between each inverter leg's output and corresponding mid-point is mid-point phase voltage, and voltage difference between mid-points of the two power sources is mid-point voltage difference. As shown in Figure 2, V_{dc1} and V_{dc2} are the DC-bus voltages of primary power source and secondary power source respectively. u_{Am}, u_{Bm} and u_{Cm} are the mid-point phase voltages of inverter1. u_{Xn}, u_{Yn} and u_{Zn} are the mid-point phase voltages of inverter2. u_{mn} is the mid-point voltage difference between the two power sources.

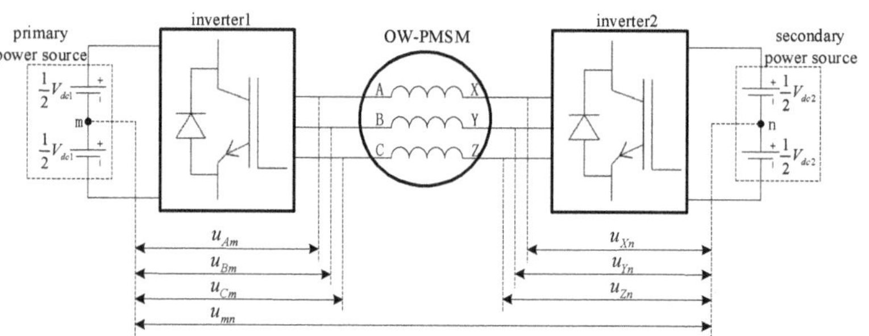

Figure 2. Schematic diagram of mid-point voltage.

The phase voltages of OW-PMSM are affected by both inverters' switching states. Based on Kirchhoff's law, the phase voltages of OW-PMSM are determined by Equation (1):

$$\begin{cases} u_{AX} = u_{Am} - u_{Xn} + u_{mn} \\ u_{BY} = u_{Bm} - u_{Yn} + u_{mn} \\ u_{CZ} = u_{Cm} - u_{Zn} + u_{mn} \end{cases} \tag{1}$$

In motor stator plane, the stator voltage vector of OW-PMSM $\vec{u}_s$ is synthesized by the three phase voltages as follow:

$$\vec{u}_s = \sqrt{\frac{2}{3}}(u_{AX}e^{j0} + u_{BY}e^{j\frac{2\pi}{3}} + u_{CZ}e^{j\frac{4\pi}{3}}) \tag{2}$$

where e^{j0}, $e^{j\frac{2\pi}{3}}$ and $e^{j\frac{4\pi}{3}}$ are spatial operators, pointing the directions of the three phases' axes respectively. $\sqrt{\frac{2}{3}}$ is the coefficient of equivalent power conversions. Similarly, the output voltage vector of the two inverters $\vec{u}_{s1}$ and $\vec{u}_{s2}$ expressed by mid-point phase voltages are given as:

$$\begin{cases} \vec{u}_{s1} = \sqrt{\frac{2}{3}}(u_{Am}e^{j0} + u_{Bm}e^{j\frac{2\pi}{3}} + u_{Cm}e^{j\frac{4\pi}{3}}) \\ \vec{u}_{s2} = \sqrt{\frac{2}{3}}(u_{Xn}e^{j0} + u_{Yn}e^{j\frac{2\pi}{3}} + u_{Zn}e^{j\frac{4\pi}{3}}) \end{cases} \tag{3}$$

This indicates $\vec{u}_{s1}$ and $\vec{u}_{s2}$ are only determined by switching states of the corresponding inverter, and not affected by the other side. Thus, $\vec{u}_{s1}$ and $\vec{u}_{s2}$ can be generated by two independent space vector pulse width modulation (SVPWM) controllers.

Substituting Equations (1) and (3) into (2) results in an expression for $\vec{u}_s$ in terms of $\vec{u}_{s1}$ and $\vec{u}_{s2}$, shown by Equation (4):

$$\vec{u}_s = \vec{u}_{s1} - \vec{u}_{s2} + \sqrt{\frac{2}{3}}u_{mn}(e^{j0} + e^{j\frac{2\pi}{3}} + e^{j\frac{4\pi}{3}}) = \vec{u}_{s1} - \vec{u}_{s2} \tag{4}$$

Because the two power sources are electrically isolated, the mid-point voltage difference u_{mn} is floating and varying along with the switching states of dual inverter. However, when synthesizing the motor stator voltage vector $\vec{u}_s$, the floating u_{mn} is eliminated as shown in Equation (4). That is because the three phase voltages contain identical component of u_{mn}, which just counteract each other in motor stator plane. In essence, when the loads of three phase windings are completely symmetrical, the synthesized motor stator voltage vector should be free of zero-sequence component, which means the following equation holds:

$$u_{AX} + u_{BY} + u_{CZ} = 0 \tag{5}$$

By substituting (1) into (5), the value of u_{mn} can be obtained.

2.2. Principle of Power Flow

From Equation (4), we know that $\vec{u}_s$ can be synthesized by $\vec{u}_{s1}$ and $\vec{u}_{s2}$, where $\vec{u}_{s1}$ and $\vec{u}_{s2}$ are generated by two independent SVPWM controllers separately. Thus, when $\vec{u}_s$ is determined and V_{dc1}, V_{dc2} are measured, the feasible region of voltage vector distribution can be obtained. As shown in Figure 3, O_1 is the origin of vector $\vec{u}_s$ and O_2 is the end of $\vec{u}_s$. Two hexagons indicating the modulation range of the two inverters can be acquired. Hexagon $A_1B_1C_1D_1E_1F_1$ centering on O_1, having a side length of $\sqrt{\frac{2}{3}}V_{dc1}$, gives the modulation range of inverter1. In the same way, hexagon $A_2B_2C_2D_2E_2F_2$ centering on O_2 with a side length of $\sqrt{\frac{2}{3}}V_{dc2}$, gives the modulation range of inverter2. Vector $\vec{u}_{s1}$ starts at O_1 and has to end within the range of hexagon $A_1B_1C_1D_1E_1F_1$, similarly vector $\vec{u}_{s2}$ starts at O_2 and has to end within the range of hexagon $A_2B_2C_2D_2E_2F_2$. According to Equation (4), the end of $\vec{u}_{s1}$ and the end of $\vec{u}_{s2}$ must coincide at one point, assuming it is I. The overlapping region of the two hexagons $A_2B_2GD_1E_1H$ corresponds to the feasible region of I, which also indicates the feasible region of voltage vector distribution.

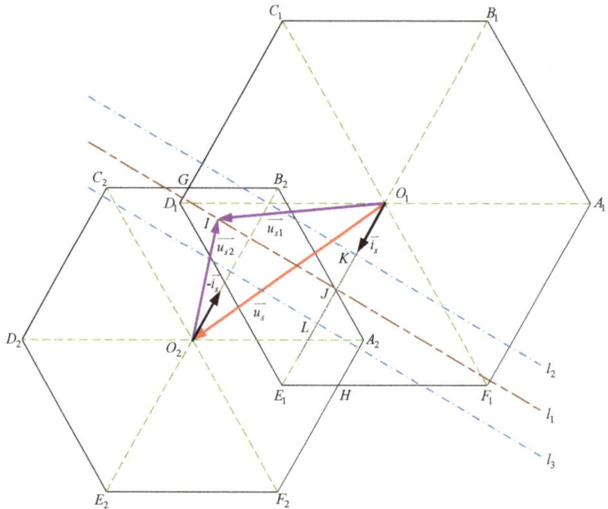

Figure 3. Schematic diagram of voltage vector distribution and power sharing.

If the motor stator currents i_{AX}, i_{BX} and i_{CX} are measured, the stator current vector of OW-PMSM $\vec{i}_s$ can be obtained as follow:

$$\vec{i}_s = \sqrt{\frac{2}{3}}\left(i_{AX}e^{j0} + i_{BY}e^{j\frac{2\pi}{3}} + i_{CZ}e^{j\frac{4\pi}{3}}\right) \tag{6}$$

Then the power sharing of dual inverter can be determined as shown in Equation (7):

$$\begin{cases} P_{inv1} = \vec{u}_{s1} \cdot \vec{i}_s \\ P_{inv2} = \vec{u}_{s2} \cdot -\vec{i}_s = -\vec{u}_{s2} \cdot \vec{i}_s \end{cases} \tag{7}$$

where P_{inv1} and P_{inv2} are the output power of inverter1 and inverter2 respectively, positive value means the corresponding inverter is outputting active power and vice versa. It is important to notice the minus sign in the expression of P_{inv2}. That is because the defined motor phase current polarities are flowing from inverter1 to inverter2, thus the direction of vector $\vec{i}_s$ on inverter2's point of view is opposite to inverter1's. In Figure 3, P_{inv1} can be obtained from vector dot product of $\vec{u}_{s1}$ and $\vec{i}_s$ at point O_1, similarly P_{inv2} can be obtained from vector dot product of $\vec{u}_{s2}$ and $-\vec{i}_s$ at point O_2. Taking inverter1 for example, we have $P_{inv1} = \vec{u}_{s1} \cdot \vec{i}_s = |\vec{i}_s||O_1J|$, where $|O_1J|$ is the projection of $\vec{u}_{s1}$ on the direction of $\vec{i}_s$. Thus all the available $\vec{u}_{s1}$ having the same projection $|O_1J|$ obtain the same value of P_{inv1}, which is indicated by line l_1 through point J and orthogonal to $\vec{i}_s$. That means all the $\vec{u}_{s1}$ starting at O_1 and ending on line l_1 gain the same P_{inv1} at current $\vec{i}_s$. Assuming we want the output power of inverter1 limited to the range of $P_{inv1} \pm \Delta P$, we can draw two lines parallel to line l_1 and having a distance of $\frac{\Delta P}{|i_s|}$ to l_1 on different side. Those are line l_2 through point K and line l_3 through point L, where $|KJ| = |LJ| = \frac{\Delta P}{|i_s|}$. Thus in prevailing circumstance, the feasible region of point I, also known as the feasible region of voltage vector distribution, is the intersection set of area $A_2B_2GD_1E_1H$ and the area between line l_2 and line l_3.

3. Desired Power Sharing Calculation Method

Because an energy conversion device is used as primary power source, we must take its traits for consideration. One is the efficiency characteristic. For example, an engine generator has a best efficiency point and a maximum efficiency curve, operating on which makes the engine under the best economy. The other is the lagging characteristic in response. For example, there is a time lag between opening the throttle and the increasing of engine's output power, or between increasing the hydrogen supply and power uprating of fuel cell. Since the secondary power source is an energy storage device and it does not need additional controls, we set it as a power buffer, to compensate the lacking power or absorb the redundant power outputted by primary power source.

Considering the two traits of the primary power source mentioned above, a desired power sharing calculation method based on a first order inertial element is proposed, as shown in Figure 4.

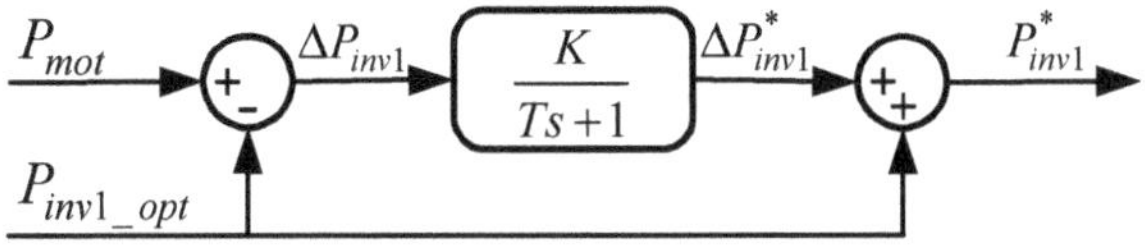

Figure 4. Calculation diagram of desired power of inverter1.

This calculator has two inputs, which are P_{mot} and P_{inv1_opt}. Where P_{mot} is the input power of OW-PMSM, identical to the summation of two inverters' output power, and can be determined as follow:

$$P_{mot} = \vec{u}_s \cdot \vec{i}_s \tag{8}$$

P_{inv1_opt} is the optimal output power of inverter1, under which the primary power source obtains maximum efficiency. P_{inv1_opt} can be set as a constant, corresponding to the best efficiency point; or set as a variable, corresponding to the maximum efficiency curve, following the changes of its output power. Then the power deviation of inverter1 ΔP_{inv1} can be obtained by subtracting P_{inv1_opt} from P_{mot}:

$$\Delta P_{inv1} = P_{mot} - P_{inv1\_opt} \tag{9}$$

ΔP_{inv1} indicates the difference between the total desired power and the ideal power output of inverter1. Thus, we use ΔP_{inv1} as the input of the first order inertial element, to generate the compensating power of inverter1. In the frequency domain, we have:

$$\Delta P^*_{inv1}(s) = \frac{K}{Ts+1}\Delta P_{inv1}(s) \tag{10}$$

where ΔP^*_{inv1} is the compensating power of inverter1, K and T are the gain and the time constant of the inertial element, respectively. In the time domain, ΔP^*_{inv1} and ΔP_{inv1} satisfy the following differential equation:

$$T\frac{d}{dt}\Delta P^*_{inv1}(t) + \Delta P^*_{inv1}(t) = K\Delta P_{inv1}(t) \tag{11}$$

In a discrete system, the current ΔP^*_{inv1} can be obtained by Equation (12), where $\Delta P^*_{inv1}(t)$ is the valve of ΔP^*_{inv1} in current sample step and ΔT is the step size of the system.

$$\Delta P^*_{inv1}(t) = \frac{K\Delta P_{inv1}(t) - \Delta P^*_{inv1}(t - \Delta T)}{T}\Delta T + \Delta P^*_{inv1}(t - \Delta T) \tag{12}$$

After ΔP^*_{inv1} is obtained, the desired output power of inverter1 P^*_{inv1} can be determined:

$$P^*_{inv1} = P_{inv1\_opt} + \Delta P^*_{inv1} \tag{13}$$

By adjusting K and T, different power sharing characteristics can be achieved. We have $K \in [0,1]$, and it affects the response amplitude of ΔP^*_{inv1}. The larger K is, the greater amplitude ΔP^*_{inv1} responds to ΔP_{inv1}. For example, when $K = 0$, we have $P^*_{inv1} = P_{inv1\_opt}$ at all time, totally regardless of the influence of ΔP_{inv1}. In this circumstance, the efficiency of the primary power source is maximized, but the secondary power source will also take the largest power fluctuation. When $K = 1$, P^*_{inv1} is completely following P_{mot} except the lag caused by the inertial element. In this circumstance, the power fluctuation of the primary power source will rise, the operating time in high efficiency area and the average efficiency will drop, but the secondary power source is not expected to supply any active power regardless of the influence of the inertial element. The time constant T affects speed and sensitivity of ΔP^*_{inv1} responding to ΔP_{inv1}. The larger T is, the more lag and inertia is in response.

4. Voltage Vector Distribution Method

This section deals with three different voltage vector distribution methods and their selection strategy. Because all these voltage vector distribution methods need to judge whether the inverter desired voltage vector is in the modulation range of the corresponding inverter, the judgmental algorithm will be introduced first.

4.1. Voltage Vector Over Range Judgmental Algorithm

The voltage vector over range judgmental algorithm is derived from the SVPWM strategy. It is used to calculate the proportions of adjacent basic vector for synthesizing desired voltage vector of a single inverter [25]. The calculation result can also indicate whether the desired voltage vector is in the modulation range, which is what we need here.

The desired voltage vector is given in static two-phase coordinate, which are D-axis component u^*_D and Q-axis component u^*_Q. As shown in Figure 5a, axis J, K and L are distributed in circular uniform. The projection of the desired voltage vector on axis J, K and L, defined as j, k and l respectively, can be obtained by Equation (14):

$$\begin{cases} j = \dfrac{u^*_Q}{2} \\ k = \dfrac{\sqrt{3}u^*_D}{2} - \dfrac{u^*_Q}{2} \\ l = -\dfrac{\sqrt{3}u^*_D}{2} - \dfrac{u^*_Q}{2} \end{cases} \tag{14}$$

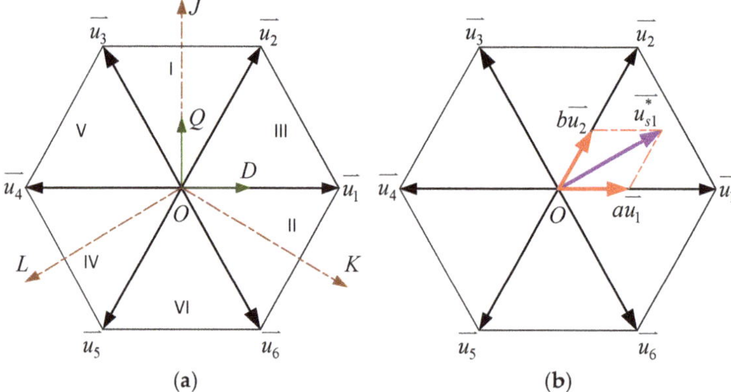

Figure 5. Sector partition and voltage vector synthesizing of space vector pulse width modulation (SVPWM). (**a**) Sector partition; (**b**) voltage vector synthesis.

Then the sector number N can be determined as follow:

$$N = sign(j) + 2sign(k) + 4sign(l) \tag{15}$$

where $sign$ is the sign function, we have $sign(x) = 1$ if $x > 0$, otherwise we have $sign(x) = 0$. There are six sectors in total, and the sector number is represented by roman numerals shown in Figure 5a. The proportion parameters x, y and z can be calculated using u_D^* and u_Q^* by Equation (16), where V_{dc} is the DC-bus voltage of the corresponding inverter.

$$\begin{cases} x = \dfrac{\sqrt{2}u_Q^*}{V_{dc}} \\ y = \dfrac{2u_D^* + \sqrt{3}u_Q^*}{\sqrt{6}V_{dc}} \\ z = \dfrac{-2u_D^* + \sqrt{3}u_Q^*}{\sqrt{6}V_{dc}} \end{cases} \tag{16}$$

Afterwards, the proportions of adjacent basic vectors to synthesize desired voltage vector, defined as a and b, can be obtained by distributing the three proportion parameters according to the sector number, which is shown in Table 1.

Table 1. Relation table for a, b and N.

N	0	1	2	3	4	5	6
a	0	z	y	−z	−x	x	−y
b	0	y	−x	x	z	−y	−z

As shown in Figure 5b, taking inverter1 as an example, the desired voltage vector $\vec{u}_{s1}^*$ lies in Sector III, and is synthesized by the adjacent basic vectors $\vec{u}_1$ and $\vec{u}_2$. We have:

$$\vec{u}_{s1}^* = a\vec{u}_1 + b\vec{u}_2 \tag{17}$$

$a + b \le 1$ indicates the desired voltage vector is within the range of the hexagon and can be modulated. Otherwise when $a + b > 1$, it means the desired voltage vector is out of the modulation range and cannot be generated integrally.

4.2. Low Switching Frequency Method

In a SVPWM control period, there are two switch commutations of each inverter leg normally. Thus, there are six switch commutations of all three legs in a single inverter. When using the dual inverter configuration, the number of switch commutations is doubled. This can cause considerable inverter switching loss and lower the inverter efficiency. However, if one inverter is forced to output only basic voltage vectors or zero vector, there is no switch commutation during the modulation period because the desired voltage vector does not need to be synthesized by two adjacent basic vectors. Thus, the switching frequency of this clamped inverter can be lower to zero if the switch commutations at the junction of modulation periods are ignored. By clamping inverter1 to output only basic voltage vectors and zero vector, the low switching frequency method can reduce the total inverter switching frequency by nearly a half, thus the inverter switching loss is significantly reduced. The desired voltage vector of inverter1 $\vec{u}_{s1}^*$ will be chosen among zero vector and basic voltage vectors lying in the feasible region of voltage vector distribution mentioned in Section 2.2, the one makes inverter1's output power closest to the desired value P_{inv1}^* will be selected.

The specific algorithm will be introduced as follow. First, matrix M_{us1} with a size of 4×7 is introduced. Each row of M_{us1} represents a candidate of voltage vectors, including six basic voltage vectors and one zero vector. The first two lines of M_{us1} are assigned as follow:

$$\begin{cases} M_{us1}(1,:) = \sqrt{\tfrac{2}{3}}U_{dc1} \circ [0, 1, \tfrac{1}{2}, -\tfrac{1}{2}, -1, -\tfrac{1}{2}, \tfrac{1}{2}] \\ M_{us1}(2,:) = \sqrt{\tfrac{2}{3}}U_{dc1} \circ [0, 0, \tfrac{\sqrt{3}}{2}, \tfrac{\sqrt{3}}{2}, 0, -\tfrac{\sqrt{3}}{2}, -\tfrac{\sqrt{3}}{2}] \end{cases} \tag{18}$$

where $M_{us1}(n,:)$ means the nth line of matrix M_{us1}, and "∘" is the operator of Hadamard product, meaning the multiplication of each corresponding element of the two participant matrixes. $M_{us1}(1,:)$ indicates the D-axis components of each candidate vector, while $M_{us1}(2,:)$ indicates the Q-axis components. Then the third line of M_{us1} can be calculated, which indicates inverter1's output power of the corresponding candidate vector:

$$M_{us1}(3,:) = M_{us1}([1,2],:)^T \times [i_D, i_Q]^T \tag{19}$$

where i_D and i_Q are D-axis component and Q-axis component of $\vec{i}_s$ respectively. The difference between inverter1's output power of each candidate vector and the desired value P^*_{inv1} can be obtained by Equation (20), assigned to the fourth line of M_{us1}:

$$M_{us1}(4,:) = |M_{us1}(3,:) - P^*_{dc1}| \tag{20}$$

Then these power difference values of inverter1 are sorted in ascending sequence. Afterwards the corresponding candidate vectors are checked respectively in the sorted sequence to verify whether it is in the feasible region until we get a positive result.

The specific procedure is introduced as follow. First, obtain inverter2's voltage vector of the corresponding candidate vector of inverter1 by $\vec{u}^*_{s2} = \vec{u}^*_{s1} - \vec{u}^*_s$, which is modified from Equation (4). Then whether this voltage vector is within the modulation range of inverter2 or not is checked by voltage vector over range judgmental algorithm introduced in sector 4.1. If the result is positive, the current candidate vector of inverter1 is valid. Otherwise the next candidate vector in the sorted sequence will be checked until the voltage vector distribution is confirmed valid.

Then the present candidate vector is assigned to inverter1's desired voltage vector, denoted by $\vec{u}^*_{s1\_LF}$, where the subscript LF means Low Frequency. The difference between inverter1's output power corresponding to $\vec{u}^*_{s1\_LF}$ and the desired value P^*_{inv1} can be obtained from the corresponding element in $M_{us1}(4,:)$, recorded as D_{Pinv1_LF}, which is necessary in the selection strategy of voltage vector distribution methods. D_{Pinv1_LF} indicates inverter1's power following deviation of the low switching frequency method. If all the candidate vectors are verified beyond the feasible region, the flag of low switching frequency method F_{LF} will be set to 0, meaning the low switching frequency method is unavailable in current situation; otherwise F_{LF} will be set to 1, indicating the low switching frequency method is valid.

There are two examples shown in Figure 6. The meanings of lines, points, and vectors are the same with Figure 3. The numbers next to vertexes of the hexagon centered on O_1 indicate the rankings of the corresponding candidate vectors in the sorted sequence, respectively. In the first situation, the first candidate vector in the sequence, whose endpoint is closest to line l_1, is within the feasible region, so it is chosen to be $\vec{u}^*_{s1\_LF}$. However, in the second situation, the first two candidate vectors indicated by dashed lines are both invalid, thus the third candidate vector in the sequence is chosen to be $\vec{u}^*_{s1\_LF}$, making inverter1's power following deviation bigger than the first situation.

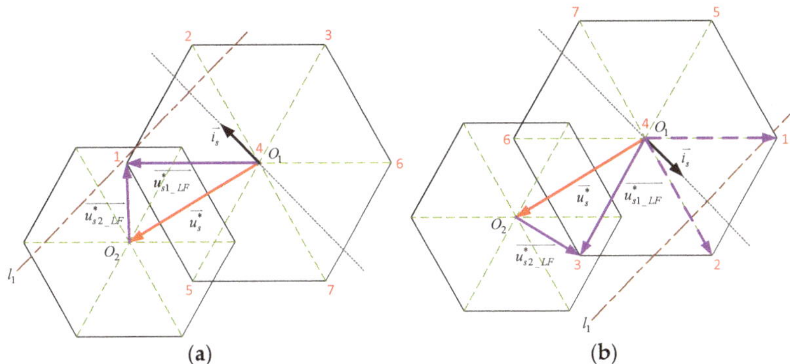

Figure 6. Voltage vector distribution examples of low switching frequency method. (**a**) First situation; (**b**) second situation.

4.3. Accurate Power Following Method

Because low switching frequency method only uses basic voltage vectors and zero vector, the difference of inverter1's output power between actual and desired value, known as verter1's power following deviation, is not minimized. The accurate power following method is proposed to solve this problem when this deviation of low switching frequency method is beyond the tolerance range. The principle of this method is simple: to generate inverter1's desired voltage vector in phase with $\vec{i_s}$, and its amplitude is determined by the desired output power of inverter1, making inverter1's power following deviation fully eliminated. The collinearity of $\vec{u_{s1}}$ and $\vec{i_s}$ could also make inverter1 free of bearing any reactive power.

The specific algorithm will be introduced as follow. First, the amplitude of $\vec{i_s}$ is obtained using its D-axis and Q-axis components by Equation (21):

$$|\vec{i_s}| = \sqrt{i_D^2 + i_Q^2} \tag{21}$$

Then the amplitude of inverter1's desired voltage vector can be determined as follow, which makes the corresponding output power of inverter1 equal to the desired value P_{inv1}^*.

$$|\vec{u_{s1}^*}| = \frac{P_{inv1}^*}{|\vec{i_s}|} \tag{22}$$

Therefore, $\vec{u_{s1}^*}$ can be obtained by Equation (23) to be in phase with $\vec{i_s}$:

$$\vec{u_{s1}^*} = \frac{\vec{i_s}}{|\vec{i_s}|}|\vec{u_{s1}^*}| \tag{23}$$

Afterwards the voltage vector over range judgmental algorithm is executed to check whether the current calculated $\vec{u_{s1}^*}$ is within the modulation range of inverter1. If the result is positive, the current calculated $\vec{u_{s1}^*}$ will be assigned to inverter1's desired voltage vector $\vec{u_{s1\_AF}^*}$, where the subscript AF means Accurate Following. Otherwise, the current calculated $\vec{u_{s1}^*}$ need to be shortened to the boundary of inverter1's modulation range by the following procedure.

We define the proportions of adjacent basic vectors for synthesizing the current calculated $\vec{u_{s1}^*}$ as a_1' and b_1', which are the results from the voltage vector over range judgmental algorithm, and we have $a_1' + b_1' > 1$ so the current calculated $\vec{u}_{s1}^*$ is beyond the modulation range. To make $a_1 + b_1 = 1$, we have:

$$\begin{cases} a_1 = \frac{a_1'}{a_1' + b_1'} \\ b_1 = \frac{b_1'}{a_1' + b_1'} \end{cases} \tag{24}$$

where a_1 and b_1 are the proportions of adjacent basic vectors for the shortened $\vec{u}_{s1}^*$, which is on the boundary of inverter1's modulation range. Then the amplitude of this shortened u_{s1}^* can be obtained by Equation (25):

$$|u_{s1}^*| = \sqrt{\frac{2}{3}} U_{dc1} \sqrt{1 - a_1 b_1} \tag{25}$$

Afterwards, the shortened $\vec{u}_{s1}^*$ can be obtained by using Equation (23) again. Then the shortened $\vec{u}_{s1}^*$ can be assigned to $u_{s1\_AF}^*$. In addition, the difference between inverter1's output power corresponding to $u_{s1\_AF}^*$ and the desired value P_{inv1}^*, denoted by D_{Pinv1_AF}, needs to be calculated by Equation (26), which is necessary in the selection strategy of voltage vector distribution methods.

$$D_{Pinv1\_AF} = |\vec{u}_{s1\_AF}^* \cdot \vec{i}_s - P_{inv1}^*| \tag{26}$$

Finally, obtain $\vec{u}_{s2}^*$ by $\vec{u}_{s2}^* = \vec{u}_{s1}^* - \vec{u}_s^*$ from Equation (4). Then check whether it is within the modulation range of inverter2. If not, the flag of accurate power following method F_{AF} will be set to 0, meaning the accurate power following method is unavailable in current situation; otherwise F_{AF} will be set to 1, indicating the calculated $\vec{u}_{s1\_AF}^*$ is valid.

There are two examples shown in Figure 7. In the first situation, line l_1 intersects $\vec{i}_s$'s extension line inside inverter1's modulation range, so inverter1's power following deviation D_{Pinv1_AF} is completely eliminated. However, in the second situation, the intersection point of line l_1 and $\vec{i}_s$'s extension line is beyond inverter1's modulation range. Thus, $\vec{u}_{s1}^*$ must be shortened to the boundary of this hexagon. Unfortunately, after being shortened, $\vec{u}_{s1}^*$ is also not in the feasible range of voltage vector distribution because the corresponding $\vec{u}_{s2}^*$ is out of inverter2's modulation range. Thus, in this situation, the accurate power following method is unavailable.

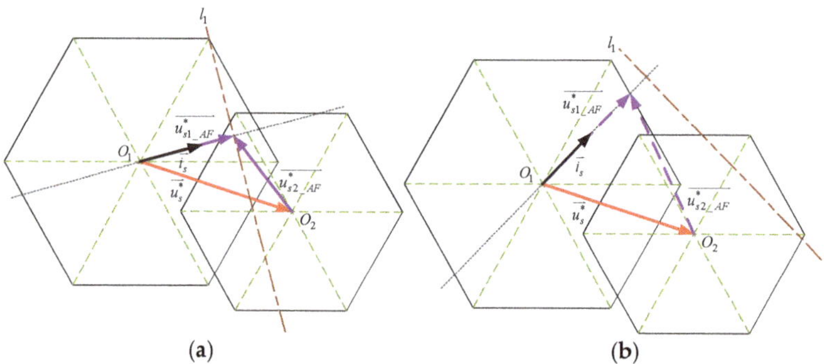

Figure 7. Voltage vector distribution examples of accurate power following method; (**a**) first situation; (**b**) second situation.

4.4. Linear Partition Method

From Equation (4) we know, when the output voltage vectors of two inverters are $180°$ out of phase, meaning $\vec{u}_{s1}$ and $\vec{u}_{s2}$ are collinear but in the opposite direction, the synthesized $\vec{u}_s$ has the maximum amplitude. Thus, if we assign $\vec{u}_{s1}^*$ and $\vec{u}_{s2}^*$ by linearly partition $\vec{u}_s$ in the proper position, the possibility of $\vec{u}_{s1}^*$ and $\vec{u}_{s2}^*$ in the feasible region can be maximized. This is the strategy of the linear partition method. If both the two methods above have failed to distribute the voltage vectors efficiently, the linear partition method is the last chance. Even if it fails to generate the desired stator voltage vector $\vec{u}_s^*$ integrally with the limitation of two inverters' DC-bus voltages, in which situation the two hexagons centered on O_1 and O_2 have no overlapping region, the linear partition method can provide a $\vec{u}_s$ having the same direction and the nearest amplitude with $\vec{u}_s^*$.

The specific algorithm will be introduced as follow. First, the amplitude of $\vec{u}_s^*$ is calculated using its D-axis and Q-axis components as follow:

$$|\vec{u}_s^*| = \sqrt{u_D^{*2} + u_Q^{*2}} \tag{27}$$

Then the desired amplitude of $\vec{u}_{s1}^*$ can be obtained by Equation (28), making the corresponding output power of inverter1 follows P_{inv1}^* accurately.

$$|\vec{u}_{s1}^*| = \frac{P_{inv1}^*}{\vec{u}_s^* \cdot \vec{i}_s}|\vec{u}_s^*| \tag{28}$$

Similar to Equation (23), $\vec{u}_{s1}^*$ can be obtained as follow to be in phase with $\vec{u}_s^*$:

$$\vec{u}_{s1}^* = \frac{\vec{u}_s^*}{|\vec{u}_s^*|}|\vec{u}_{s1}^*| \tag{29}$$

Then the voltage vector over range judgmental algorithm is executed to check whether the current calculated $\vec{u}_{s1}^*$ is within the modulation range of inverter1. If the answer is negative, the current calculated $\vec{u}_{s1}^*$ need to be shortened to the boundary of inverter1's modulation range by the procedure introduced in sector 4.3, that is using Equations (24), (25) and (29) in turn.

Afterwards, obtain $\vec{u}_{s2}^*$ by $\vec{u}_{s2}^* = \vec{u}_{s1}^* - \vec{u}_s^*$, which is modified from Equation (4). Then check whether $\vec{u}_{s2}^*$ is within the modulation range of inverter2. If positive, the latest calculated $\vec{u}_{s1}^*$ will be assigned to inverter1's desired voltage vector $\vec{u}_{s1\_LP}^*$, where the subscript LP means Linear Partition. Otherwise, the current calculated $\vec{u}_{s2}^*$ need to be shortened to the boundary of inverter2's modulation range by the same way with $\vec{u}_{s1}^*$. After $\vec{u}_{s2}^*$ is shortened, we need to recalculate the corresponding $\vec{u}_{s1}^*$ by $\vec{u}_{s1}^* = \vec{u}_s^* + \vec{u}_{s2}^*$, and check whether it is within the modulation range again. If not, shorten it to the boundary of its modulation range. Finally the recalculated $\vec{u}_{s1}^*$ can be assigned to $\vec{u}_{s1\_LP}^*$. Similarly, the difference between inverter1's output power corresponding to $\vec{u}_{s1\_LP}^*$ and the desired value P_{inv1}^*, denoted by D_{Pinv1_LP}, needs to be calculated for voltage vector distribution method selection strategy. Equation (26) is also suitable for this occasion.

There are four examples shown in Figure 8. In the first situation, line l_1 intersects $\vec{u}_s^*$ right inside the feasible range of voltage vector distribution, thus the first calculated $\vec{u}_{s1}^*$ is valid. Under this circumstance the output power of inverter1 follows P_{inv1}^* accurately. In the second situation, P_{inv1}^* is too big so line l_1 is completely out of inverter1's modulation range. Thus, $\vec{u}_{s1}^*$ has to be shortened to the boundary of the hexagon centered on O_1. In the third situation, the first calculated $\vec{u}_{s1}^{*}$, which makes the output power of inverter1 equals to P_{inv1}^{*}, is out of feasible range of voltage vector distribution because the corresponding $\vec{u}_{s2}^{*}$ is beyond its modulation range. Thus, after $\vec{u}_{s2}^{*}$ is shortened to the hexagon centered on O_2, $\vec{u}_{s1}^{*}$ is recalculated to compensate the shortage. In the fourth situation, the amplitude of $\vec{u}_{s}^{*}$ is too big so that the two hexagons centered on O_1 and O_2 have no overlapping region. Thus, $\vec{u}_{s1}^{*}$ and $\vec{u}_{s2}^{*}$ both reached their maximum amplitude collinearly with $\vec{u}_{s}^{*}$ in the opposite direction to synthesize $\vec{u}_{s}$, which is the best available result.

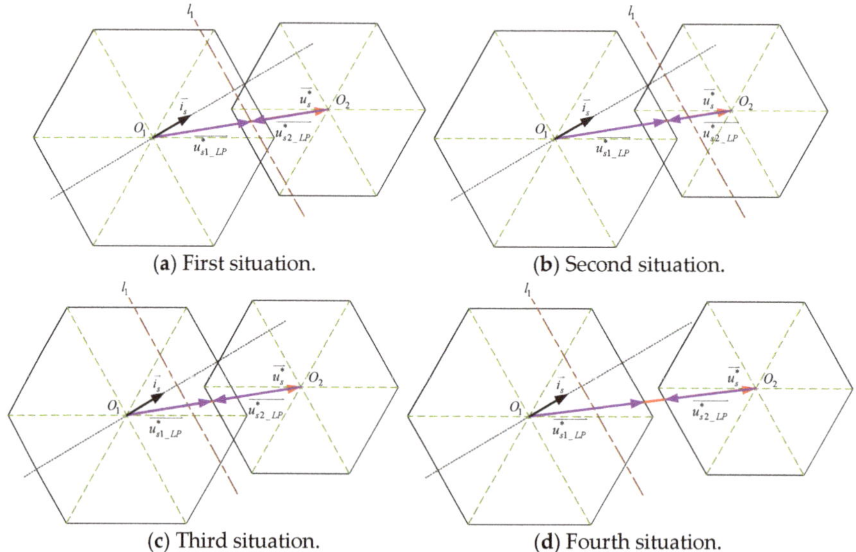

(**a**) First situation.

(**b**) Second situation.

(**c**) Third situation.

(**d**) Fourth situation.

Figure 8. Voltage vector distribution examples of linear partition method.

4.5. Voltage Vector Distribution Method Selection Strategy

Each of these three voltage vector distribution methods introduced above has worked out a result to distribute voltage vectors generated by dual inverter to synthesize the desired stator voltage vector of OW-PMSM. The voltage vector distribution method selection strategy deals with the results from these voltage vector distribution methods by selecting the optimal one as the final result of voltage vector distribution.

The general idea of this selection strategy is expounded as follow. We have three evaluation indexes, which are the accurate synthesizing of desired stator voltage vector, the following of inverter1's desired output power in acceptable accuracy, and the reduction of inverter switching frequency, ranked from highest priority to lowest. In other words, by priority, first we need to ensure the voltage vector distribution result can generate the desired stator voltage vector or as integrally as possible. Based on that, we try to make inverter1's output power follow the desired value in a specific tolerance range. After these two conditions are met, we can think about lowering the inverter switching frequency. The best situation is these three conditions are all satisfied. Otherwise, we discard the reduction of inverter switching frequency. Then we give up the accurate power following of inverter1. At least the synthesizing of desired voltage vector has to be ensured.

For the second condition mentioned above, we need to define the inverter1's maximum power following deviation, denoted by D_{Pinv1_max}. If inverter1's output power is within the range of $[P_{inv1}^{*} - D_{Pinv1\_max}, P_{inv1}^{*} + D_{Pinv1\_max}]$, we consider inverter1's output power is following the desired value within tolerance range. D_{Pinv1_max} can be adjusted according to the control requirements of the primary power source.

The specific algorithm of this selection strategy will be introduced as follow. First, the flag of low switching frequency method F_{LF} is checked. If $F_{LF} = 1$, meaning the low switching frequency method is available, the selection will be made between the low switching frequency method and the accurate power following method. Then three conditions are checked, which are $F_{AF} = 0$, $D_{Pinv1\_LF} \leq D_{Pinv1\_AF}$, and $D_{Pinv1\_LF} \leq D_{Pinv1\_max}$. If anyone of these conditions is met, indicating the low switching frequency method meets inverter1's accurate power following condition or has a better performance in inverter1's power following, the low switching frequency method will be selected. If all these three conditions are not met, the accurate power following method will be selected. However, if $F_{LF} = 0$, meaning the low switching frequency method is invalid in current situation, the selection will have to be made between the accurate power following method and the linear partition method. When $F_{AF} = 1$ and $D_{Pinv1\_AF} \leq D_{Pinv1\_LP}$ are both met, indicating the accurate power following method is available and performs better in inverter1's power following, the accurate power following method will be selected. Otherwise, the linear partition method will be selected to synthesize the desired stator voltage vector as integrally as possible. The flowchart of the voltage vector distribution method selection strategy is shown in Figure 9.

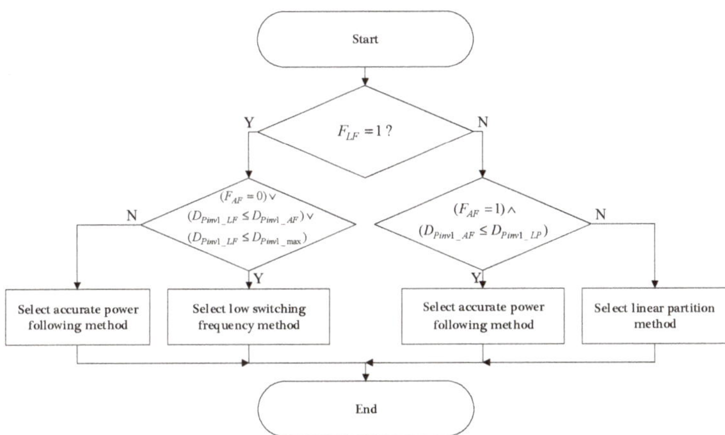

Figure 9. Flowchart of the voltage vector distribution method selection strategy.

5. Drive System Simulation

In this sector the overall configuration of the proposed system is summarized and the simulation results of the proposed methods are analyzed.

5.1. Overall Configuration

The overall configuration of the proposed system is shown in Figure 10. It consists of two parts, which are the drive system circuit and the drive system controller. The drive system circuit has the same structure as the system shown in Figure 1, equipped with current sensors measuring the phase currents of OW-PMSM, and voltage sensors measuring DC-bus voltages of dual inverter. The capacitors C1 and C2 are used to filter the voltage fluctuation of the power sources and provide reactive power needed.

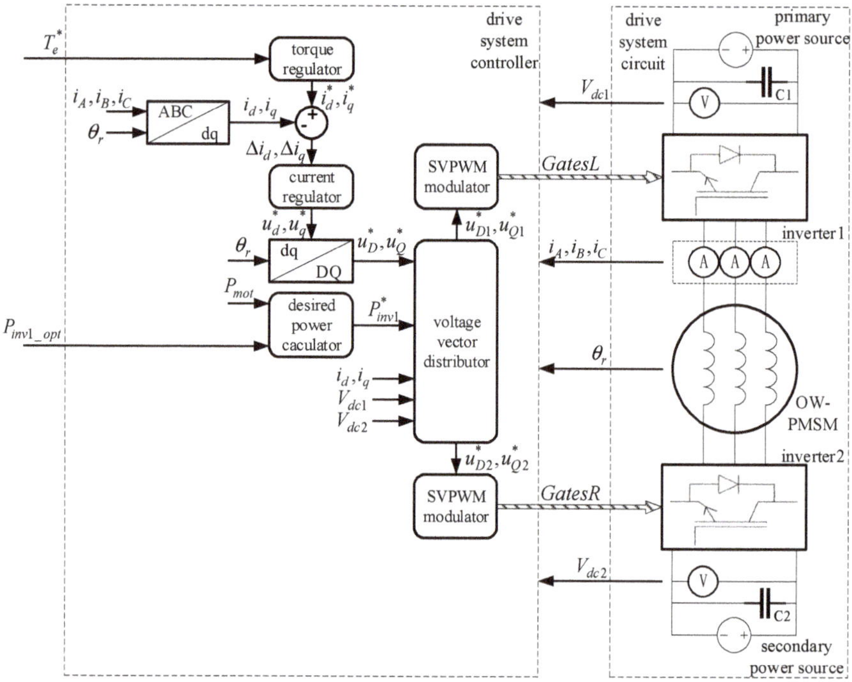

Figure 10. Overall configuration of the proposed system.

In the drive system controller, the desired torque T_e^* and the optimal output power of inverter1 P_{inv1_opt} are inputted from vehicle control unit. In torque regulator, the desired stator current $\vec{i}_s^*$ is calculated by MTPA (maximum torque per ampere) method in constant torque region and CBE (constant back EMF) method in flux weakening region [26,27]. Then the desired stator voltage vector $\vec{u}_s^*$ is generated by the current regulator, which is actual a PI controller. The desired power sharing calculation method introduced in sector 3 is executed in the desired power calculator, outputting desired output power of inverter1 P_{inv1}^*. Afterwards, the voltage vector distribution method expounded in sector 4 is carried out in the voltage vector distributor. After the desired stator voltage vector is partitioned, the desired voltage vectors of inverter1 and inverter2, known as $\vec{u}_{s1}^*$ and $\vec{u}_{s2}^*$, are sent to the corresponding SVPWM modulators respectively. Where *GatesL* and *GatesR* are the gate control signals for inverter1 and inverter2 respectively.

5.2. Simulation Results

To validate to proposed control methods, we ran a simulation of the OW-PMSM drive system on Matlab/Simulink platform. The basic parameters of the drive system circuit are shown in Table 2 and the controller parameters are shown in Table 3. It is important to note that an extra motor speed controller, which is actual a PI controller, was used to generate the desired torque T_e^* inputted to the torque regulator mentioned above to make the motor speed follow the preset value. This motor speed controller is not necessary in an electric vehicle and not shown in Figure 10.

Table 2. Parameters of the drive system circuit. OW-PMSM: open-end winding permanent magnet synchronous motor.

Modules	Items	Parameters
Solver	Solve type	Discrete
	Time step T_s/s	5×10^{-7}
OW-PMSM	Motor type	Interior open-end winding PMSM
	Number of pole pairs p_0	4
	Stator resistance R_s/Ω	0.1
	Fundamental amplitude and third harmonic amplitude of permanent magnet flux linkage $[\psi_f, \psi_{f3}]$/Wb	[0.2, 0.01]
	d-axis and q-axis inductance $[L_d, L_q]$/F	[0.0012, 0.0015]
	Zero sequence inductance L_0/F	0.0003
	Rotational inertia of rotor J_m/kg m^{-2}	0.011
	Cullen and viscous resistance coefficient	[0.001, 0.0005]
Inverter devices	On-resistance R_{on}/Ω	0.01
	Forward voltage drop V_f/V	0.8
	Current fall time and tailing time $[T_f, T_t]$/s	$[1, 1.5] \times 10^{-6}$
	Current capacity of each phase i_{max}/A	160
Power Sources	DC-bus voltages of primary and secondary power source $[V_{dc1}, V_{dc2}]$/V	[300, 200]

Table 3. Parameters of the drive system controller. SVPWM: space vector pulse width modulation.

Modules	Items	Parameters
Motor speed controller	Sampling time T_{s_SC}/s	1×10^{-4}
	Proportionality coefficient P_{SC}	0.2
	Integral coefficient I_{SC}	2
Torque regulator	Sampling time T_{s_TR}/s	1×10^{-4}
	Voltage saturation coefficient k_u	0.95
Current regulator	Sampling time T_{s_CR}/s	1×10^{-4}
	Proportionality coefficient P_{CR}	2
	Integral coefficient I_{CR}	120
Desired power calculator	Sampling time T_{s_PC}/s	1×10^{-4}
	Gain of the inertial element K_{PC}	0.5
	Time constant of the inertial element T_{PC}/s	0.05
Voltage vector distributor	Sampling time T_{s_VD}/s	1×10^{-4}
	Maximum power difference of inverter1 D_{Pinv1_max}/kW	3
SVPWM modulator	Sampling time T_{s_SVM}/s	1×10^{-4}
	Control period T_{c_SVM}/s	1×10^{-4}

This simulation's duration was 0.9 s. The expected motor speed linearly increased from 0 to 6000 r/min in 0–0.3 s and stayed at 6000 r/min till 0.6 s. Then the speed linearly dropped to 0 in 0.6–0.9 s. The load torque jumped from 0 to 60 N·m at 0.05 s and remained until simulation finished.

Curves of motor's expected and actual rotational speed are shown in Figure 11a, curves of motor's desired torque, electromagnetic torque, and load torque are shown in Figure 11b.

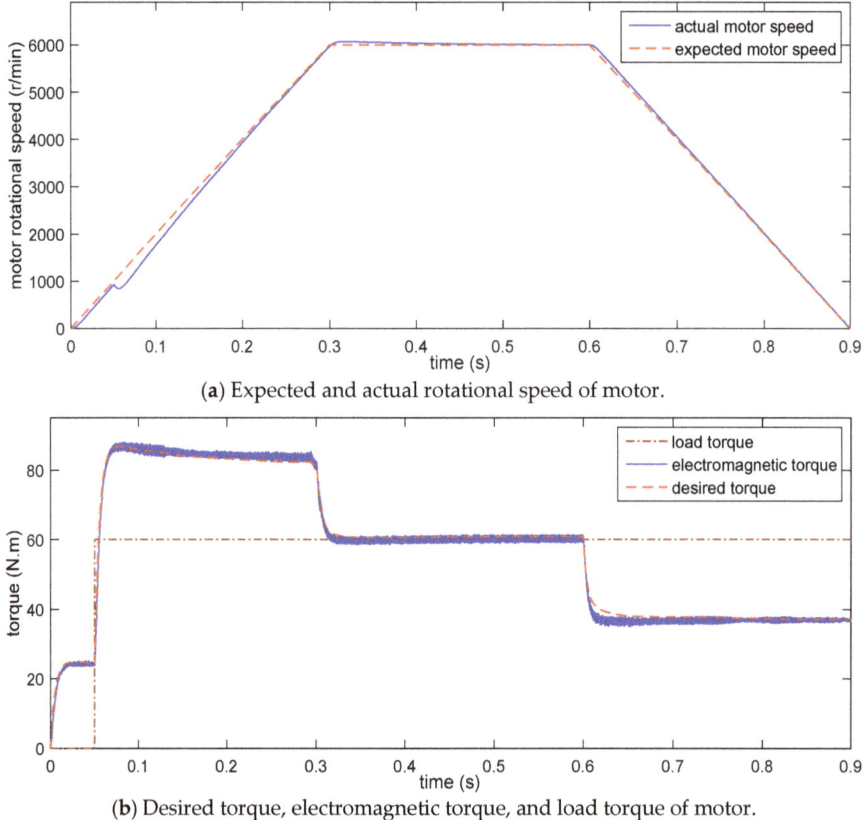

(**a**) Expected and actual rotational speed of motor.

(**b**) Desired torque, electromagnetic torque, and load torque of motor.

Figure 11. Curves of motor speed and torque.

We can observe from Figure 11a that the motor rotary speed could smoothly and swiftly follow the preset value and only had a slight fluctuation at 0.05 s when the load torque jumped. Figure 11b also indicates that motor's electromagnetic torque followed the desired torque well, the amplitude its fluctuation was limited within 3 N·m under dual inverter's modulation by SVPWM strategy with an operating frequency of 10 kHz. The torque's fluctuation was reduced by about 50% compared to the multi-level current hysteresis modulation we proposed in [24].

Local curves of phase A's voltage and current are shown in Figure 12a,b respectively, with a time range of 0.5–0.51 s.

Theoretically, there are five levels in motor's phase voltage of a single inverter system under SVPWM strategy. In a dual inverter system, if the two DC-bus voltages are equal, there are nine levels in motor's phase voltage. Furthermore, there are even more phase voltage levels with unequal DC-bus voltages. These additional levels made the waveform of phase A's voltage shown in Figure 12a more close to the sine wave, compared to a single inverter system. Figure 12b also shows the current ripple was reduced significantly, compared to the system under the multi-level current hysteresis modulation [24].

In addition, we made a harmonic analysis of phase A's voltage in the time range of 0.35–0.55 s, while the motor was working in a steady state at the speed of 6000 r/min. The result is shown in Figure 13.

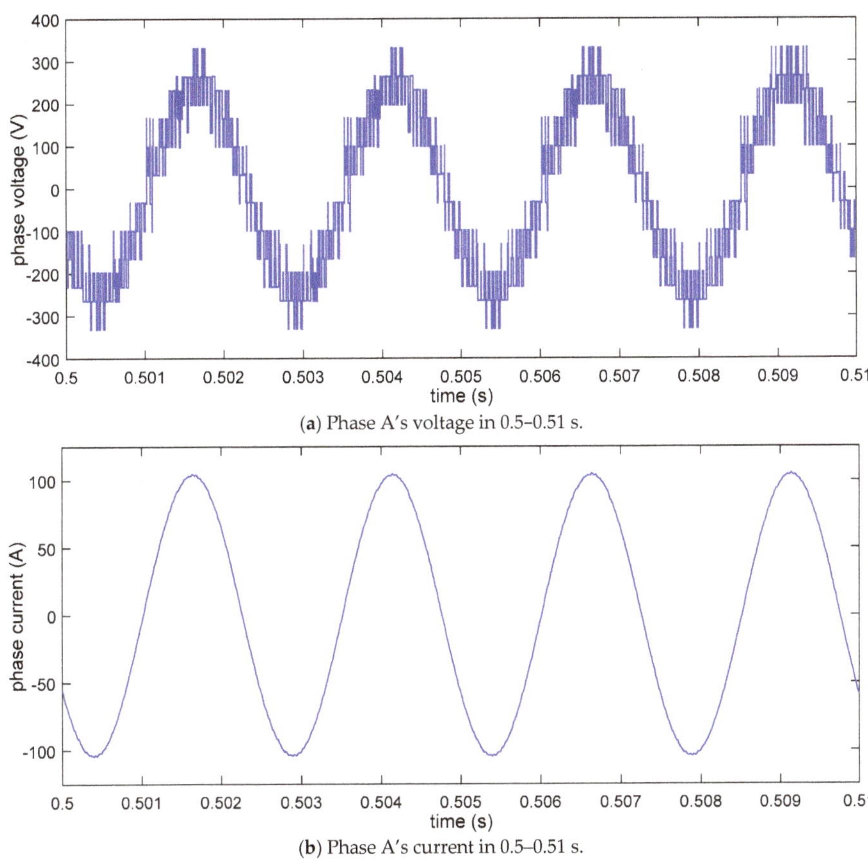

(**a**) Phase A's voltage in 0.5–0.51 s.

(**b**) Phase A's current in 0.5–0.51 s.

Figure 12. Local curves of phase A's voltage and current and harmonic component.

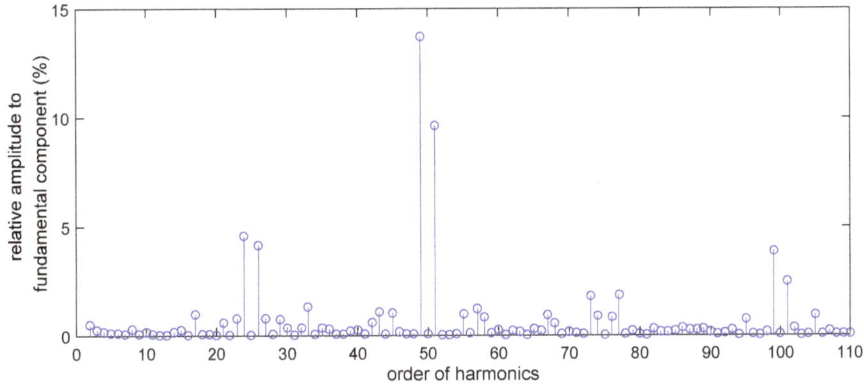

Figure 13. Relative amplitude to fundamental component of phase A's voltage in 0.35–0.55 s.

From Figure 13 we can see, the proportion of the harmonic components are very low and mainly concentrate around 25th harmonic and its multiples. The components around 50th harmonic are relatively large, with amplitudes above 10% of the fundamental amplitude. Considering the fundamental frequency is 400 Hz (rotor rotate speed multiplies number of pole-pairs), the frequency of the 50th harmonic is 20 kHz, which is just twice as the SVPWM control frequency.

Curves in regard to inverter1's output power following effect, and curves of two inverters' output power are shown in Figure 14a,b respectively.

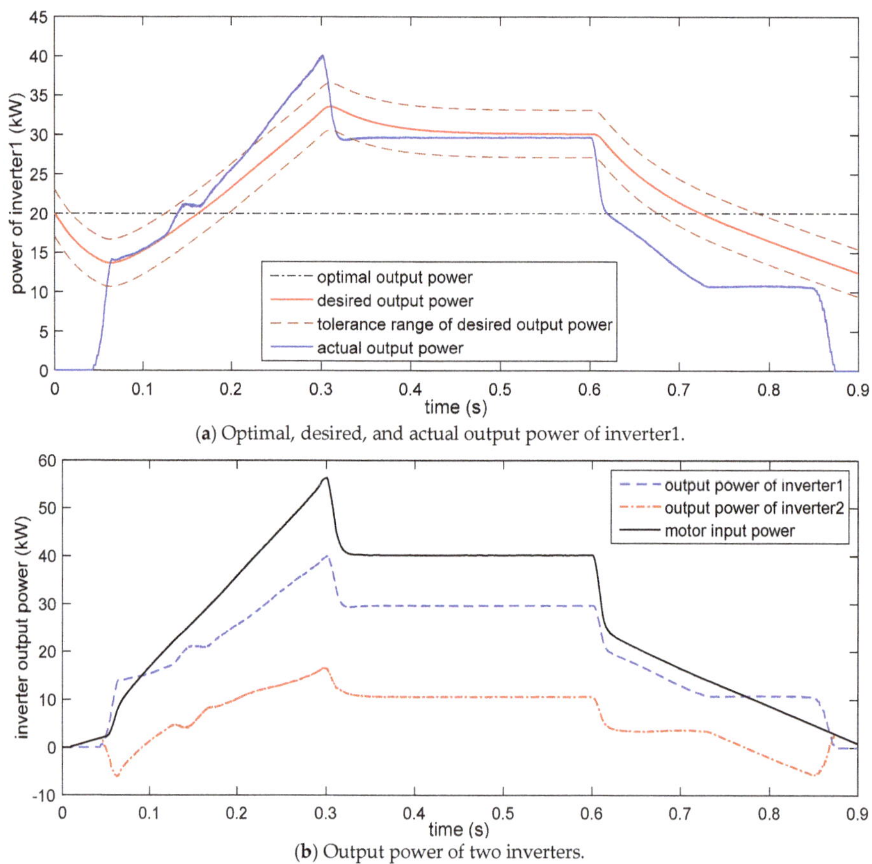

(**a**) Optimal, desired, and actual output power of inverter1.

(**b**) Output power of two inverters.

Figure 14. Curves of inverter output power.

From Figure 14a we can see, while inverter1's optimal output power P_{inv1_opt} was fixed at 20 kW as a constant, the desired output power P^*_{inv1} had some fluctuations to approach the motor input power shown in Figure 13b, which relieved the power outputting stress on inverter2. Due to the inertial element in the desired power calculator, the curve of P^*_{inv1} never went too far from P_{inv1_opt} and had a moderate fluctuation, which had a positive effect on the primary power source's efficiency performance. The tolerance range of inverter1's power following, which is determined by $[P^*_{inv1} - D_{Pinv1\_max}, P^*_{inv1} + D_{Pinv1\_max}]$, is indicated by the dashed lines. Within this range, inverter1's output power was considered following P^*_{inv1} accurately enough, thus the low switching frequency method was selected in priority to reduce inverter1's switching frequency. Most of the time, inverter1's actual output power stayed in the tolerance range as expected. However, in first 0.05 s and after the motor speed started going down, inverter1's actual output power couldn't follow P^*_{inv1} well. Because during these periods, motor's load torque was too low to demand a stator current vector with an enough amplitude for the power distribution in a big ratio or power transfer between the two power sources. The stator current is the medium of power transfer through the motor, thus a stator current vector with a considerable amplitude is necessary for the power distribution in a wide range or power transfer between the two power sources. In other words, the range of power distribution ratio or power transfer between the two power sources is limited by the amplitude of motor's stator current vector, which is also determined by motor's load torque. This limitation was discussed in [23], but it's not completely suitable for the circumstance here.

As shown in Figure 14b, the motor's input power was determined by motor's speed demand and the load torque, and was the summation of the two inverters' output power. While inverter1's output power was controlled to follow P^*_{inv1}, inverter2's output power was made to compensate the lacking power or absorb the redundant power. During some periods where inverter1's output power was more than motor's requirement, inverter2's output power was negative, meaning inverter2 was absorbing power. In other words, the primary power source was charging the secondary power source through the motor.

We define the voltage vector distribution mode number to specify the status of voltage vector distribution. Relations between the mode number and the specific circumstances of voltage vector distribution are listed in Table 4. And the curve of voltage vector distribution mode number is shown in Figure 15.

Table 4. Meanings of voltage vector distribution mode number.

Voltage Vector Distribution Method	Specific Circumstance	Mode Number
Low switching frequency method	The nth candidate vector is selected (the mode number indicates the ranking of selected vector in the sorted sequence)	1~7
Accurate power following method	inverter1's power following deviation D_{Pinv1_AF} equals to 0	0
	inverter1's power following deviation D_{Pinv1_AF} is not equal to 0	−1
Linear partition method	inverter1's power following deviation D_{Pinv1_LP} equals to 0	−2
	inverter1's power following deviation D_{Pinv1_LP} is not equal to 0, but $\vec{u}^*_s$ can be integrally synthesized	−3
	$\vec{u}^*_s$ cannot be integrally synthesized	−4

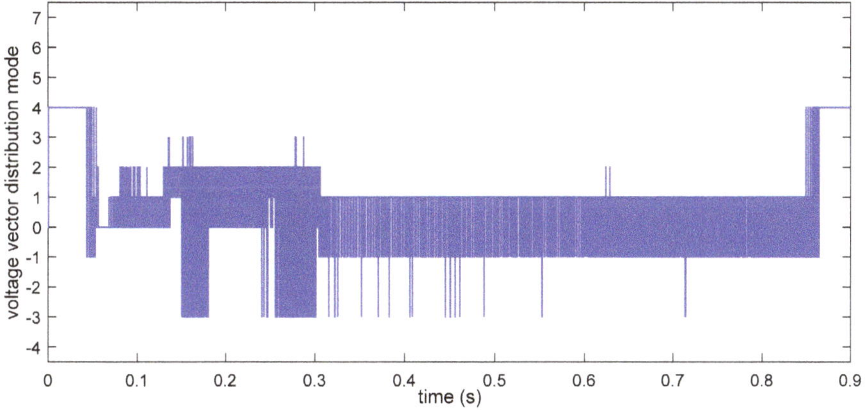

Figure 15. Curve of voltage vector distribution mode number.

As shown in Figure 15, some values of the mode number never appeared, such as −4, 5, 6, and 7. The absence of −4 is because in this simulation, the amplitude of $\vec{u}_s^*$ was not too high and the DC-bus voltages of the two power sources were enough, thus $\vec{u}_s^*$ could always be integrally synthesized. In low switching frequency method, inverter1's output power values corresponding to the candidate voltage vectors ranked behind were not in the tolerance range of inverter1's power following. Thus the absence of 5, 6, 7 is mostly decided by this tolerance range. The narrower this range is, the fewer the candidate voltage vectors can be selected.

The voltage vector distribution of inverter1 and inverter2 in static DQ plane are shown in Figure 16a,b respectively, with a sampling step of 0.002 s. The curves of inverter switching frequency and inverter losses are shown in Figure 17a,b respectively.

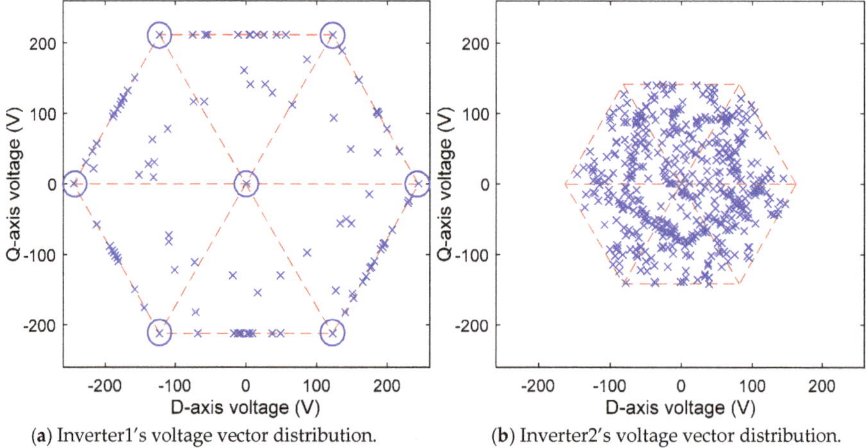

(**a**) Inverter1's voltage vector distribution.　　(**b**) Inverter2's voltage vector distribution.

Figure 16. Inverter voltage vector distribution in static DQ plane.

From Figure 16 we can see, all the sampled voltage vectors, noted as "×", were within the modulation range of the corresponding inverter, indicated by the dashed hexagon, as expected. However, the distribution characteristics of these voltage vectors are different between inverter1 and inverter2. In Figure 16a, many sampled voltage vectors overlapped at the circled positions, which are the vertexes and the center of inverter1's hexagon, where the candidate voltage vectors of low switching frequency method are lying. That is because during most of the simulation time, the low switching frequency method was selected, which can be discovered from Figure 15. For this reason, the sampled voltage vectors of inverter1 appeared sparser than inverter2's shown in Figure 16b, while their amount were actually the same. We can also discover that there was a certain amount of sampled voltage vectors on the boundary of inverter1's modulation range. That is because sometimes inverter1's output power was hard to stay in its tolerance range, on these occasions inverter1 had to generate the voltage vector with highest amplitude to make its output power closer to the tolerance range. Relatively, as Figure 16b shows, the sampled voltage vectors of inverter2 were almost uniformly distributed due to inverter2's role of a compensator.

In a large proportion of the simulation time inverter1 was outputting basic voltage vectors and saturated voltage vectors, of which the switch commutations in a SVPWM period are fewer than regular voltage vectors. This resulted in the switching frequency of inverter1 much lower than inverter2, especially in the steady state operation when motor's speed and electromagnetic torque were not changing, as shown in Figure 17a. The inverter losses consist of the on-state loss and the switching loss, in which the on-state loss is only determined by the current through the inverter, while the switching loss is in direct proportion to the inverter switching frequency. From Figure 17b we can see most of the time the switching loss was lower than the on-state loss due to inverter1's low switching frequency, resulting in the improvement of inverter efficiency.

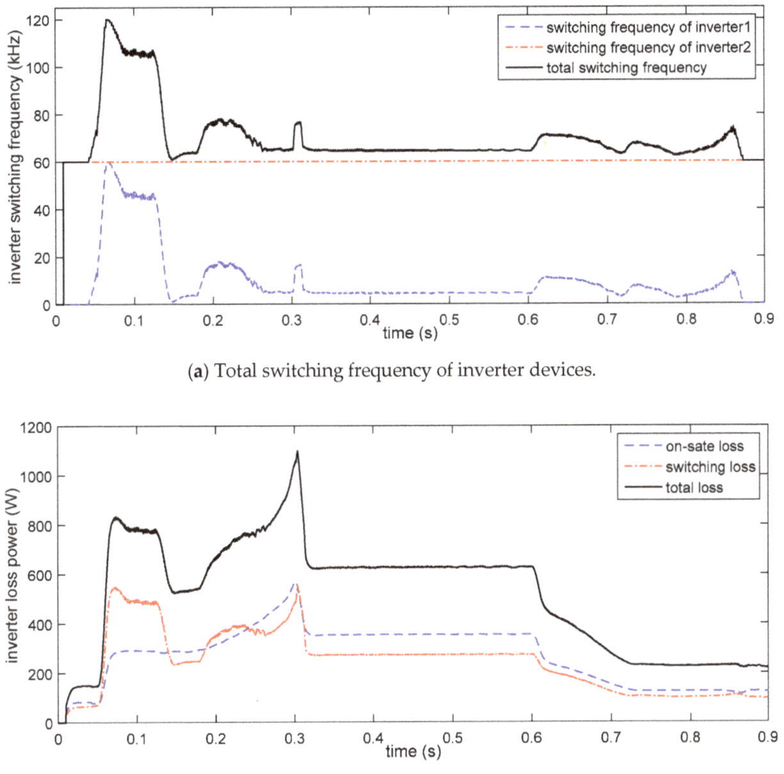

(**a**) Total switching frequency of inverter devices.

(**b**) On-state loss and switching loss power of inverter.

Figure 17. Curves of inverter switching frequency and inverter loss power.

6. Conclusions

This paper, aims at an OW-PMSM drive system fed by dual inverter for electric vehicles, in which each inverter is powered by an independent power source. The primary power source is an energy conversion device while the secondary power source consists of an energy storage element. A desired power sharing calculation method has been proposed. By using a first order inertial element, it could manage the power sharing between the two power sources to optimize the energy efficiency of the primary power source. Furthermore, three different voltage vector distribution methods with various advantages have been proposed, and their selection strategy could select the optimal one according to the operating conditions. Based on the integral synthesizing of the desired stator voltage vector, the proposed voltage vector distribution method could reduce the inverter switching frequency while making the primary power source follow its desired output power. Finally, a simulation of the drive system on Matlab/Simulink platform has been executed to validate the proposed methods.

The proposed system is suitable for electric vehicles with a single motor driving and powered by two power sources, so that the power flow can be handled by the dual inverter and OW-PMSM without a DC/DC converter and the DC voltage is lowered. The desired power sharing calculation method and voltage vector distribution method could improve the primary power source's energy efficiency by regulating its output power and lessen inverter1's switching loss by reducing its switching frequency, respectively. By controlling the optimal output power of inverter1 inputted to the system, they could also handle the energy management between the two power sources. The proposed methods provide a theoretical basis and implementation scheme a for dual inverter OW-PMSM drive system with two isolated power sources in electric vehicles.

Future research will be directed towards finding an energy management method matched with this system for electric vehicles to maximize the overall efficiency and driving range, and extending the ranges of power sharing or power transfer between two power sources when the load torque is not high enough. After solving the existing practical issues, an experimental verification of the proposed system will also be executed in an electric vehicle.

Acknowledgments: This work was supported by the China Postdoctoral Science Foundation (2014M561290), the Energy Administration of Jilin Province [2016]35, the Jilin Province Science and Technology Development Fund (20150520115JH), the Jilin Province Science and Technology Development Fund (20180101062JC), and Energy Conservation and Ecological Protection of Special Funds Project of Jilin Province (2017).

Author Contributions: Nan Xu and Yi-Fan Jia conceived the control method and wrote the full manuscript; Yu-Kuan Li and Di Zhao performed the simulation and revised the manuscript; Xin Tang analyzed the data; Liang Chu analyzed and evaluated the simulation results and gave valuable suggestions.

Conflicts of Interest: The authors declare no conflict of interest.

References

1. Park, J.W.; Koo, D.H.; Kim, J.M.; Kim, H.G. Improvement of control characteristics of interior permanent magnet synchronous motor for electric vehicle. *IEEE Trans. Ind. Appl.* **2001**, *37*, 1754–1760. [CrossRef]
2. Boazzo, B.; Pellegrino, G. Model-based direct flux vector control of permanent-magnet synchronous motor drives. *IEEE Trans. Ind. Appl.* **2015**, *51*, 3126–3136. [CrossRef]
3. Park, C.S.; Lee, J.S. A torque error compensation algorithm for surface mounted permanent magnet synchronous machines with respect to magnet temperature variations. *Energies* **2017**, *10*, 1365. [CrossRef]
4. Nam, K.-T.; Kim, H.; Lee, S.-J.; Kuc, T.-Y. Observer-Based Rejection of Cogging Torque Disturbance for Permanent Magnet Motors. *Appl. Sci.* **2017**, *7*, 867. [CrossRef]
5. Wu, H.X.; Li, L.Y.; Kou, B.Q.; Ping, Z. The research on energy regeneration of permanent magnet synchronous motor used for hybrid electric vehicle. In Proceedings of the Vehicle Power and Propulsion Conference (VPPC '08), Harbin, China, 3–5 September 2008; pp. 1–4. [CrossRef]
6. Howlader, A.M.; Urasaki, N.; Senjyu, T.; Yona, A.; Saber, A.Y. Optimal pam control for a buck boost dc-dc converter with a wide-speed-range of operation for a PMSM. *J. Power Electron.* **2010**, *10*, 477–484. [CrossRef]
7. Athavale, A.; Sasaki, K.; Gagas, B.S.; Kato, T.; Lorenz, R. Variable flux permanent magnet synchronous machine (VF-PMSM) design methodologies to meet electric vehicle traction requirements with reduced losses. *IEEE Trans. Ind. Appl.* **2017**, 1–8. [CrossRef]
8. Howlader, A.M.; Urasaki, N.; Senjyu, T.; Yona, A. Wide-Speed-Range optimal PAM control for permanent magnet synchronous motor. In Proceedings of the IEEE International Conference on Electrical Machines and Systems, Tokyo, Japan, 15–18 November 2009; Volume 2007, pp. 1–5.
9. Li, Z.; Onar, O.; Khaligh, A.; Schaltz, E. Design and control of a multiple input DC/DC converter for battery/ultra-capacitor based electric vehicle power system. In Proceedings of the IEEE Applied Power Electronics Conference and Exposition, Washington, DC, USA, 15–19 February 2009; pp. 591–596.
10. Gualous, H.; Gustin, F.; Berthon, A.; Dakyo, B. DC/DC converter design for supercapacitor and battery power management in hybrid vehicle applications—Polynomial control strategy. *IRE Trans. Ind. Electron.* **2010**, *57*, 587–597. [CrossRef]
11. An, Q.; Liu, J.; Peng, Z.; Sun, L.; Sun, L. Dual-space vector control of open-end winding permanent magnet synchronous motor drive fed by dual inverter. *IEEE Trans. Power Electron.* **2016**, *31*, 8329–8342. [CrossRef]
12. Nguyen, N.K.; Semail, E.; Meinguet, F.; Sandulescu, P.; Kestelyn, X.; Aslan, B. Different virtual stator winding configurations of open-end winding five-phase PM machines for wide speed range without flux weakening operation. In Proceedings of the IEEE European Conference on Power Electronics and Applications, Lille, France, 2–6 September 2013; pp. 1–8.
13. Sandulescu, P.; Meinguet, F.; Kestelyn, X.; Semail, E.; Bruyere, A. Control strategies for open-end winding drives operating in the flux-weakening region. *IEEE Trans. Power Electron.* **2014**, *29*, 4829–4842. [CrossRef]

14. Kwak, M.S.; Sul, S.K. Flux weakening control of an open winding machine with isolated dual inverters. In Proceedings of the IEEE Industry Applications Conference 2007, New Orleans, LA, USA, 23–27 September 2007; pp. 251–255.
15. Welchko, B.A. A double-ended inverter system for the combined propulsion and energy management functions in hybrid vehicles with energy storage. In Proceedings of the IEEE Industrial Electronics Society, Raleigh, NC, USA, 6–10 November 2005; p. 6.
16. Machiya, H.; Haga, H.; Kondo, S. High efficiency drive method of an open-winding induction machine driven by dual inverter using capacitor across dc bus. *Electr. Eng. Jpn.* **2016**, *197*, 62–72. [CrossRef]
17. Drisya, V.; Samina, T. Supply voltage boosting using a floating capacitor bridge in a 3 level space vector modulated inverter system for an open-end winding induction motor drive. In Proceedings of the IEEE International Conference on Control Communication & Computing, Trivandrum, India, 19–21 November 2015; pp. 165–169.
18. Sivakumar, K.; Das, A.; Ramchand, R.; Patel, C.; Gopakumar, K. A hybrid multilevel inverter topology for an open-end winding induction-motor drive using two-level inverters in series with a capacitor-fed h-bridge cell. *IRE Trans. Ind. Electron.* **2010**, *57*, 3707–3714. [CrossRef]
19. Pan, D.; Huh, K.K.; Lipo, T.A. Efficiency improvement and evaluation of floating capacitor open-winding PM motor drive for EV application. In Proceedings of the IEEE Energy Conversion Congress and Exposition, Pittsburgh, PA, USA, 14–18 September 2014; pp. 837–844.
20. Park, J.S.; Nam, K. Dual Inverter Strategy for High Speed Operation of HEV Permanent Magnet Synchronous Motor. In Proceedings of the IEEE Industry Applications Conference, Tampa, FL, USA, 8–12 October 2006; Volume 1, pp. 488–494.
21. Sun, D.; Zheng, Z.; Lin, B.; Zhou, W.; Chen, M. A hybrid PWM based field weakening strategy for hybrid-inverter driven open-winding PMSM system. *IEEE Trans. Power Appar. Syst.* **2017**, 2912–2918. [CrossRef]
22. Griva, G.; Oleschuk, V. Flexible PWM control of three-level inverters for fuel cell/battery supplied aeromobile drive. In Proceedings of the IEEE Industrial Electronics, Porto, Portugal, 3–5 November 2009; pp. 3735–3740.
23. Casadei, D.; Grandi, G.; Lega, A.; Rossi, C. Multilevel operation and input power balancing for a dual two-level inverter with insulated DC sources. *IEEE Trans. Ind. Appl.* **2008**, *44*, 1815–1824. [CrossRef]
24. Chu, L.; Jia, Y.F.; Chen, D.-S.; Xu, N.; Wang, Y.-W.; Tang, X.; Xu, Z. Research on Control Strategies of an Open-End Winding Permanent Magnet Synchronous Driving Motor (OW-PMSM)-Equipped Dual Inverter with a Switchable Winding Mode for Electric Vehicles. *Energies* **2017**, *10*, 616. [CrossRef]
25. Hendawi, E.; Khater, F.; Shaltout, A. Analysis, simulation and implementation of space vector pulse width modulation inverter. In Proceedings of the WSEAS International Conference on Applications of Electrical Engineering, World Scientific and Engineering Academy and Society (WSEAS), Penang, Malaysia, 23–25 March 2010; pp. 124–131.
26. Sun, T.; Wang, J.; Chen, X. Maximum torque per ampere (MTPA) control for interior permanent magnet synchronous machine drives based on virtual signal injection. *IEEE Trans. Power Electron.* **2015**, *30*, 5036–5045. [CrossRef]
27. Hu, D.; Zhu, L.; Xu, L. Maximum Torque per Volt operation and stability improvement of PMSM in deep flux-weakening Region. In Proceedings of the IEEE Energy Conversion Congress and Exposition, Raleigh, NC, USA, 15–20 September 2012; pp. 1233–1237.

MDPI

St. Alban-Anlage 66

4052 Basel

Switzerland

Tel. +41 61 683 77 34

Fax +41 61 302 89 18

www.mdpi.com

Applied Sciences Editorial Office

E-mail: applsci@mdpi.com

www.mdpi.com/journal/applsci